CMOS ELECTRONICS

IEEE Press
445 Hoes Lane
Piscataway, NJ 08855

IEEE Press Editorial Board
Stamatios V. Kartalopoulos, *Editor in Chief*

M. Akay	M. E. El-Hawary	F. M. B. Periera
J. B. Anderson	R. Leonardi	C. Singh
R. J. Baker	M. Montrose	S. Tewksbury
J. E. Brewer	M. S. Newman	G. Zobrist

Kenneth Moore, *Director of Book and Information Services (BIS)*
Catherine Faduska, *Senior Acquisitions Editor*
Christina Kuhnen, *Associate Acquisitions Editor*

IEEE Computer Society, *Sponsor*
C-S Liaison to IEEE Press, Michael Williams

Technical Reviewers
Cecilia Metra, University of Bologna, Italy
Antonio Rubio, Polytechnical University of Catalonia, Barcelona, Spain
Robert Aitken, Artisan Corp., San Jose, California
Robert Madge, LSI Logic Corp., Gresham, Oregon
Joe Clement, William Filler, Duane Bowman, and David Monroe, Sandia National Laboratories, Albuquerque, New Mexico
Jose Luis Rosselló and Sebastia Bota, University of the Balearic Islands, Spain
Harry Weaver and Don Neamen, Univerisity of New Mexico, Albuquerque, New Mexico
Manoj Sachdev, University of Waterloo, Ontario, Canada

CMOS ELECTRONICS
HOW IT WORKS, HOW IT FAILS

JAUME SEGURA
Universitat de les Illes Balears

CHARLES F. HAWKINS
University of New Mexico

IEEE Computer Society, *Sponsor*

IEEE PRESS

A JOHN WILEY & SONS, INC., PUBLICATION

Copyright © 2004 by the Institute of Electrical and Electronics Engineers, Inc. All rights reserved.

Published simultaneously in Canada.

No part of this publication may be reproduced, stored in a retrieval system or transmitted in any form or by any means, electronic, mechanical, photocopying, recording, scanning or otherwise, except as permitted under Section 107 or 108 of the 1976 United States Copyright Act, without either the prior written permission of the Publisher, or authorization through payment of the appropriate per-copy fee to the Copyright Clearance Center, Inc., 222 Rosewood Drive, Danvers, MA 01923, (978) 750-8400, fax (978) 646-8600, or on the web at www.copyright.com. Requests to the Publisher for permission should be addressed to the Permissions Department, John Wiley & Sons, Inc., 111 River Street, Hoboken, NJ 07030, (201) 748-6011, fax (201) 748-6008.

Limit of Liability/Disclaimer of Warranty: While the publisher and author have used their best efforts in preparing this book, they make no representation or warranties with respect to the accuracy or completeness of the contents of this book and specifically disclaim any implied warranties of merchantability or fitness for a particular purpose. No warranty may be created or extended by sales representatives or written sales materials. The advice and strategies contained herein may not be suitable for your situation. You should consult with a professional where appropriate. Neither the publisher nor author shall be liable for any loss of profit or any other commercial damages, including but not limited to special, incidental, consequential, or other damages.

For general information on our other products and services please contact our Customer Care Department within the U.S. at 877-762-2974, outside the U.S. at 317-572-3993 or fax 317-572-4002.

Wiley also publishes its books in a variety of electronic formats. Some content that appears in print, however, may not be available in electronic format.

Library of Congress Cataloging-in-Publication Data is available.

ISBN 0-471-47669-2

Printed in the United States of America.

10 9 8 7 6 5 4 3 2 1

A Mumara, a la seva memòria, i al meu Pare.
To Patricia, Pau, and Andreu for love, for sharing life, and for happiness
—Jaume Segura

To Jan, who has shared the mostly ups, the sometimes downs, and the eternal middle road of life. To our children, Andrea, David, and Shannon, their spouses, and the grandchildren who are following, I hope this work will be a small record of an intense period in which all of you were close to my thoughts.
—Charles F. Hawkins

CONTENTS

Foreword — xiii

Preface — xv

PART I CMOS FUNDAMENTALS

1 Electrical Circuit Analysis — 3
- 1.1 Introduction — 3
- 1.2 Voltage and Current Laws — 3
  - 1.2.1 Kirchhoff's Voltage Law (KVL) — 5
  - 1.2.2 Kirchhoff's Current Law (KCL) — 7
- 1.3 Capacitors — 17
  - 1.3.1 Capacitor Connections — 19
  - 1.3.2 Capacitor Voltage Dividers — 20
  - 1.3.3 Charging and Discharging Capacitors — 22
- 1.4 Diodes — 24
  - 1.4.1 Diode Resistor Circuits — 25
  - 1.4.2 Diode Resistance — 28
- 1.5 Summary — 29
- Bibliography — 29
- Exercises — 29

2 Semiconductor Physics — 37

- 2.1 Semiconductor Fundamentals — 37
  - 2.1.1 Metals, Insulators, and Semiconductors — 37
  - 2.1.2 Carriers in Semiconductors: Electrons and Holes — 39
  - 2.1.3 Determining Carrier Population — 41
- 2.2 Intrinsic and Extrinsic Semiconductors — 42
  - 2.2.1 n-Type Semiconductors — 42
  - 2.2.2 p-Type Semiconductors — 44
- 2.3 Carrier Transport in Semiconductors — 44
  - 2.3.1 Drift Current — 45
  - 2.3.2 Diffusion Current — 46
- 2.4 The pn Junction — 46
- 2.5 Biasing the pn Junction: I–V Characteristics — 48
  - 2.5.1 The pn Junction under Forward Bias — 49
  - 2.5.2 The pn Junction under Reverse Bias — 49
- 2.6 Parasitics in the Diode — 50
- 2.7 Summary — 51
- Bibliography — 51
- Exercises — 52

3 MOSFET Transistors — 53

- 3.1 Principles of Operation: Long-Channel Transistors — 53
  - 3.1.1 The MOSFET as a Digital Switch — 54
  - 3.1.2 Physical Structure of MOSFETs — 55
  - 3.1.3 Understanding MOS Transistor Operation: A Descriptive Approach — 56
  - 3.1.4 MOSFET Input Characteristics — 59
  - 3.1.5 nMOS Transistor Output Characteristics — 60
  - 3.1.6 pMOS Transistor Output Characteristics — 70
- 3.2 Threshold Voltage in MOS Transistors — 77
- 3.3 Parasitic Capacitors in MOS Transistors — 79
  - 3.3.1 Non-Voltage-Dependent Internal Capacitors — 79
  - 3.3.2 Voltage-Dependent Internal Capacitors — 80
- 3.4 Device Scaling: Short-Channel MOS Transistors — 81
  - 3.4.1 Channel Length Modulation — 83
  - 3.4.2 Velocity Saturation — 83
  - 3.4.3 Putting it All Together: A Physically Based Model — 85
  - 3.4.4 An Empirical Short-Channel Model for Manual Calculations — 87
  - 3.4.5 Other Submicron Effects — 91
- 3.5 Summary — 94

		References	94
		Exercises	94

4 CMOS Basic Gates — 99

4.1	Introduction	99
4.2	The CMOS Inverter	100
	4.2.1 Inverter Static Operation	100
	4.2.2 Dynamic Operation	107
	4.2.3 Inverter Speed Property	109
	4.2.4 CMOS Inverter Power Consumption	113
	4.2.5 Sizing and Inverter Buffers	116
4.3	NAND Gates	118
4.4	NOR Gates	120
4.5	CMOS Transmission Gates	122
4.6	Summary	123
	Bibliography	123
	Exercises	123

5 CMOS Basic Circuits — 127

5.1	Combinational logic	127
	5.1.1 CMOS Static Logic	127
	5.1.2 Tri-State Gates	131
	5.1.3 Pass Transistor Logic	132
	5.1.4 Dynamic CMOS Logic	133
5.2	Sequential Logic	137
	5.2.1 Register Design	137
	5.2.2 Semiconductor Memories (RAMs)	139
5.3	Input–Output (I/O) Circuitry	142
	5.3.1 Input Circuitry: Protecting ICs from Outside Environment	143
	5.3.2 Input Circuitry: Providing "Clean" Input Levels	144
	5.3.3 Output Circuitry, Driving Large Loads	145
	5.3.4 Input–Output Circuitry: Providing Bidirectional Pins	146
5.4	Summary	146
	References	147
	Exercises	148

PART II FAILURE MODES, DEFECTS, AND TESTING OF CMOS ICs

6 Failure Mechanisms in CMOS IC Materials — 153

6.1	Introduction	153
6.2	Materials Science of IC Metals	154

6.3	Metal Failure Modes		159
	6.3.1	Electromigration	159
	6.3.2	Metal Stress Voiding	169
	6.3.3	Copper Interconnect Reliability	176
6.4	Oxide Failure Modes		178
	6.4.1	Oxide Wearout	180
	6.4.2	Hot-Carrier Injection (HCI)	185
	6.4.3	Defect-Induced Oxide Breakdown	191
	6.4.4	Process-Induced Oxide Damage	191
	6.4.5	Negative Bias Temperature Instability (NBTI)	191
6.5	Conclusion		192
	Acknowledgments		192
	Bibliography		192
	Exercises		195

7 Bridging Defects — 199

7.1	Introduction		199
7.2	Bridges in ICs: Critical Resistance and Modeling		200
	7.2.1	Critical Resistance	200
	7.2.2	Fault models for Bridging Defects on Logic Gate Nodes (BF)	207
7.3	Gate Oxide Shorts (GOS)		208
	7.3.1	Gate Oxide Short Models	209
7.4	Bridges in Combinational Circuits		214
	7.4.1	Nonfeedback Bridging Faults	214
	7.4.2	Feedback Bridging Faults	214
7.5	Bridges in Sequential Circuits		217
	7.5.1	Bridges in Flip-Flops	217
	7.5.2	Semiconductor Memories	218
7.6	Bridging Faults and Technology Scaling		218
7.7	Conclusion		219
	References		219
	Exercises		220

8 Open Defects — 223

8.1	Introduction		223
8.2	Modeling Floating Nodes in ICs		224
	8.2.1	Supply–Ground Capacitor Coupling in Open Circuits	224
	8.2.2	Effect of Surrounding Lines	226
	8.2.3	Influence of the Charge from MOSFETs	228
	8.2.4	Tunneling Effects	229

		8.2.5	Other Effects	232
	8.3	Open Defect Classes		232
		8.3.1	Transistor-On Open Defect	232
		8.3.2	Transistor Pair-On and Transistor Pair-On / Off	234
		8.3.3	The Open Delay Defect	236
		8.3.4	CMOS Memory Open Defect	237
		8.3.5	Sequential Circuit Opens	238
	8.4	Summary		239
		References		239
		Exercises		240

9 Parametric Failures — 243

- 9.1 Introduction — 243
- 9.2 Intrinsic Parametric Failures — 246
  - 9.2.1 Transistor Parameter Variation — 246
  - 9.2.2 Impact on Device Intrinsic Electrical Properties — 249
  - 9.2.3 Line Interconnect Intrinsic Parameter Variation — 252
  - 9.2.4 Temperature Effect — 255
- 9.3 Intrinsic Parametric Failure Impact on IC Behavior — 256
  - 9.3.1 Interconnect Models — 257
  - 9.3.2 Noise — 259
  - 9.3.3 Delay — 264
- 9.4 Extrinsic Parametric Failure — 271
  - 9.4.1 Extrinsic Weak Interconnect Opens — 271
  - 9.4.2 Extrinsic Resistive Vias and Contacts — 271
  - 9.4.3 Extrinsic Metal Mousebites — 272
  - 9.4.4 Extrinsic Metal Slivers — 275
- 9.5 Conclusion — 275
- References — 276
- Exercises — 279

10 Defect-Based Testing — 281

- 10.1 Introduction — 281
- 10.2 Digital IC Testing: The Basics — 282
  - 10.2.1 Voltage-Based Testing — 283
  - 10.2.2 Speed Testing — 288
  - 10.2.3 Current-Based Testing — 289
  - 10.2.4 Comparative Test Methods Studies — 292
- 10.3 Design for Test — 293
  - 10.3.1 Scan Design — 294

	10.3.2	BIST	296
	10.3.3	Special Test Structures for Next Levels of Assembly	299
10.4	Defect-Based Testing (DBT)		300
	10.4.1	Bridge Defects	301
	10.4.2	Opens	304
	10.4.3	DBT Summary	307
10.5	Testing Nanometer ICs		309
	10.5.1	Nanometer Effects on Testing	311
	10.5.2	Voltage-Based Testing	314
	10.5.3	Current-Based Testing	315
	10.5.4	Delay Testing Using Statisitical Timing Analysis	318
	10.5.5	Low-V_{DD} Testing	318
	10.5.6	Multiparameter Testing	320
10.6	Conclusions		322
	Bibliography		322
	References		322
	Exercises		326

Appendix A Solutions to Self-Exercises   331

A.1	Chapter 1	331
A.2	Chapter 3	334
A.3	Chapter 4	335
A.4	Chapter 5	336
A.5	Chapter 6	336
A.6	Chapter 7	337
A.8	Chapter 8	337
A.8	Chapter 10	337

Index   339

About the Authors   347

FOREWORD

Advances in electronics have followed Moore's Law, by scaling feature sizes every generation, doubling transistor integration capacity every two years, resulting in complex chips of today—a treadmill that we take for granted. No doubt that design complexity has grown tremendously and therefore gets tremendous attention, but often overlooked is the technology underlying reliability, and cost-effective test and product engineering of these complex chips. This is becoming especially crucial as we transition from yesterday's micro-electronics to today's nano-electronics.

This book, *CMOS Electronics: How It Works, How It Fails,* written by Professor Segura and Professor Hawkins, addresses just that—the technology underlying failure analysis, testing, and product engineering. The book starts with fundamental device physics, describes how MOS transistors work, how logic circuits are built, and then eases into failure mechanisms of these circuits. Thus the reader gets a very clear picture of failure mechanisms, how to detect them, and how to avoid them. The book covers the latest advances in failure analysis and test and product engineering, such as defects due to bridges, opens, and parametrics, and formulates test strategies to observe these defects in defect-based testing.

As technology progresses with even smaller geometries, you will have to comprehend test and product engineering upfront in the design. That is why this book is a refreshing change as it introduces design and test together.

I would like to thank Professor Segura and Professor Hawkins for giving me an opportunity to read a draft of this book; I surely enjoyed it, learned a lot from it, and I am sure that you the readers will find it rewarding, too.

SHEKHAR BORKAR
Intel Fellow
Director, Circuit Research

PREFACE

If you find the mysteries and varieties of integrated circuit failures challenging, then this book is for you. It is also for you if you work in the CMOS integrated circuits (IC) industry, but admit to knowing little about the electronics itself or the important electronics of failure. The goal of this book is knowledge of the electronic behavior that failing CMOS integrated circuits exhibit to customers and suppliers. The emphasis is on electronics at the transistor circuit level, to gain a deeper understanding of why failing circuits act as they do.

There are two audiences for this book. The first are those in the industry who have had little or no instruction in CMOS electronics, but are surrounded by the electronic symptoms that permeate a CMOS customer or supplier. You may have a physics, chemical engineering, chemistry, or biology education and need to understand the circuitry that you help manufacture. Part I of the book is designed to bring you up to speed in preparation for Part II, which is an analysis of the nature of CMOS failure mechanisms. The second audience is the electrical engineering professional or student who will benefit from a systematic description of the electronic behavior mechanisms of abnormal circuits. Part II describes the material reliability failures, bridge and open circuit defects, and the subtle parametric failures that are plaguing advanced technology ICs. The last chapter assembles this information to implement a defect-based detection strategy used in IC testing. Five key groups should benefit from this approach: test engineers, failure analysts, reliability engineers, yield improvement engineers, and designers. Managers and others who deal with IC abnormalities will also benefit.

CMOS manufacturing environments are surrounded with symptoms that can indicate serious test, design, yield, or reliability problems. Knowledge of how CMOS circuits work and how they fail can rapidly guide you to the nature of a problem. Is the problem related to test, design, reliability, failure analysis, yield–quality issues, or problems that may occur during characterization of a product? Is the symptom an outcome of random defects, or is it systematic, with a common failure signature? Is the defect a bridging problem, an open circuit problem, or a subtle speed-related problem?

We know how bridge and open circuit defects cause performance degradation or failure, and we are closing in on a complete picture of the IC failures that depend upon parametric properties of the circuit environment. These properties include temperature and power supply values. There are cost savings when you can shorten the time to diagnosis and intelligently plan test strategies that minimize test escapes. Rapid insights into abnormal electronic behavior shorten the time to diagnosis and also allow more deterministic planning of test strategies. Determination of root cause of failure is much more efficient if you can isolate one of the failure modes from their electronic properties early in the process.

Circuit abnormalities are a constant concern, and the ability to rapidly estimate the defect type is the first step: namely, what are we looking for? Or when planning test strategies, what test methods should we use to match against particular defects? The effect of the many varieties of defects on circuit operation is analyzed in the book with numerical examples. The early portion of the book assumes little knowledge of circuits and transistors, but builds to a mature description of the electronic properties of defects in Part II. The book adopts a self-learning style with many problems, examples, and self-exercises

Part I (Chapters 1–5) describes how defect-free CMOS circuits work and Part II (Chapters 6–10) describes CMOS circuit failure properties and mechanisms. Chapter 1 begins with Ohm's and Kirchhoff's laws, emphasizing circuit analysis by inspection, and concludes with an analysis of diodes and capacitors. Chapter 2 introduces semiconductor physics and how that knowledge leads to transistor operation in Chapter 3. Chapter 3 has abundant problem-solving examples to gain confidence with transistor operation and their interplay when connected to resistors. Chapter 4 builds to transistor circuit operation with two and four transistors such as CMOS inverters, NAND and NOR logic gates, and transmission gates. Chapter 5 shows how CMOS transistor circuits are synthesized from Boolean algebra equations, and also compares different design styles.

Part II builds on the foundation of how good, or defect-free, CMOS circuits operate and extends that to circuit failures. Chapter 6 examines why IC metal and oxide materials fail in time, producing the dreaded reliability failures. Chapters 7–9 analyze the electronic failure properties of bridge, open, and parametric variation to formulate a test strategy that matches suspected defects to a detection method sensitive to that defect type. Chapter 10 brings all of this together in a test engineering approach called defect-based testing (DBT). The links to test, failure analysis, and reliability are emphasized.

The math used in CMOS electronics varies from simple algebraic expressions to complex physical and timing models whose solutions are suitable only for computers. Fortunately, algebraic equations serve most needs when we manually analyze a circuit's current and voltage response in the presence of a defect. A few simple calculus equations appear, particularly for introducing timing and power relations. A goal is to understand the subtle pass/fail operation or loss of noise margin for defective circuits.

The rapid pace of CMOS technologies influenced these electronic descriptions. Virtually all modern IC transistors are now short-channel devices with different model equations than their long-channel transistor predecessors. Although first-order models for short-channel transistors may at first seem simpler than long-channel equivalents, we found that this simplicity bred inaccuracy and clumsiness when used for manual transistor circuit analysis. We describe these problems and one approach for analyzing short-channel devices in Chapter 3, but chose long-channel transistor models to illustrate how to calculate voltages and currents. It was the only expedient way to give readers the intuitive insights into the nature of transistors when these devices are subjected to various voltage biases.

CMOS field effect transistors do not exist in isolation in the IC. Parasitic bipolar transistors have a small but critical role in the destructive CMOS latchup condition, in electrostatic discharge (ESD) protection circuits, and in some parasitic defect structures. However, we chose to contain the size of the book by only including a brief description of bipolar transistors in sections where they are mentioned.

The knowledge poured into this book came from several sources. The authors teach electronics at their universities, and much data and knowledge were taken from collaborative work done at Sandia National Labs and at Intel Corporation. We are indebted to many persons in our defect electronics education and particularly acknowledge those with whom we worked closely. These include Jerry Soden and Ed Cole of Sandia National Labs, Antonio Rubio of the Polytechnical University of Catalonia, Jose Luis Rosselló of the University of the Balearic Islands, Alan Righter of Analog Devices, and Ali Keshavarzi, who hosted each of us on university sabbaticals at the Intel facilities in Portland, Oregon and Rio Rancho, New Mexico.

Each chapter had from one to three reviewers. All reviewers conveyed a personal feeling of wanting to make this a better book. These persons are: Sebastià Bota of the University of Barcelona, Harry Weaver and Don Neamen of the University of New Mexico, Jose Luis Rosselló of the University of the Balearic Islands, Antonio Rubio of the Polytechnic University of Catalonia, Manoj Sachdev of the University of Waterloo, Joe Clement, Dave Monroe, and Duane Bowman of Sandia National Labs, Cecilia Metra of the University of Bologna, Bob Madge of LSI Logic Corp., and Rob Aitken of Artisan Corp. We also thank editing and computer support from Gloria Ayuso of the University of the Balearic Islands and Francesc Segura from DMS, Inc.

The seminal ideas and original drafts for the book began in the Spanish city of Palma de Mallorca on the Balearic Islands. Over the next four years, about two-thirds of the book was written in Mallorca with the remainder at the University of New Mexico in Albuquerque in the United States. Countless editorial revisions occurred, with memorable ones on long flights across the Atlantic, in the cafes of the ancient quarter of Palma, and in the coffee shops around the University of New Mexico, where more than four centuries earlier, Spanish farmers grew crops.

The book is intended to flow easily for those without an EE degree and can be a one-semester course in a university. We found from class teaching with this material that the full book is suitable for senior or graduate students with non-EE backgrounds. The EE students can skip or review Chapters 1–2, and go directly to Chapters 3–10. Chapters 3–5 are often taught at the undergraduate EE level, but we found that the focus on CMOS electronics here is more than typically taught. The style blends the descriptive portions of the text with many examples and exercises to encourage self-study. The learning tools are a pad of paper, pen, pocket calculator, isolated time, and motivation to learn. The rewards are insights into the deep mysteries of CMOS IC behavior. For additional material related to this book, visit http://omaha.uib.es/cmosbook/index.html

JAUME SEGURA
CHARLES F. HAWKINS

January 2004

PART I

CMOS FUNDAMENTALS

CHAPTER 1

ELECTRICAL CIRCUIT ANALYSIS

1.1 INTRODUCTION

We understand complex integrated circuits (ICs) through simple building blocks. CMOS transistors have inherent parasitic structures, such as diodes, resistors, and capacitors, whereas the whole circuit may have inductor properties in the signal lines. We must know these elements and their many applications since they provide a basis for understanding transistors and whole-circuit operation.

Resistors are found in circuit speed and bridge-defect circuit analysis. Capacitors are needed to analyze circuit speed properties and in power stabilization, whereas inductors introduce an unwanted parasitic effect on power supply voltages when logic gates change state. Transistors have inherent diodes, and diodes are also used as electrical protective elements for the IC signal input/output pins. This chapter examines circuits with resistors, capacitors, diodes, and power sources. Inductance circuit laws and applications are described in later chapters. We illustrate the basic laws of circuit analysis with many examples, exercises, and problems. The intention is to learn and solve sufficient problems to enhance one's knowledge of circuits and prepare for future chapters. This material was selected from an abundance of circuit topics as being more relevant to the later chapters that discuss how CMOS transistor circuits work and how they fail.

1.2 VOLTAGE AND CURRENT LAWS

Voltage, current, and resistance are the three major physical magnitudes upon which we will base the theory of circuits. Voltage is the potential energy of a charged particle in an electric field, as measured in units of volts (V), that has the physical units of Newton ·

m/coulomb. Current is the movement of charged particles and is measured in coulombs per second or amperes (A). Electrons are the charges that move in transistors and interconnections of integrated circuits, whereas positive charge carriers are found in some specialty applications outside of integrated circuits.

Three laws define the distribution of currents and voltages in a circuit with resistors: Kirchhoff's voltage and current laws, and the volt–ampere relation for resistors defined by Ohm's law. Ohm's law relates the current and voltage in a resistor as

$$V = R \times I \tag{1.1}$$

This law relates the voltage drop (V) across a resistor R when a current I passes through it. An electron loses potential energy when it passes through a resistor. Ohm's law is important because we can now predict the current obtained when a voltage is applied to a resistor or, equivalently, the voltage that will appear at the resistor terminals when forcing a current.

An equivalent statement of Ohm's law is that the ratio of voltage applied to a resistor to subsequent current in that resistor is a constant $R = V/I$, with a unit of volts per ampere called an ohm (Ω). Three examples of Ohm's law in Figure 1.1 show that any of the three variables can be found if the other two are known. We chose a rectangle as the symbol for a resistor as it often appears in CAD (computer-aided design) printouts of schematics and it is easier to control in these word processing tools.

The ground symbol at the bottom of each circuit is necessary to give a common reference point for all other nodes. The other circuit node voltages are measured (or calculated) with respect to the ground node. Typically, the ground node is electrically tied through a building wire called the common to the voltage generating plant wiring. Battery circuits use another ground point such as the portable metal chassis that contains the circuit. Notice that the current direction is defined by the positive charge with respect to the positive terminal of a voltage supply, or by the voltage drop convention with respect to a positive charge. This seems to contradict our statements that all current in resistors and transistors is due to negative-charge carriers. This conceptual conflict has historic origins. Ben Franklin is believed to have started this convention with his famous kite-in-a-thunderstorm experiment. He introduced the terms positive and negative to describe what he called electrical fluid. This terminology was accepted, and not overturned when we found out later that current is actually carried by negative-charge carriers (i.e., electrons). Fortunately, when we calculate voltage, current, and power in a circuit, a positive-charge hypothesis gives the same results as a negative-charge hypothesis. Engineers accept the positive convention, and typically think little about it.

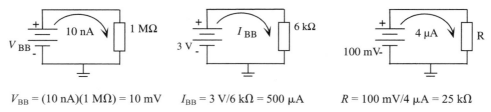

$V_{BB} = (10 \text{ nA})(1 \text{ M}\Omega) = 10 \text{ mV}$ $I_{BB} = 3 \text{ V}/6 \text{ k}\Omega = 500 \text{ }\mu\text{A}$ $R = 100 \text{ mV}/4 \text{ }\mu\text{A} = 25 \text{ k}\Omega$

Figure 1.1. Ohm's law examples. The battery positive terminal indicates where the positive charge exits the source. The resistor positive voltage terminal is where positive charge enters.

An electron loses energy as it passes through a resistance, and that energy is lost as heat. Energy per unit of time is power. The power loss in an element is the product of voltage and current, whose unit is the watt (W):

$$P = VI \tag{1.2}$$

1.2.1 Kirchhoff's Voltage Law (KVL)

This law states that "the sum of the voltage drops across elements in a circuit loop is zero." If we apply a voltage to a circuit of many serial elements, then the sum of the voltage drops across the circuit elements (resistors) must equal the applied voltage. The KVL is an energy conservation statement allowing calculation of voltage drops across individual elements: energy input must equal energy dissipated.

Voltage Sources. An ideal voltage source supplies a constant voltage, no matter the amount of current drawn, although real voltage sources have an upper current limit. Figure 1.2 illustrates the KVL law where V_{BB} represents a battery or bias voltage source. The polarities of the driving voltage V_{BB} and resistor voltages are indicated for the clockwise direction of the current.

Naming V_1 the voltage drop across resistor R_1, V_2 that across resistor R_2, and, subsequently, V_5 for R_5, the KVL states that

$$V_{BB} = V_1 + V_2 + V_3 + V_4 + V_5 \tag{1.3}$$

Note that the resistor connections in Figure 1.2 force the same current I_{BB} through all resistors. When this happens, i.e., when the same current is forced through two or more resistors, they are said to be connected in series. Applying Ohm's law to each resistor of Figure 1.2, we obtain $V_i = R_i \times I_{BB}$ (where i takes any value from 1 to 5). Applying Ohm's law to each voltage drop at the right-hand side of Equation (1.3) we obtain

$$\begin{aligned} V_{BB} &= R_1 I_{BB} + R_2 I_{BB} + R_3 I_{BB} + R_4 I_{BB} + R_5 I_{BB} \\ &= (R_1 + R_2 + R_3 + R_4 + R_5) I_{BB} \\ &= R_{eq} I_{BB} \end{aligned} \tag{1.4}$$

where $R_{eq} = R_1 + R_2 + R_3 + R_4 + R_5$. The main conclusion is that when a number of resistors are connected in series, they can be reduced to an equivalent single resistor whose value is the sum of the resistor values connected in series.

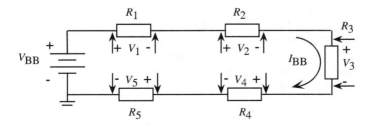

Figure 1.2. KVL seen in series elements.

6 CHAPTER 1 ELECTRICAL CIRCUIT ANALYSIS

■ **EXAMPLE 1.1**

Figure 1.3(a) shows a 5 V source driving two resistors in series. The parameters are referenced to the ground node. Show that the KVL holds for the circuit.

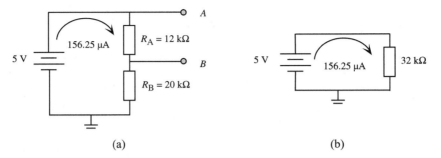

Figure 1.3. (a) Circuit illustrating KVL. (b) Equivalent circuit. The power supply cannot tell if the two series resistors or their equivalent resistance are connected to the power source terminals.

The voltage drop across R_A is

$$V_A = R_A \times I = 12\ \text{k}\Omega \times 156.25\ \mu\text{A} = 1.875\ \text{V}$$

Similarly, the voltage across R_B is 3.125 V. The applied 5 V must equal the series drops across the two resistors or 1.875 V + 3.125 V = 5 V. The last sentence is a verification of the KVL. ■

The current in Figure 1.3(a) through the 12 kΩ in series with a 20 kΩ resistance is equal to that of a single 32 kΩ resistor (Figure 1.3(b)). The voltage across the two resistors in Figure 1.3(a) is 5 V and, when divided by the current (156.25 μA), gives an equivalent series resistance of (5 V/156.25 μA) = 32 kΩ. Figure 1.3(b) is an equivalent reduced circuit of that in Figure 1.3(a).

Current Sources. We introduced voltage power sources first since they are more familiar in our daily lives. We buy voltage batteries in a store or plug computers, appliances, or lamps into a voltage socket in the wall. However, another power source exists, called a current source, that has the property of forcing a current out of one terminal that is independent of the resistor load. Although not as common, you can buy current power sources, and they have important niche applications.

Current sources are an integral property of transistors. CMOS transistors act as current sources during the crucial change of logic state. If you have a digital watch with a microcontroller of about 200k transistors, then about 5% of the transistors may switch during a clock transition, so 10k current sources are momentarily active on your wrist.

Figure 1.4 shows a resistive circuit driven by a current source. The voltage across the current source can be calculated by applying Ohm's law to the resistor connected between the current source terminals. The current source as an ideal element provides a fixed current value, so that the voltage drop across the current source will be determined by the element or elements connected at its output. The ideal current source can supply an infinite voltage, but real current sources have a maximum voltage limit.

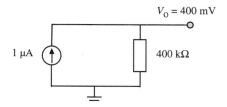

Figure 1.4. A current source driving a resistor load.

1.2.2 Kirchhoff's Current Law (KCL)

The KCL states that "the sum of the currents at a circuit node is zero." Current is a mass flow of charge. Therefore the mass entering the node must equal the mass exiting it. Figure 1.5 shows current entering a node and distributed to three branches. Equation (1.5) is a statement of the KCL that is as essential as the KVL in Equation (1.3) for computing circuit variables. Electrical current is the amount of charge (electrons) Q moving in time, or dQ/dt. Since current itself is a flow (dQ/dt), it is grammatically incorrect to say that "current flows." Grammatically, charge flows, but current does not.

$$I_0 = I_1 + I_2 + I_3 \tag{1.5}$$

The voltage across the terminals of parallel resistors is equal for each resistor, and the currents are different if the resistors have different values. Figure 1.6 shows two resistors connected in parallel with a voltage source of 2.5 V. Ohm's law shows a different current in each path, since the resistors are different, whereas all have the same voltage drop.

$$I_A = \frac{2.5 \text{ V}}{100 \text{ k}\Omega} = 25 \text{ }\mu\text{A}$$
$$I_B = \frac{2.5 \text{ V}}{150 \text{ k}\Omega} = 16.675 \text{ }\mu\text{A} \tag{1.6}$$

Applying Equation (1.5), the total current delivered by the battery is

$$I_{BB} = I_A + I_B = 25 \text{ }\mu\text{A} + 16.67 \text{ }\mu\text{A} = 41.67 \text{ }\mu\text{A} \tag{1.7}$$

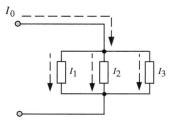

Figure 1.5. KCL and its current summation at a node.

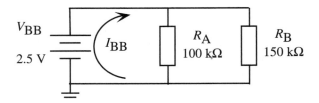

Figure 1.6. Parallel resistance example.

and

$$\frac{V_{BB}}{I_{BB}} = \frac{2.5 \text{ V}}{41.67 \text{ μA}} = 60 \text{ kΩ} \quad (1.8)$$

Notice that the sum of the currents in each resistor branch is equal to the total current from the power supply, and that the resistor currents will differ when the resistors are unequal. The equivalent parallel resistance R_{eq} in the resistor network in Figure 1.6 is $V_{BB} = R_{eq}(I_A + I_B)$. From this expression and using Ohm's law we get

$$\frac{V_{BB}}{R_{eq}} = I_A + I_B$$

$$I_A = \frac{V_{BB}}{R_A} \quad (1.9)$$

$$I_B = \frac{V_{BB}}{R_B}$$

and

$$\frac{V_{BB}}{R_{eq}} = \frac{V_{BB}}{R_A} + \frac{V_{BB}}{R_B}$$

$$\frac{1}{R_{eq}} = \frac{1}{R_A} + \frac{1}{R_B} \quad (1.10)$$

$$R_{eq} = \frac{1}{\frac{1}{R_A} + \frac{1}{R_B}} = \frac{1}{\frac{1}{100 \text{ kΩ}} + \frac{1}{150 \text{ kΩ}}} = 60 \text{ kΩ}$$

This is the expected result from Equation (1.8). R_{eq} is the equivalent resistance of R_A and R_B in parallel, which is notationly expressed as $R_{eq} = R_A \| R_B$. In general, for n resistances in parallel,

$$R_{eq} = (R_1 \| R_2 \| \cdots \| R_n) = \frac{1}{\frac{1}{R_1} + \frac{1}{R_2} + \cdots + \frac{1}{R_n}} \quad (1.11)$$

The following examples and self-exercises will help you to gain confidence on the concepts discussed.

1.2 VOLTAGE AND CURRENT LAWS

Self-Exercise 1.1

Calculate V_o and the voltage drop V_p across the parallel resistors in Figure 1.7. *Hint:* Replace the 250 kΩ and 180 kΩ resistors by their equivalent resistance and apply KVL to the equivalent circuit.

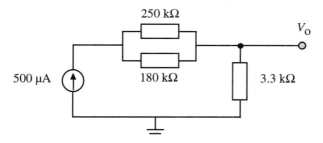

Figure 1.7.

Self-Exercise 1.2

Calculate R_3 in the circuit of Figure 1.8.

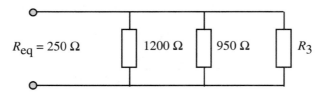

Figure 1.8. Equivalent parallel resistance.

When the number of resistors in parallel is two, Equation (1.11) reduces to

$$R_p = \frac{R_A \times R_B}{R_A + R_B} \qquad (1.12)$$

■ **EXAMPLE 1.2**

Calculate the terminal resistance of the resistors in Figures 1.9(a) and (b).

$$R_{eq} = 1 \text{ M}\Omega \| 2.3 \text{ M}\Omega$$
$$= \frac{(10^6)(2.3 \times 10^6)}{10^6 + 2.3 \times 10^6}$$
$$= 697 \text{ k}\Omega$$

10 CHAPTER 1 ELECTRICAL CIRCUIT ANALYSIS

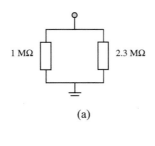

(a)

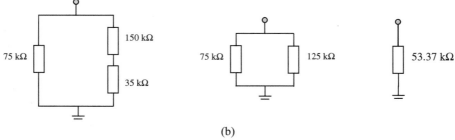

(b)

Figure 1.9. Parallel resistance calculations and equivalent circuits.

The equivalent resistance at the network in Figure 1.9(b) is found by combining the series resistors to 185 kΩ and then calculating the parallel equivalent:

$$R_{eq} = 75 \text{ k}\Omega \| 185 \text{ k}\Omega$$
$$= \frac{(75 \times 10^3)(185 \times 10^3)}{75 \times 10^3 + 185 \times 10^3}$$
$$= 53.37 \text{ k}\Omega$$

■

Self-Exercise 1.3

Calculate the resistance at the voltage source terminals R_{in}, I_{BB}, and V_0 at the terminals in Figure 1.10. If you are good, you can do this in your head.

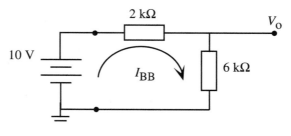

Figure 1.10.

Self-Exercise 1.4

Use Equations (1.11) or (1.12) and calculate the parallel resistance for circuits in Figures 1.11(a)–(d). Estimates of the terminal resistances for circuits in (a) and (b) should be done in your head. Circuits in (c) and (d) show that the effect of a large parallel resistance becomes negligible.

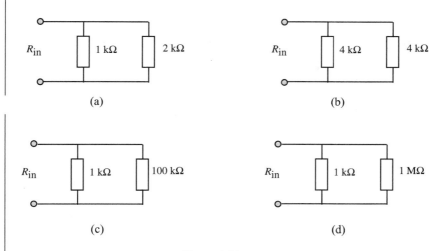

Figure 1.11.

Self-Exercise 1.5

Calculate R_{in}, I_{BB}, and V_0 in Figure 1.12. Estimate the correctness of your answer in your head.

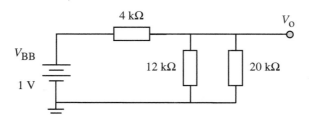

Figure 1.12.

Self-Exercise 1.6

(a) In Figure 1.13, find I_3 if $I_0 = 100$ μA, $I_1 = 50$ μA, and $I_2 = 10$ μA. (b) If $R_3 = 50$ kΩ, what are R_1 and R_2?

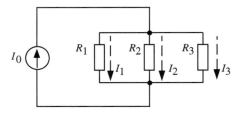

Figure 1.13.

Self-Exercise 1.7

Calculate R_1 and R_2 in Figure 1.14.

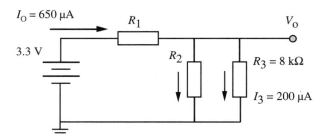

Figure 1.14.

Self-Exercise 1.8

If the voltage across the current source is 10 V in Figure 1.15, what is R_1?

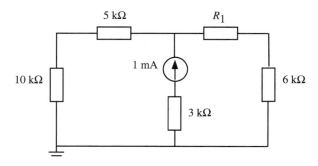

Figure 1.15.

Resistance Calculations by Inspection. A shorthand notation for the terminal resistance of networks allows for quick estimations and checking of results. The calculations are defined before computing occurs. Some exercises below will illustrate this. Solutions are given in the Appendix.

1.2 VOLTAGE AND CURRENT LAWS

Self-Exercise 1.9

Write the shorthand notation for the terminal resistance of the circuits in Figures 1.16(a) and (b).

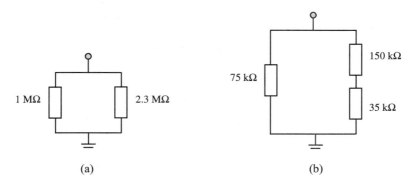

Figure 1.16.

Self-Exercise 1.10

Write the shorthand notation for the terminal resistance of the three circuits in Figure 1.17.

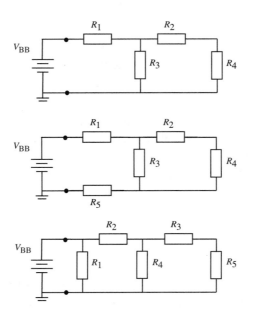

Figure 1.17. Terminal resistance using shorthand notation.

Self-Exercise 1.11

In the lower circuit of Figure 1.17, $R_1 = 20$ kΩ, $R_2 = 15$ kΩ, $R_3 = 25$ kΩ, $R_4 = 8$ kΩ, $R_5 = 5$ kΩ. Calculate R_{eq} for these three circuits.

Dividers. Some circuit topologies are repetitive and lend themselves to analysis by inspection. Two major inspection techniques use *voltage divider* and *current divider* concepts that take their analysis from the KVL and KCL. These are illustrated below with derivations of simple circuits followed by several examples and exercises. The examples have slightly more elements, but they reinforce previous examples and emphasize analysis by inspection.

Figure 1.18 shows a circuit with good visual voltage divider properties that we will illustrate in calculating V_3.

The KVL equation is

$$V_{BB} = I_{BB}(R_1) + I_{BB}(R_2) + I_{BB}(R_3) = I_{BB}(R_1 + R_2 + R_3)$$

$$I_{BB} = \frac{V_{BB}}{R_1 + R_2 + R_3} = \frac{V_3}{R_3} \tag{1.13}$$

and

$$V_3 = \frac{R_3}{R_1 + R_2 + R_3} V_{BB} \tag{1.14}$$

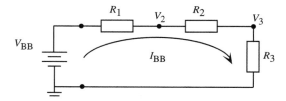

Figure 1.18. Voltage divider circuit.

Equation (1.14) is a shorthand statement of the voltage divider. It is written by inspection, and calculations follow. The voltage dropped by each resistor is proportional to their fraction of the whole series resistance. Figure 1.18 is very visual, and you should be able to write the voltage divider expression by inspection for any voltage drop. For example, the voltage from node V_2 to ground is

$$V_2 = \frac{R_2 + R_3}{R_1 + R_2 + R_3} V_{BB} \tag{1.15}$$

Self-Exercise 1.12

Use inspection and calculate the voltage at V_0 (Figure 1.19). Verify that the sum of the voltage drops is equal to V_{BB}. Write the input resistance R_{in} by inspection and calculate the current I_{BB}.

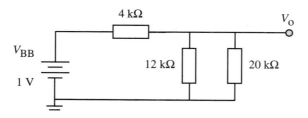

Figure 1.19. Voltage divider analysis circuit.

Self-Exercise 1.13

Write the expression for R_{in} at the input terminals, V_0, and the power supply current (Figure 1.20).

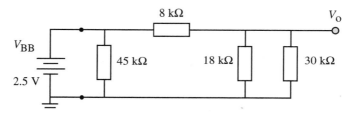

Figure 1.20.

Current divider expressions are visual, allowing you to see the splitting of current as it enters branches. Figure 1.21 shows two resistors that share total current I_{BB}.
KVL gives

$$V_{BB} = (R_1 \| R_2)I_{BB} = \frac{R_1 \times R_2}{R_1 + R_2} I_{BB} = (I_1)(R_1) = (I_2)(R_2) \tag{1.16}$$

then

$$I_1 = \frac{R_2}{R_1 + R_2} I_{BB} \quad \text{and} \quad I_2 = \frac{R_1}{R_1 + R_2} I_{BB} \tag{1.17}$$

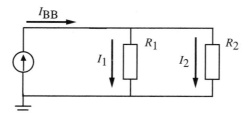

Figure 1.21. Current divider.

16 CHAPTER 1 ELECTRICAL CIRCUIT ANALYSIS

Currents divide in two parallel branches by an amount proportional to the opposite leg resistance divided by the sum of the two resistors. This relation should be memorized, as was done for the voltage divider.

Self-Exercise 1.14

Write the current expression by inspection and solve for currents in the 12 kΩ and 20 kΩ paths in Figure 1.22.

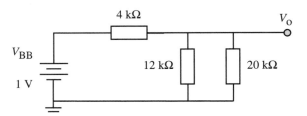

Figure 1.22.

Self-Exercise 1.15

(a) Write the current expression by inspection and solve for currents in all resistors in Figure 1.23, where $I_{BB} = 185.4$ µA. (b) Calculate V_{BB}.

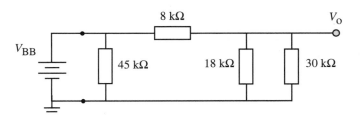

Figure 1.23.

Self-Exercise 1.16

(a) Solve for current in all resistive paths in Figure 1.24 using the technique of inspection. (b) Calculate a new value for the 20 kΩ resistor so that its current is 5 µA.

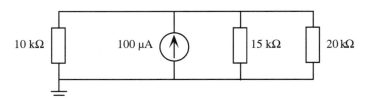

Figure 1.24.

Self-Exercise 1.17

In Figure 1.25, calculate V_0, I_2, and I_9.

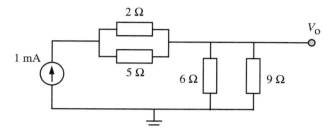

Figure 1.25.

Self-Exercise 1.18

(a) Write R_{in} between the battery terminals by inspection and solve (Figure 1.26).
(b) Write the $I_{1.5k}$ expression by inspection and solve. This is a larger circuit, but it presents no problem if we adhere to the shorthand style. We write R_{in} between battery terminals by inspection, and calculate $I_{1.5k}$ by current divider inspection.

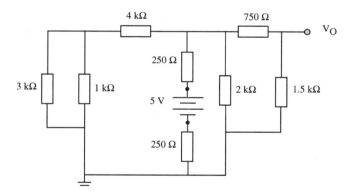

Figure 1.26.

1.3 CAPACITORS

Capacitors appear in CMOS digital circuits as parasitic elements intrinsic to transistors or with the metals used for interconnections. They have an important effect on the time for a transistor to switch between on and off states, and also contribute to propagation delay between gates due to interconnection capacitance. Capacitors also cause a type of noise called cross-talk. This appears especially in high-speed circuits, in which the voltage at one interconnection line is affected by another interconnection line that is isolated but located close to it. Cross-talk is discussed in later chapters.

The behavior and structure of capacitors inherent to interconnection lines are significantly different from the parasitic capacitors found in diodes and transistors. We introduce ideal parallel plate capacitors that are often used to model wiring capacitance. Capacitors inherent to transistors and diodes act differently and are discussed later.

A capacitor has two conducting plates separated by an insulator, as represented in Figure 1.27(a). When a DC voltage is applied across the conducting plates (terminals) of the capacitor, the steady-state current is zero since the plates are isolated by the insulator. The effect of the applied voltage is to store charges of opposite sign at the plates of the capacitor.

The capacitor circuit symbol is shown in Figure 1.27(b). Capacitors are characterized by a parameter called capacitance (C) that is measured in Farads. Strictly, capacitance is defined as the charge variation ∂Q induced in the capacitor when voltage is changed by a quantity ∂V, i.e.,

$$C = \frac{\partial Q}{\partial V} \tag{1.18}$$

This ratio is constant in parallel plate capacitors, independent of the voltage applied to the capacitor. Capacitance is simply the ratio between the charge stored and the voltage applied, i.e., $C = Q/V$, with units of Coulombs per Volt called a Farad. This quantity can also be computed from the geometry of the parallel plate and the properties of the insulator used to construct it. This expression is

$$C = \frac{\varepsilon_{ins} A}{d} \tag{1.19}$$

where ε_{ins} is an inherent parameter of the insulator, called permittivity, that measures the resistance of the material to an electric field; A is the area of the plates used to construct the capacitor; and d the distance separating the plates.

Although a voltage applied to the terminals of a capacitor does not move net charge through the dielectric, it can displace charge within it. If the voltage changes with time, then the displacement of charge also changes, causing what is known as displacement current, that cannot be distinguished from a conduction current at the capacitor terminals. Since this current is proportional to the rate at which the voltage across the capacitor changes with time, the relation between the applied voltage and the capacitor current is

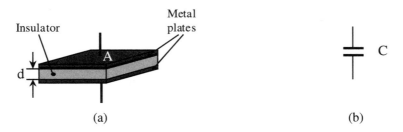

Figure 1.27. (a) Parallel plate capacitor. (b) Circuit symbol.

$$i = C\frac{dV}{dt} \qquad (1.20)$$

If the voltage is DC, then $dV/dt = 0$ and the current is zero. An important consequence of Equation (1.20) is that the voltage at the terminals of a capacitor cannot change instantaneously, since this would lead to an infinite current. That is physically impossible. In later chapters, we will see that any logic gate constructed within an IC has a parasitic capacitor at its output. Therefore, the transition from one voltage level to another will always have a delay time since the voltage output cannot change instantaneously. Trying to make these output capacitors as small as possible is a major goal of the IC industry in order to obtain faster circuits.

1.3.1 Capacitor Connections

Capacitors, like resistors, can be connected in series and in parallel. We will show the equivalent capacitance calculations when they are in these configurations.

Capacitors in parallel have the same terminal voltage, and charge distributes according to the relative capacitance value differences (Figure 1.28(a)). The equivalent capacitor is equal to the sum of the capacitors:

$$C_1 = \frac{Q_1}{V}, \qquad C_2 = \frac{Q_2}{V}$$

$$C_1 + C_2 = \frac{Q_1}{V} + \frac{Q_2}{V} = \frac{Q_1 + Q_2}{V} \qquad (1.21)$$

$$C_{eq} = \frac{Q_{eq}}{V}$$

where $C_{eq} = C_1 + C_2$, and $Q_{eq} = Q_1 + Q_2$. Capacitors connected in parallel simply add their values to get the equivalent capacitance.

Capacitors connected in series have the same charge stored, whereas the voltage depends on the relative value of the capacitor (Figure 1.28(b)). In this case, the expression for the equivalent capacitor is analogous to the expression obtained when connecting resistors in parallel:

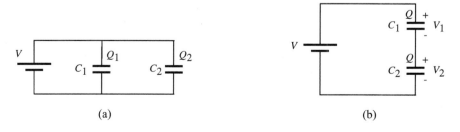

Figure 1.28. Capacitance interconnection. (a) Parallel. (b) Series.

CHAPTER 1 ELECTRICAL CIRCUIT ANALYSIS

$$C_1 = \frac{Q}{V_1}, \quad C_2 = \frac{Q}{V_2}$$

$$\frac{1}{C_1} + \frac{1}{C_2} = \frac{V_1}{Q} + \frac{V_2}{Q} = \frac{V_1 + V_2}{Q} \tag{1.22}$$

$$C_{eq} = \frac{Q}{V_{eq}}$$

where $V_{eq} = V_1 + V_2$, and

$$C_{eq} = \frac{C_1 C_2}{C_1 + C_2}$$

■ EXAMPLE 1.3

In Figures 1.28(a) and (b), $C_1 = 20$ pF and $C_2 = 60$ pF. Calculate the equivalent capacitance seen by the voltage source.

(a) $\quad C_{eq} = C_1 + C_2 = 20$ pF $+ 60$ pF $= 80$ pF

(b) $\quad C_{eq} = \dfrac{1}{\dfrac{1}{C_1} + \dfrac{1}{C_2}} = \dfrac{1}{1/20 \text{ pF} + 1/60 \text{ pF}} = 15$ pF

■

Self-Exercise 1.19

Calculate the terminal equivalent capacitance for the circuits in Figure 1.29.

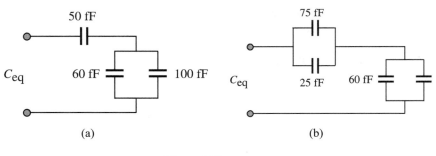

Figure 1.29.

1.3.2 Capacitor Voltage Dividers

There are open circuit defect situations in CMOS circuits in which capacitors couple voltages to otherwise unconnected nodes. This simple connection is a capacitance voltage di-

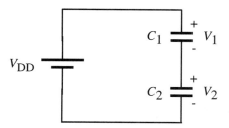

Figure 1.30. Capacitance voltage divider.

vider circuit (Figure 1.30). The voltage across each capacitor is a fraction of the total voltage V_{DD} across both terminals.

■ **EXAMPLE 1.4**

Derive the relation between the voltage across each capacitor C_1 and C_2 in Figure 1.30 to the terminal voltage V_{DD}.

The charge across the plates of the series capacitors is equal so that $Q_1 = Q_2$. The capacitance relation $C = Q/V$ allows us to write

$$Q_1 = Q_2 \quad C_1 V_1 = C_2 V_2$$

or

$$V_2 = \frac{C_1}{C_2} V_1$$

Since

$$V_{DD} = V_1 + V_2$$

then

$$V_2 = V_{DD} - V_1 = \frac{C_1}{C_2} V_1$$

Solve for

$$V_1 = \frac{C_2}{C_1 + C_2} V_{DD}$$

and get

$$V_2 = \frac{C_1}{C_1 + C_2} V_{DD}$$

The form of the capacitor divider is similar to the resistor voltage divider except the numerator term differs. ■

Self-Exercise 1.20

Solve for V_1 and V_2 in Figure 1.31.

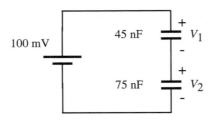

Figure 1.31.

Self-Exercise 1.21

If $V_2 = 700$ mV, what is the driving terminal voltage V_D in Figure 1.32?

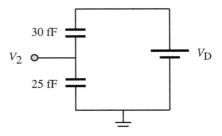

Figure 1.32.

1.3.3 Charging and Discharging Capacitors

So far we have discussed the behavior of circuits with capacitors in the steady state, i.e., when DC sources drive the circuit. In these cases, the analysis of the circuit is done assuming that it reached a stationary state. Conceptually, these cases are different from situations in which the circuit source makes a sudden transition, or a DC source is applied to a discharged capacitor through a switch. In these situations, there is a period of time during which the circuit is not in a stationary state but in a transient state. These cases are important in digital CMOS ICs, since node state changes in ICs are transient states determining the timing and power characteristics of the circuit. We will analyze charge and discharge of capacitors with an example.

EXAMPLE 1.5

In the circuit of Figure 1.33, draw the voltage and current evolution at the capacitor with time starting at $t = 0$ when the switch is closed. Assume $V_{in} = 5$ V and that the capacitor is initially at 0 V.

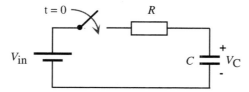

Figure 1.33.

The Kirchoff laws for current and voltage can be applied to circuits with capacitors as we did with resistors. Thus, once the switch is closed, the KVL must follow at any time:

$$V_{in} = V_R + V_C$$

The Kirchoff current law applied to this circuit states that the current through the resistor must be equal to the current through the capacitor, or

$$\frac{V_R}{R} = C\frac{dV_C}{dt}$$

Using the KVL equation, we can express the voltage across the resistor in terms of the voltage across the capacitor, obtaining

$$\frac{V_{in} - V_C}{R} = C\frac{dV_C}{dt}$$

This equation relates the input voltage to the voltage at the capacitor. The solution gives the time evolution of the voltage across the capacitor,

$$V_C = V_{in}(1 - e^{-t/RC})$$

The current through the capacitor is

$$I_C = I_R = \frac{V_{in} - V_C}{R}$$

$$I_C = \frac{V_{in}}{R}e^{-t/RC}$$

At $t = 0$, the capacitor voltage is zero (it is discharged) and the current is maximum (the voltage drop at the resistor is maximum), whereas in DC (for $t \to \infty$) the capacitor voltage is equal to the source voltage and the current is zero. This example shows that the voltage evolution is exponential when charging a capaci-

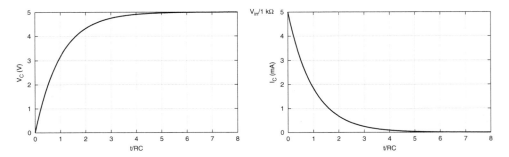

Figure 1.34.

tor through a resistor. The time constant is defined for $t = RC$, that is, the time required to charge the capacitor to $(1 - e^{-1})$ of its final value, or 63%. This means that the larger the value of the resistor or capacitor, the longer it takes to charge/discharge it (Figure 1.34). ∎

1.4 DIODES

A circuit analysis of the semiconductor diode is presented below; later chapters discuss its physics and role in transistor construction. Diodes do not act like resistors; they are nonlinear. Diodes pass significant current at one voltage polarity and near zero current for the opposite polarity. A typical diode nonlinear current–voltage relation is shown in Figure 1.35(a) and its circuit symbol in Figure 1.35(b). The positive terminal is called the anode, and the negative one is called the cathode. The diode equation is

$$I_D = I_S(e^{\frac{qV_D}{kT}} - 1) \tag{1.23}$$

where k is the Boltzmann constant ($k = 1.38 \times 10^{-23}$ J/K), q is the charge of the electron ($q = 1.6 \times 10^{-19}$ C), and I_S is the reverse biased current. The quantity kT/q is called the thermal voltage (V_T) whose value is 0.0259 V at T = 300 K; usually, we use $V_T = 26$ mV at that

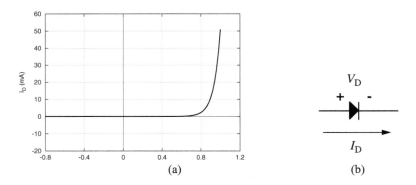

Figure 1.35. (a) Diode I–V characteristics and (b) symbol.

temperature. When the diode applied voltage is positive and well beyond the thermal voltage ($V_D \gg V_T = kT/q$), Equation (1.23) becomes

$$I_D = I_S e^{\frac{qV_D}{kT}} \tag{1.24}$$

The voltage across the diode can be solved from Equation (1.23) as

$$V_D = \frac{kT}{q} \ln\left(\frac{I_D}{I_S} + 1\right) \tag{1.25}$$

For forward bias applications $I_D/I_S \gg 1$ and this reduces to

$$V_D = \frac{kT}{q} \ln \frac{I_D}{I_S} \tag{1.26}$$

> *Self-Exercise 1.22*
>
> (a) Calculate the forward diode voltage if $T = 25°C$, $I_D = 200$ nA, and $I_S = 1$ nA. Compute from Equation (1.25). (b) At what current will the voltage drop be 400 mV?

Diode Equations (1.23)–(1.26) are useful in their pure form only at the temperature at which I_S was measured. These equations predict that I_D will exponentially drop as temperature rises which is not so. I_S is more temperature-sensitive than the temperature exponential and doubles for about every 10°C rise. The result is that diode current markedly increases as temperature rises.

1.4.1 Diode Resistor Circuits

Figure 1.36 shows a circuit that can be solved for all currents and node element voltages if we know the reverse bias saturation current I_S.

■ **EXAMPLE 1.6**

If $I_S = 10$ nA at room temperature, what is the voltage across the diode in Figure 1.36 and what is I_D? Let $kT/q = 26$ mV.

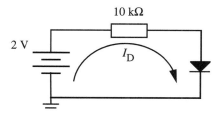

Figure 1.36. Forward-biased diode analysis.

Write KVL using the diode voltage expression:

$$2\text{ V} = I_D(10\text{ k}\Omega) + (26\text{ mV})\ln\left(\frac{I_D}{I_S} + 1\right)$$

This equation has one unknown (I_D), but it is difficult to solve analytically, so an iterative method is easiest. Values of I_D are substituted into the equation, and the value that balances the LHS and RHS is a close approximation. A starting point for I_D can be estimated from the upper bound on I_D. If $V_D = 0$, then $I_D = 2\text{ V}/10\text{ k}\Omega = 200\text{ }\mu\text{A}$. I_D cannot be larger than 200 µA. A close solution is $I_D = 175\text{ }\mu\text{A}$.

The diode voltage is

$$V_D = \frac{kT}{q}\ln\frac{I_D}{I_S}$$

$$= 26\text{ mV} \times \ln\frac{175\text{ uA}}{10\text{ nA}} = 244.2\text{ mV}$$

∎

Self-Exercise 1.23

Estimate I_D and V_D in Figure 1.37 for $I_S = 1$ nA.

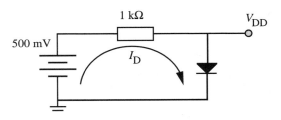

Figure 1.37.

■ EXAMPLE 1.7

Figure 1.38 shows two circuits with the diode cathode connected to the positive terminal of a power supply ($I_S = 100$ nA). What is V_o in both circuits?

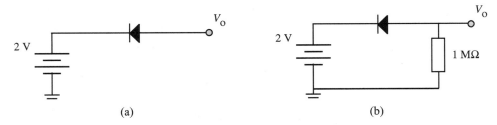

Figure 1.38.

Figure 1.38(a) has a floating node at V_0 so there is no current in the diode. Since $I_D = 0$, the diode voltage drop $V_D = 0$ and

$$V_0 = V_D + 2\text{ V} = 2\text{ V}$$

Figure 1.38(b) shows a current path to ground. The diode is reversed-biased and $I_{BB} = -I_D = 100$ nA. Then

$$V_0 = I_{BB} \times 1\text{ M}\Omega = 100\text{ nA} \times 1\text{ M}\Omega = 100\text{ mV}$$

■

Both problems in Example 1.7 can be analyzed using Equations (1.23) to (1.26) or observing the process in the *I–V* curve of Figure 1.35. In Figure 1.38(a), the operating point is at the origin. In Figure 1.38(b), it has moved to the left of the origin.

Self-Exercise 1.24

The circuit in Figure 1.39 is similar to IC protection circuits connected to the input pins of an integrated circuit. The diodes protect the logic circuit block when input pin (pad) voltages are accidentally higher than the power supply voltage (V_{DD}) or lower than the ground voltage. If $V_{PAD} > 5$ V, then diode D_2 turns on and bleeds charge away from the input pin. The same process occurs through diode D_1 if the input pad voltage becomes less than ground (0 V). An integrated circuit tester evaluates the diodes by forcing current (100 μA) and measuring the voltage. If the protection circuit is damaged, an abnormal voltage is usually read at the damaged pin.

(a) If diode reverse bias saturation current is $I_S = 100$ nA, what is the expected input voltage measured if the diodes are good and R_1 and R_2 are small? Apply ±100 μA to assess both diodes.

(b) If the upper diode has a dead short across it, what is V_{IN} when the test examines the upper diode?

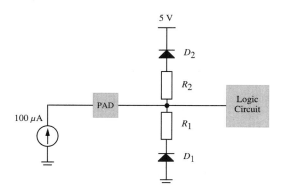

Figure 1.39.

Self-Exercise 1.25

Calculate V_0 and V_{D1} in Figure 1.40, where $I_S = 100$ μA and $T = 25$ °C.

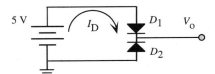

Figure 1.40.

1.4.2 Diode Resistance

Although diodes do not obey Ohm's law, a small signal variation in the forward bias can define a resistance from the slope of the I–V curve. The distinction with linear elements is important as we cannot simply divide a diode DC voltage by its DC current. That result is meaningless.

The diode curve is repeated and enlarged in Figure 1.41. If a small signal variation $v(t)$ is applied in addition to the DC operating voltage V_{DC}, then the exponential current/voltage characteristic can be approximated to a line [given that $v(t)$ is small enough] and an equivalent resistance for that bias operation can be defined. Bias point 1 has a current change with voltage that is larger than that of bias point 2; therefore, the dynamic resistance of the diode is smaller at bias point 1. Note that this resistance value depends strongly on the operating voltage bias value. Each point on the curve has a slope in the forward bias. Dividing the DC voltage by the current, V_1/I_1, is not the same as V_2/I_2. Therefore, these DC relations are meaningless. However, dynamic resistance concepts are important in certain transistor applications.

This concept is seen in manipulation of the diode equation, where the forward-biased dynamic resistance r_d is

$$\frac{1}{r_d} = \frac{dI_D}{dV_D} = \frac{d[I_S e^{qV_D/kT}]}{dV_D} = \frac{qI_S e^{qV_D/kT}}{kT} = \frac{qI_D}{kT} \qquad (1.27)$$

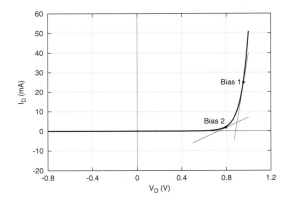

Figure 1.41.

The diode dynamic resistance is

$$r_d = \frac{dV_D}{dI_D} = \frac{kT}{qI_D} \qquad (1.28)$$

or at room temperature

$$r_d \approx \frac{26 \text{ mV}}{I_D} \qquad (1.29)$$

> *Self-Exercise 1.26*
>
> Find the diode dynamic resistance at room temperature for $I_D = 1$ μA, 100 μA, 1 mA, and 10 mA.

How do we use the concept of dynamic diode resistance? A diode can be biased at a DC current, and small changes about that operating point have a resistance. A small sinusoid voltage (v_D) causes a diode current (i_D) change equal to v_D/r_D. The other important point is that you cannot simply divide DC terminal voltage by DC terminal current to calculate resistance. This is true for diodes and also for transistors, as will be seen later.

1.5 SUMMARY

This chapter introduced the basic analysis of circuits with power supplies, resistors, capacitors, and diodes. Kirchhoff's current and voltage laws were combined with Ohm's law to calculate node voltages and element currents for a variety of circuits. The technique of solving for currents and voltages by inspection is a powerful one because of the rapid insight into the nature of circuits it provides. Finally, the section on diodes illustrated analysis with a nonlinear element. The exercises at the end of the chapter should provide sufficient drill to prepare for subsequent chapters, which will introduce the MOSFET transistor and its simple configurations.

BIBLIOGRAPHY

1. R. C. Dorf and J. A. Svoboda, *Introduction to Electric Circuits,* 4th ed., Wiley, 1998.
2. D. E. Johnson, J. R. Johnson, J. L. Hilburn, and P. Scott, *Electric Circuit Analysis,* 3rd ed., Prentice-Hall, 1989.
3. J. W. Nilsson and S. A. Riedel, *Electric Circuits,* 6th ed., Prentice-Hall, 2000.
4. A. J. Rosa and R. E. Thomas, *The Analysis and Design of Linear Circuits,* 4th ed., Wiley, 2003.

EXERCISES

1.1. Write the shorthand expression for R_{eq} at the open terminals in Figure 1.42.

1.2. Write the shorthand expression for R_{eq} at the open terminals in Figure 1.43.

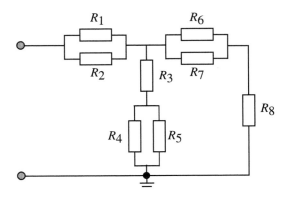

Figure 1.42.

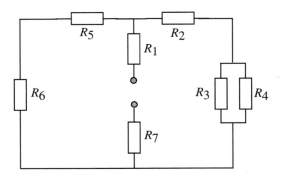

Figure 1.43.

1.3. For the circuit in Figure 1.44, (a) calculate V_0; (b) calculate I_{2M}.

1.4. Calculate V_0 by first writing a voltage divider expression and then solving for V_0 (Figure 1.45a and b).

1.5. Write the shorthand notation for current I_2 in resistor R_2 in Figure 1.46 as a function of driving current I.

1.6. For the circuit in Figure 1.47, (a) solve for V_0 using a voltage divider expression; (b) solve for I_{2K}; (c) solve for I_{900}.

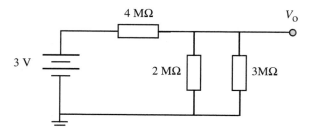

Figure 1.44.

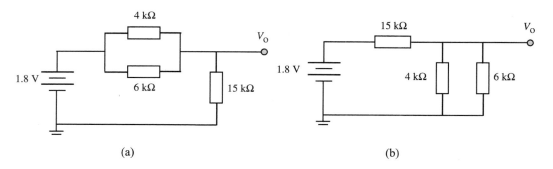

(a) (b)

Figure 1.45.

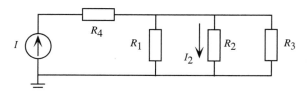

Figure 1.46.

32 CHAPTER 1 ELECTRICAL CIRCUIT ANALYSIS

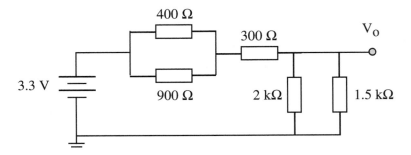

Figure 1.47.

1.7. Use the circuit analysis technique by inspection, and write the shorthand expression to calculate I_{2K} for Figure 1.48.

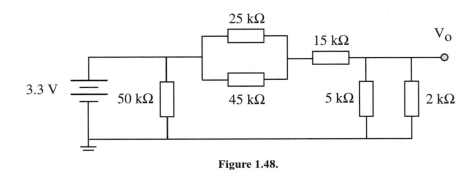

Figure 1.48.

1.8. Given the circuit in Figure 1.49, (a) write the expression for I_{450} and solve; (b) write the expression for V_{800}; (c) show that $I_{800} + I_{400} = 2$ mA.

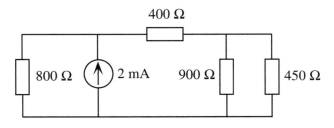

Figure 1.49.

1.9. Find I_{6k} in Figure 1.50. *Hint:* when we have two power supplies and a linear (resistive) network, we solve in three steps.

1. Set one power supply to 0 V and calculate current in the 6 kΩ resistor from the nonzero power supply.
2. Reverse the role and recalculate I_{6k}.
3. The final answer is the sum of the two currents.

This is known as the superposition theorem and can be applied only for linear elements.

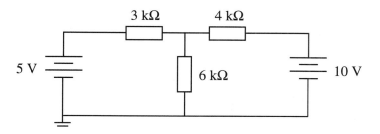

Figure 1.50.

1.10. Find the equivalent capacitance at the input nodes in Figure 1.51.

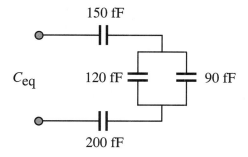

Figure 1.51.

1.11. Find C_1 in Figure 1.52.

1.12. Solve for I_D and V_D in Figure 1.53, where the diode has the value $I_S = 1$ μA.

1.13. Calculate V_0 in Figure 1.54, given that the reverse-bias saturation current $I_S = 1$ nA, and you are at room temperature.

34 CHAPTER 1 ELECTRICAL CIRCUIT ANALYSIS

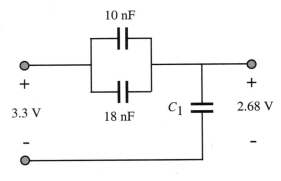

Figure 1.52.

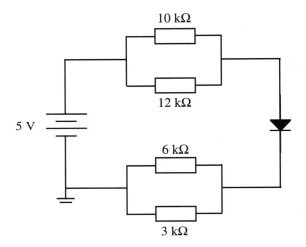

Figure 1.53.

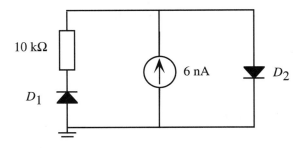

Figure 1.54.

1.14. Diode D_1 in Figure 1.55 has a reverse-bias saturation current of $I_{01} = 1$ nA, and diode D_2 has $I_{02} = 4$ nA. At room temperature, what is V_0?

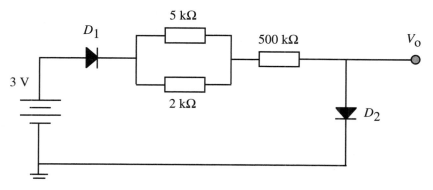

Figure 1.55.

1.15. Calculate the voltage across the diodes in Figure 1.56, given that the reverse-bias saturation current in D_1 is $I_{01} = 175$ nA and $I_{02} = 100$ nA.

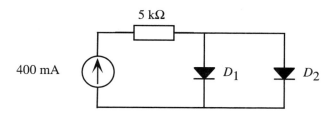

Figure 1.56.

CHAPTER 2

SEMICONDUCTOR PHYSICS

2.1 SEMICONDUCTOR FUNDAMENTALS

2.1.1 Metals, Insulators, and Semiconductors

Materials are classified by their physical properties. Conductivity measures the amount of current through a material when a voltage is applied, and materials are classified into three conductivity groups: metals, insulators, and semiconductors. Metals present practically no resistance to carrier flow, whereas insulators allow virtually no electrical carrier flow under an applied voltage. Semiconductors are unique, and can behave as conductors or insulators. This chapter introduces the principles of semiconductor physics, using only a few equations and numerical examples. Emphasis is on providing a common understanding of these principles and a basic description of how the devices work. The language and visual and mathematical models of semiconductor physics permeate CMOS manufacturing.

The classic explanation for conduction differences between these materials uses the energy band model of solids that derives from quantum mechanics. Neils Bohr found that electrons in an atom could not have arbitrary energy values, but had defined, discrete (quantum) energy values. A basic principle of quantum mechanics is that energy is distributed in quantum packets, and cannot take continuous values. Electrons orbit at discrete distances from the nucleus. Figure 2.1(a) shows the allowed energy levels of a hydrogen atom electron. These energy levels are those that the electron can take that are discrete. The s and p energy level symbols are taken from quantum mechanical convention.

In a two-atom system, each energy level in the single atom system splits into two sublevels, as shown in Figure 2.1(b). When more atoms are added to construct a crystalline solid, the energy levels successively split, leading to the picture in Figure 2.1(c), where

CMOS Electronics: How It Works, How It Fails. By Jaume Segura and Charles F. Hawkins
ISBN 0-471-47669-2 © 2004 Institute of Electrical and Electronics Engineers

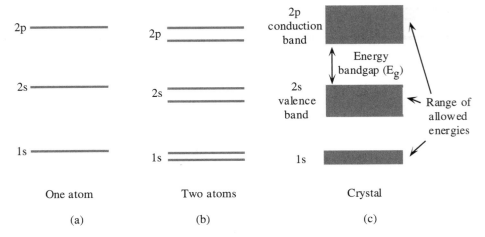

Figure 2.1. (a) Energy levels in a single atom, (b) two atoms, and (c) a solid.

energy bands separated by gaps of forbidden energies (called band gaps) replace single energy levels. The band gap width depends on the type of atom used to build the solid, and it determines the conductive properties of the material.

Energy bands have different conduction properties. The outermost energy band is called the conduction band, and the next-lower one is called the valence band or outer shell. An electron having an energy corresponding to the conduction band is not tied to any atom, and can move "freely" through the solid. Such an electron contributes to current when a voltage is applied. An electron in the valence band has an energy that is attached to an atom of the solid, and is not "free" to move within the solid when a voltage is applied.

Energy bands help us more easily understand the conductive properties of different materials. Figure 2.2(a) shows the energy bands of a metal; the lowest energy value of the conduction band is below the maximum energy of the valence band. This means that the

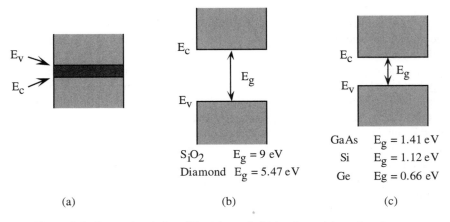

Figure 2.2. Energy bands in solids: (a) metal, (b) insulator, (c) semiconductor.

metal conduction band has an abundance of electrons that are available for conduction. It takes very little additional energy to move an electron from the valence to the conduction band since the bands are merged.

The energy band structure of an insulator is shown in Figure 2.2(b). Insulators have a large energy gap between the valence and conduction bands. The thermal energy needed for an electron to go from the valence to the conduction band is so high that only a few electrons within the material can acquire such an energy and jump over the gap. A voltage applied to the material will cause almost no current since virtually all electrons are tied to atoms in the valence band.

Semiconductors are the third class of conducting material, and show an intermediate behavior. The valence and conduction bands are not merged, but the energy gap is small enough so that some electrons are energetic enough to jump across it. The energy needed for electrons to jump across the gap comes from the ambient temperature or photon energy. We will deal with thermal energy since most integrated circuits are sealed and admit no light from the environment. The process of gaining enough thermal energy and jumping from the valence to the conduction band is inherently statistical. This means that electrons in a solid are continuously moving up and down between the valence and conduction bands. However, at a given temperature there is a population of electrons in the conduction band that contribute to the current.

Since the energy used by an electron to jump the gap is thermal, the population of electrons in the conduction band depends on the temperature. At absolute zero temperature, there is no thermal energy in a pure semiconductor, so that no electron has enough energy to jump across the gap. As temperature increases, the number of conducting electrons increases.

The differences in gap energies between insulators and semiconductors are related to how electrons are arranged within atoms. Electrons are grouped into layers around the atomic nucleus, and electrons in the internal layers cannot be separated from the nucleus. Only electrons from the outside valence layer may jump from their bounded valence state to the conducting state (free from the attractive forces of the atomic nucleus). Atoms of conducting materials have several layers of orbiting electrons. The number of electrons required to fill a given layer remains constant and independent of the atomic element. An atom having all layers completely filled will have all electrons (even those at the outmost layer) strongly "tied" to the nucleus. A large amount of energy is needed to break such a layer and take one electron out of the atom. These atoms are known as noble gases, since they do not react with other elements, as their electrons are closely packed. Atoms in which the external layer is not "closed" (more electrons are required to completely fill such layers) have electrons more lightly attached to the atom. As a result, only a small amount of energy is needed to separate an external electron from the atom. This is the case in metals; the outside electrons belong to the solid instead of being attached to some nucleus.

2.1.2 Carriers in Semiconductors: Electrons and Holes

The only contribution to current in a metal is from electrons in the conducting band. In contrast, semiconductor current has two contributions: one from electrons in the conduction band and the other from electron vacancies in the valence band caused by electrons that jumped into the conduction band. The vacancy of an electron in the valence band leaves an empty local charge space of a value equal to the electron charge, but of opposite sign. Therefore, it has a net charge $+q$ (q being the charge value of an electron; $q = -1.6 \times 10^{-19}$ Coulomb).

The most common semiconductor material for ICs is silicon, that has four electrons in its outer energy band (Figure 2.3(a)). When silicon is crystalline, each electron of the valence layer is shared with one electron of a neighbor atom so that by sharing, each Si atom has eight outer shell electrons. If a valence electron gains enough thermal energy to jump into the conduction band, it leaves a vacancy position. Such a position is available to another valence electron to move into and leave a vacancy at its original site. This process can now be repeated for a third valence electron moving into this last vacancy, and so on. It is important to note that this process does not require the moving valence electron to go to the conduction band to move to such a vacancy. The electron vacancy can be seen as a "particle" of positive charge that moves in the opposite direction to the valence electron. Such a virtual particle with associated charge $+q$ is called a "hole." When silicon is constructed as a crystal, it behaves as a semiconductor.

When the semiconductor is in equilibrium and there is no external electromagnetic field or temperature gradient, then electrons and holes move randomly in space, and no net current is observed. When an electric field is applied, the hole movement is not random, but drifts in the same direction as the field. This gives a net current contribution from holes, in addition to the current contribution from electrons in the conduction band. These dual conduction mechanisms in solids are detailed in the next section.

In a pure silicon material, electrons and holes are created in pairs. The "creation" of a free electron jumping into the conduction band creates a hole in the valence band, whereas an electron dropping from the conduction to the valence band implies that a hole disappears. When an electron jumps from the valence to the conduction band, the process is called an electron–hole pair creation, and when an electron jumps back from the conduction band to the valence band, the process is referred to as electron–hole recombination, or simply recombination. Energy is needed for electron–hole pair formation, but in the opposite process, energy is released when an electron recombines with a hole. This is illustrated in Figure 2.4 for the energy band gap model of a semiconductor and its solid-state physical representation. It emphasizes that mobile carriers are electrons in the conduction band and holes in the valence band. The process of electron–hole creation and

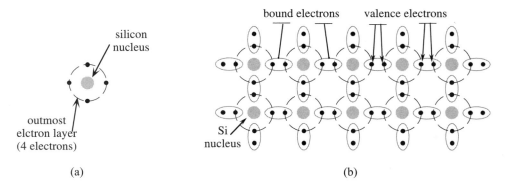

Figure 2.3. (a) Representation of a silicon atom with its four electrons at the outmost layer, and (b) picture of the silicon structure in a crystalline solid.

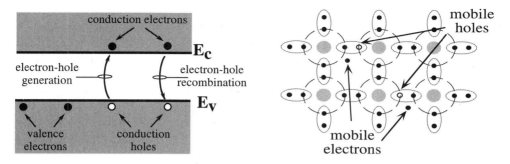

Figure 2.4. Electron-hole pair creation and recombination in the band gap model (left), and its representation in the solid (right).

recombination is inherently statistical—electron–hole pairs are continuously created and recombined in a semiconductor at a given temperature.

2.1.3 Determining Carrier Population*

We will refer to electrons as the carriers in the conduction band and holes as the carriers in the valence band. Electrons and holes are mobile carriers, and we must estimate their populations at each temperature that are statistically available in each band. Fermi–Dirac statistics accounts for the basic properties of electrons in solids, and gives us a mathematical model of the electron statistics for the concentration of carriers. Electrons have a duality property, acting as particles or as indistinguishable waves, and follow the Pauli principle of exclusion.† Since electrons are quantum elements, we describe the probability of an electron being at a given energy for a given temperature. The probability of an electron at a given energy level E is expressed with a probability function $f(E)$, which may be derived from the Fermi–Dirac statistics (see [2] or [1]), obtaining

$$f(E) = \frac{1}{1 + e^{(E-E_F)/kT}} \quad (2.1)$$

where k is the Boltzmann constant (1.38×10^{-23} J/K), T is temperature in Kelvin, and E_F is known as the Fermi energy level, an important parameter for semiconductors. The Fermi energy level is the energy at which the probability function equals 0.5 (this can be easily verified by substituting $E = E_F$ in the equation), or the energy level below which the probability function is 1 for $T = 0$ K. At absolute zero temperature, all possible energy levels are filled. If $f_e(E)$ is the probability function of an electron being at a given energy (E), then holes are the "dual" or complementary particles. The probability of a given hole being at such an energy is $f_h(E) = 1 - f_e(E)$.

Since we are describing electrons in a solid, we must account for the number of available energy states. For example, no state is available within the forbidden energy gap. The

---

*This subsection can be skipped without loss of continuity.
†The Pauli principle of exclusion states that two quantum objects (electrons in this case) cannot occupy the same energy levels. As a result, the maximum number of electrons per layer in a given atom is fixed.

formulation of the number of states $N(E)$ available for an electron at a given energy in a solid allows us to find the concentration of electrons n having an energy interval dE as

$$n = f(E)N(E)dE \tag{2.2}$$

The electron concentration within the conduction band is found by integrating such a concentration from $E = E_C$ to $E \to \infty$. Thus,

$$n_0 = \int_{E_C}^{\infty} f(E)N(E)dE \tag{2.3}$$

where n_0 stands for the number of carriers in equilibrium. Similarly, the hole concentration is

$$h_0 = \int_{-\infty}^{E_v} [1 - f(E)]N(E)dE \tag{2.4}$$

A detailed derivation for $N(E)$ and Equations (2.3) and (2.4) are beyond the scope of this work. For a detailed analysis we refer to any of the books cited at the end of the chapter.

2.2 INTRINSIC AND EXTRINSIC SEMICONDUCTORS

The previous concepts applied to "pure" semiconductors, in which all atoms are of the same type. This is referred to as an *intrinsic semiconductor*, implying that the number of electrons is equal to the number of holes, since they are generated in pairs. *Extrinsic semiconductors* are created by intentionally adding impurities* to the semiconductor to increase the concentration of one carrier type (electrons or holes) without increasing the concentration of the other, thus breaking the symmetry between the number of electrons and holes. The intentional substitution of a silicon atom by another element is called *doping*. This process depends on the type of impurity added and the number of impurities introduced in a unit volume. There are *n*-type impurities that increase electron concentration and *p*-type impurities that increase hole concentration. The number of impurities (atoms/unit volume) injected into the solid is much less than the number of silicon atoms, and the crystalline structure of the semiconductor is not globally disturbed.

2.2.1 *n*-Type Semiconductors

Silicon has four electrons in its outer layer that form bonds with neighbor atoms. We can increase the electron concentration without changing the hole concentration by replacing some silicon atoms with Periodic Table Group V atoms having five electrons in their external layer (arsenic, phosphorus, or antimony). Four of the five external electrons of the substituting atom create bonds with the neighboring silicon atoms, while the fifth one is almost free to move within the solid (Figure 2.5). At room temperature, the thermal energy is enough to activate such a fifth electron and move it into the conduction band.

---

*Any material, no matter what its quality, always has unintended impurities. The effect of such impurities can be neglected if they are kept to a minimum. Additionally, crystalline solids are not perfect crystals and may have some "irregularities" that impact the energy band structure.

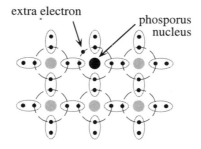

Figure 2.5. Adding a donor atom creates a mobile electron without creating a hole.

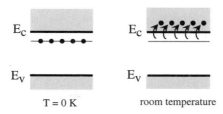

Figure 2.6. Picture of the donor effect in the band-gap energy model with change in temperature.

The electron jumps into the conducting energy band without creating a hole, so no electron–hole pairs are generated, only free electrons. In this case, each impurity atom is called a *donor*, since it implies that an extra electron is donated to the semiconductor. Silicon is a Group IV atom in the Periodic Table and donor atoms come from the Group V atoms.

When extrinsic silicon is doped with donors, the number of conducting electrons is approximately equal to the number of donor atoms injected (N_D) plus some electrons coming from electron–hole pair creation. When donor concentration greatly exceeds the normal intrinsic population of carriers, then the conducting electron concentration is essentially that of the donor concentration. By adding a specific concentration of donors, the population of electrons can be made significantly higher than that of holes. An extrinsic semiconductor doped with donor impurities is called an *n*-type semiconductor.

When a considerable number of donor impurities are added (10^{15}–10^{17} atoms per cm^3), the effect on the bandgap model is creation of an allowed energy level within the gap close to the E_c (Figure 2.6). At zero Kelvin temperature, all electrons are at this energy level and are not mobile within the solid. At room temperature, the energy required to jump into the conduction band is small, and all donor electrons are ionized and remain in the conduction band.

Doping a semiconductor does not increase its net charge, since the negative q charge excess of the fifth electron with respect to the "replaced" silicon atom is balanced by the atomic number* of the donor impurity (5 instead of 4 for silicon) giving an extra positive charge that compensates the electron charge.

*The atomic number is the number of protons of a neutral (nonionized) atom.

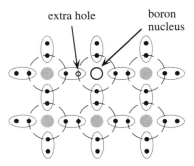

Figure 2.7. Adding acceptor atoms creates a mobile hole without creating an electron.

2.2.2 p-Type Semiconductors

We can increase the number of holes without increasing the number of electrons by replacing silicon atoms with elements from Group III in the Periodic Table that have three electrons in the outermost layer. The three electrons of the impurity bond with three of the neighbor silicon atoms, and the fourth bond is missing (Figure 2.7). Impurities having three electrons in the outermost layer are called *acceptors* since a vacancy is created that can accept free electrons. Acceptor doping in the energy-band model creates an energy level close to the silicon valence band, as shown in Figure 2.8. At room temperature, electrons in the valence band have enough energy to jump to this level. This creates a hole in the silicon without injecting electrons into the conduction band. At a given temperature, the concentration of holes in a semiconductor in which a number N_A of acceptor impurities were introduced will be N_A plus some of the holes contributed by the electron–hole pairs created at that temperature. Boron is a Group III atom and is the most popular of acceptor doping atoms.

2.3 CARRIER TRANSPORT IN SEMICONDUCTORS

The movement of electrons and holes in a semiconductor is called carrier transport. Computation of carrier transport requires knowing the carrier concentration calculated from the Fermi function and its subsequent derivations. It also requires the laws governing movement of carriers within the solid. Movement of electrons and holes within a semi-

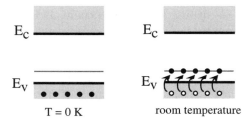

Figure 2.8. Effect of acceptor doping in the band-gap model.

conductor is different from carriers traveling in free space since collisions of carriers with the lattice impact their mobility. Temperature also plays an important role in carrier transport since it determines the carrier population and also affects carrier movement within the solid because of atomic thermal agitation. Although carriers are in constant motion within a solid, an unbiased semiconductor will not register a net current since carrier movement is random and no preferred direction is collectively chosen. A force is required for net movement of a charge to occur. The two main carrier movement mechanisms in solids are drift and diffusion.

2.3.1 Drift Current

Drift is carrier movement due to an electric or magnetic field. An electric field $\mathscr{E}$ applied to a semiconductor causes electrons in the conduction band to move in the direction opposite to the electric field, whereas holes in the valence band move in the same direction of this field. An electric field causes energy band bending, as shown in Figure 2.9(a) and the carrier movement indicated in Figure 2.9(b). Electron and hole current densities (electron hopping in the valence band) have the same direction and both contribute to current.

At relatively low electric fields, the velocity acquired by electrons and holes has a linear dependence on the applied external field. This dependence is maintained until a certain electric field is reached, and then the carrier velocity saturates. Velocity saturation occurs because carriers collide with the lattice atoms of the crystal. For electric fields beyond 10^5 V/cm, electron and hole velocities reach a maximum value in pure silicon at room temperature and no longer depend on the applied field. Modern deep-submicron transistors show velocity saturation during switching.

The relation between drift velocity and electric field for carriers is given by

$$v_d = \mu_0 \mathscr{E} \left[1 + \left(\frac{\mu_0 \mathscr{E}}{v_{\text{sat}}} \right)^\beta \right]^{-1/\beta} \quad (2.5)$$

where $\beta \approx 1$ for electrons, $\beta \approx 2$ for holes, μ_0 is the proportionality factor between the electric field $\mathscr{E}$ and the carrier velocity at low electric fields, and v_{sat} is the velocity saturation value reached for high electric fields. Note that this expression leads to $v_d = \mu_0 \mathscr{E}$ for low electric fields and $v_d = v_{\text{sat}}$ for large electric field strengths.

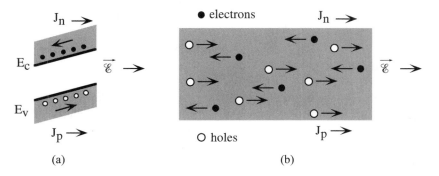

Figure 2.9. Band bending by the effect of an external applied field $\mathscr{E}$.

The current density of charge carriers J_d in an electric field is derived from a series of relations shown below, where v is the velocity of moving charge, N is the moving carrier concentration, q is carrier charge, and $\mathcal{E}$ is the electric field pushing the charge through a solid with mobility μ.

$$J_d = vqN = \mu q \mathcal{E} N \qquad (2.6)$$

2.3.2 Diffusion Current

Diffusion is a thermal mechanism that moves particles from high-density macroscopic regions to low-density ones, so that in the final situation, the particle distribution in space is uniform. Electrons and holes diffusing in a solid are moving charged particles that create a diffusion current. Diffusion motion is described by Fick's law, stating that

$$J = -D \nabla N \qquad (2.7)$$

where J is flux in particles/cm^2-sec, ∇N is the gradient of particles, and D is the diffusion coefficient. The equation states that the flux of particles is zero if there is a uniform distribution (i.e., ∇N is zero). Electrons and holes diffusing in the same direction give rise to opposite current densities since their charge signs are opposite. Expressions for electron and hole diffusion currents are

$$J_{n|\text{diff}} = qD_n \nabla n$$
$$J_{p|\text{diff}} = -qD_p \nabla p \qquad (2.8)$$

D_n and D_p are electron and hole diffusion coefficients that differ because electron and hole mobilities are different. The relation between the diffusion coefficient and the mobility is

$$\frac{D_x}{\mu_x} = \frac{kT}{q} \qquad (2.9)$$

where x must be replaced by n for electrons and by p for holes. This equation is known as the Einstein relationship. Drift is the dominant charge transport mechanism in CMOS field-effect transistors, whereas diffusion plays a secondary role. Velocity saturation of electrons and holes occurs in the drift mechanism of all modern CMOS transistors.

2.4 THE *pn* JUNCTION

Diodes are simple semiconductor devices that are the building blocks of MOS transistors. Diodes have a junction formed by joining a *p*-type and *n*-type semiconductor. This is referred to as a *pn* junction. To understand *pn* junction properties, assume an ideal case in which two pieces of semiconductors with opposite doping are initially separated [Figure 2.10(a)] and then joined [Figure 2.10(b)]. The lattice structure is not lost at the joining surface, and the doping concentration has a sharp change from the left side (*n*-doped) to the right side (*p*-doped).

2.4 THE pn JUNCTION

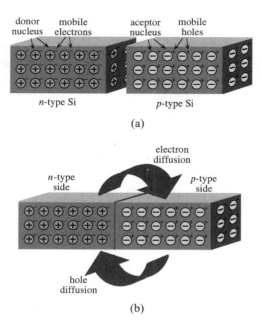

Figure 2.10. Two pieces of semiconductor materials with opposite doping: (a) separated and (b) joined.

Since the n-type and p-type semiconductor bars are in equilibrium, the total net charge in each bar is zero. At the instant when the semiconductor pieces join, there is a momentary abrupt change in electron and hole concentration at the joining surface. Strong concentration gradients exist for electrons on the n-side and holes on the p-side. This nonequilibrium condition exists for a short time during which electrons at the junction start to diffuse from the initial n-type bar into the p-type one, while holes close to the junction move away from the p-doped semiconductor into the n-type one. If electrons and holes were not charged particles, then this diffusing process would continue until electron and hole concentrations were uniform along the whole piece of joined semiconductors.

The reality is different, since electrons and holes are charged particles and their diffusion creates electric fields in the semiconductor at both sides of the junction. Carriers close to the junction are the first particles to diffuse away and recombine when they meet opposite carrier types on the opposite side of the junction. As a consequence of this carrier migration and recombination, all dopant atoms close to the joining surface are ionized. This creates a zone of net charge (Figure 2.11) around the junction (positive at the n-side and negative at the p-side) that induces an electric field pointing from the n-side to the p-side. Carriers moving by diffusion now "feel" the electric field as an opposing force when trying to diffuse. Their final motion depends on which conducting mechanism is stronger. As more carriers move by diffusion, more atoms are ionized, and the electric field strength increases. Finally, the net diffusion mechanism stops when the induced internal electric field (which increases with the number of carriers moving by diffusion) reaches a value such that its force exactly balances the force tending to diffuse carriers across the junction. The result is creation of a depletion region with all donors and acceptors ionized where the net charge and electric field are nonzero.

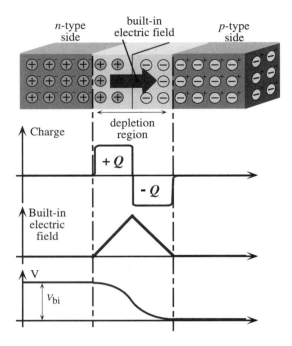

Figure 2.11. A diode in equilibrium, showing the charge, electric field, and potential internal distribution.

Figure 2.11 shows a picture of the diode junction in equilibrium (no external electric field, light, or temperature gradient are present) and the charge, electric field, and potential distribution at each point within the diode. Notice that in the regions where atoms are not ionized (out of the depletion region), there is charge neutrality, so that no electric field or potential drop exist. The net charge in the depletion region is positive in the *n*-type side and negative in the *p*-type side. As a result, the electric field increases while moving from the neutral regions to the junction site. The electric field distribution causes a voltage difference between the two oppositely doped regions that depends mainly on the doping levels. This is called the built-in junction potential (V_{bi}) shown at the bottom of Figure 2.11.

At equilibrium, a charge zone exists on both sides of the junction in which all donors and acceptor atoms are ionized. This zone is known as the depletion region or space-charge zone with a high electric field and a potential that increases when moving from the *p*-type zone to the *n*-type zone. Outside the boundaries of the space charge region within the semiconductor, the electric field, net charge, and potential gradient are all zero.

2.5 BIASING THE *pn* JUNCTION: I–V CHARACTERISTICS

The previous section showed that when two semiconductor bars of opposite doping are joined, a built-in electric field appears, preventing electrons from diffusing too far away from the *n*-side and holes from diffusing away from the *p*-side. Once the system is in equilibrium, no current exists since the junction is isolated with no external conducting path. We will now analyze the behavior of the junction when an external voltage is applied.

2.5.1 The *pn* Junction under Forward Bias

Assume that an external voltage source is connected to a diode, inducing more positive voltage at the *p*-side with respect to the *n*-side (Figure 2.12). The external voltage creates an electric field opposed to the built-in one, decreasing its strength. The built-in electric field is a barrier for electrons diffusing from the *n*-side to the *p*-side and for holes diffusing from the *p*-side to the *n*-side. An external reduction of the built-in field favors the diffusion process. Since many conducting electrons are on the *n*-side and many holes are on the *p*-side, then many electrons will diffuse into the *p*-side and many holes will diffuse into the *n*-side. This process causes a permanent current since electrons must be replaced at the *n*-type side through the voltage source, and holes are required at the *p*-type side. The larger the applied voltage bias, the higher the diode current.

2.5.2 The *pn* Junction under Reverse Bias

When the voltage applied to the diode is positive at the *n*-side with respect to the *p*-side, it induces an external electric field that increases the built-in junction electric field (Figure 2.13). This reduces electron diffusion from the *n*-side silicon into the *p*-side, and hole diffusion in the opposite direction. Only electrons or holes within the junction region itself are accelerated by the junction electric field and contribute to overall current. Since very few electrons are present in a *p*-type material, and few holes are in the *n*-type side, the reverse bias current is very small. In fact, this current, called the reverse-bias saturation current, arises from thermal generation of electron–hole pairs in the depletion region itself. The free electron and hole carriers are then rapidly swept out of the junction, forming the current at the diode terminals.

The device-level current/voltage characteristics of the diode were introduced in Chapter 1. The relation between the applied diode voltage V_D and the obtained current I_D is exponential and reproduced here:

$$I_D = I_S \left(e^{\frac{qV_D}{kT}} - 1 \right) \tag{2.10}$$

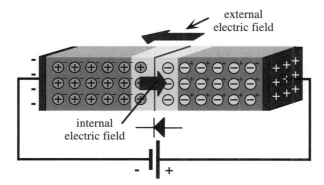

Figure 2.12. Forward-biased diode showing internal junction depletion-field reduction.

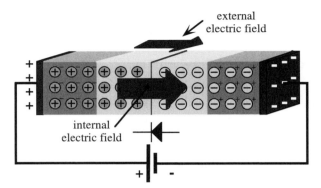

Figure 2.13. Reverse-biased diode showing internal junction depletion-field increase.

Remember that I_0 is called the reverse-bias saturation current. When V_D is large and negative, then $I_D = -I_0$. The I–V characteristic of the diode rapidly increases for positive diode voltages, whereas very little current is obtained for negative diode voltages.

2.6 PARASITICS IN THE DIODE

Diodes are useful as nonsymmetric components that allow current in one direction but not in the opposite. All diodes show a behavior that deviates from their ideality and this deviation can be modeled by so-called *parasitic elements.* Parasitic elements are undesired and can be resistance, capacitance, or inductance. The diode current leakage in the off-state (ideally not conducting) can be modeled by a high parasitic (undesired) resistance.

In CMOS technology, two parasitic diodes exist in each transistor. These diodes are always reverse-biased in ICs, and their main degradation effects at the circuit level are related to reverse current leakage or to delay through parasitic capacitors of the diode.

We saw that when a diode is reverse-biased, the external voltage increases the internal electric field strength. This widens the depletion region. The higher the reverse voltage, the wider the depletion region, and the larger the total net charge in this region. Conversely, a forward bias narrows the depletion region, and reduces the fixed charges across the *pn* junction. Therefore, a diode has an *internal capacitor* since there is a charge variation induced by a voltage variation. From Chapter 1, we know that the term capacitance is defined as the charge variation in a component due to voltage variation at its terminals, i.e.,

$$C = \frac{\partial Q}{\partial V} \qquad (2.11)$$

The parasitic capacitor inherent to the *pn* junction is different from the passive or parallel-plate capacitors seen in Chapter 1 because the charge–voltage ratio varies with the applied voltage. Since the Q/V quotient is not constant when the applied voltage changes, a fixed capacitor value cannot be assigned to the parasitic capacitor.

The effect of the diode parasitic capacitor at the circuit level is significant, since it must be charged and discharged when a gate is switching. This contributes to circuit delay

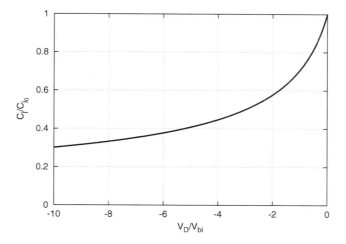

Figure 2.14. Plot of C_j versus applied voltage (V_D) for a diode.

and other secondary effects discussed later. In many cases, the exact dependence of the diode parasitic capacitor with the applied voltage is replaced by an approximate value. The value of the capacitor with the applied voltage is

$$C_j = \frac{C_{j0}}{\left(1 - \frac{V_D}{V_{bi}}\right)^{1/2}} \qquad (2.12)$$

where C_{j0} is a constant depending on the *pn* doping values, fundamental constants of the silicon, the area of the surfaces being joined, and the built-in junction potential; V_D is the reverse applied voltage, and V_{bi} is the built-in junction potential. Figure 2.14 plots the capacitance normalized to C_{j0} with respect to the reverse applied voltage. Equation (2.12) breaks down in the diode forward-bias regions for large positive values of V_D. In the next chapter, it will be shown how this parasitic device affects device operation, and how it is modeled at the circuit level.

2.7 SUMMARY

The construction of integrated circuits ultimately depends on a strong physical base. Chapter 2 provides a brief introduction to these essential concepts. Although a few modeling equations were given, the purpose is qualitative understanding of the language and physical flow of semiconductor conduction, related doping properties, and diode characteristics. These ideas permeate later chapters.

BIBLIOGRAPHY

1. N. W. Ashcroft and N. D. Mermin, *Solid State Physics,* Saunders College, 1976.
2. C. Kittel, *Introduction to Solid State Physics,* 6th ed., Wiley, 1986

3. D. A. Neamen, *Semiconductor Physics and Devices—Basic Principles,* 3rd ed., McGraw-Hill, 2003.
4. R. F. Pierret, *Semiconductor Device Fundamentals,* Addison-Wesley, 1996.

EXERCISES

2.1. Discuss the following:
   (a) What is the band-gap voltage E_g in the solid state model?
   (b) Use the band-gap voltage to distinguish between metals, semiconductors, and insulators.

2.2. Metal electrical conduction is done by electrons. How does semiconductor conduction differ?

2.3. What energy process occurs when electron–hole pairs are created, and when they recombine?

2.4. What is the Fermi Level?

2.5. Electrical conduction in a metal is done entirely by electrons. The dominant form of conduction in a semiconductor can either be by holes or electrons. How does insertion of a Group III or a Group V element into a host Group IV Si element affect the choice of dominant hole or electron injection.

2.6. Two forces dominate the net motion of carriers in a semiconductor: drift and diffusion. Describe the conditions that promote these two mechanisms.

2.7. Describe how an electric field appears across a *pn* junction and the dynamic relation of this $\mathscr{E}$-field to charge movement across the *pn* junction.

2.8. The diode reverse bias saturation current I_S originates in the depletion region with its high electric field. The sources of this current are the thermal creation of electron–hole pairs that are then swept out of the junction by this high electric field. If the reverse-bias voltage is made larger, what is the impact on I_S? Will I_S increase, decrease, or stay the same? Explain.

2.9. The diode equation (2.10) predicts an exponential increase in diode current I_D as diode voltage V_D increases. Assume that you measured a diode in the forward-bias region and found an exponential relation at the lower V_D, but the curve tended toward a straight line at higher voltages. Explain.

2.10. Chapter 1 described capacitors made of two metal plates separated by a dielectric. Describe how a *pn* junction capacitance differs from the simple metal plate dielectric capacitor.

CHAPTER 3

MOSFET TRANSISTORS

MOSFET transistors are the basic element of today's integrated circuits (ICs). Our goal here is to impart the analytical ability and transistor insights that electrical engineers use in solving IC problems. An abundance of examples and self-exercises are provided to develop intuitive responses to digital transistor circuit operation. We begin with a simple picture of transistors as switches, and evolve to more developed analytical models. We want to smoothly lead the way through these topics, providing knowledge and insight about transistors in the long- and short-channel technologies. The information in this chapter is a foundation for subsequent chapters, and is a basis for understanding the electronic aberrations of defective circuits.

3.1 PRINCIPLES OF OPERATION: LONG-CHANNEL TRANSISTORS

Transistors are the basic blocks for building electronic circuits. A major difference between transistors and passive elements (resistors, capacitors, inductors, and diodes) is that transistor current and voltage characteristics vary with the voltage (or current) on a control terminal. There are two types of transistors with different physical principles: bipolar transistors and field effect transistors (FETs). There is only one type of bipolar transistor—the bipolar junction transistor (BJT)—and two types of FET devices—the junction field effect transistor (JFET) and the metal oxide semiconductor field effect transistor (MOSFET). Today's digital ICs mainly use MOSFETs, whereas bipolars are used in specific digital technologies and more generally in analog circuits. JFETs have specific applications and are not used in digital applications. We will focus on MOSFETs since they appear in more than 90% of today's digital applications.

MOSFETs have three signal terminals: gate (G), source (S) and drain (D), plus the bulk terminal (B), to which the gate, drain, and source voltages are referenced. Figure 3.1(a) shows a MOSFET with its four terminals and its thin insulator of SiO_2 (with thickness T_{OX}) between the gate and bulk. Figure 3.1(b) shows symbols commonly used for MOSFETs, where the bulk terminal is labeled (B) or implied (not drawn). There are two types of MOSFET transistors, the nMOS transistor and the pMOS transistor, depending on the polarity of the carriers responsible for conduction. A simple description of the device will introduce basic concepts and terms.

3.1.1 The MOSFET as a Digital Switch

The simplest view of MOSFET logic operation treats the transistor as a switch. The gate terminal is analogous to the light switch on the wall. When the gate has a high voltage, the transistor closes like a switch, and the drain and source terminals are electrically connected. Just as a light switch requires a certain force level to activate it, the transistor needs a certain voltage level to connect the drain and source terminals. This voltage is called the transistor threshold voltage V_t and is a fixed voltage different for nMOS (V_{tn}) and pMOS (V_{tp}) devices in a given fabrication process. The nMOS threshold voltage V_{tn} is always positive, whereas the pMOS threshold voltage V_{tp} is always negative.

Transistors act as switches with two conducting states, on and off, depending on the control (gate) terminal voltage. An ideal transistor has a zero ohm resistance between the drain and source when it is in the on-state, and infinite resistance between these terminals in the off-state. The ideal device should also switch between on- and off-states with a zero delay time as soon as the control variable changes state.

Unfortunately, transistors are not ideal switches. MOS transistors have a small, equivalent drain–source resistance in their conducting state and a high but not infinite resistance in the off-state. Additionally, the delay of a transistor to switch between on- and off-states is not zero. Several parameters (both geometric and technology related) determine the on and off equivalent resistance and capacitance that degrade the time needed to switch between both states.

Figure 3.2 shows switch models for the nMOS and pMOS transistors. In the nMOS, the source is the reference terminal, which is always at the lower voltage, so that $V_{DS} \geq 0$, and $V_{GS} \geq 0$. Since V_{tn} is always positive, the off-state occurs when $V_{GS} < V_{tn}$ so the de-

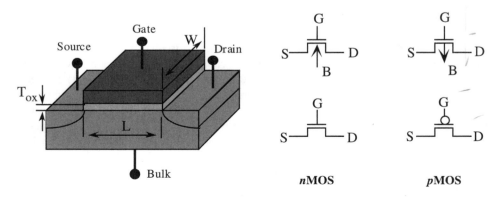

Figure 3.1. (a) MOS structure. (b) Symbols used at the circuit level.

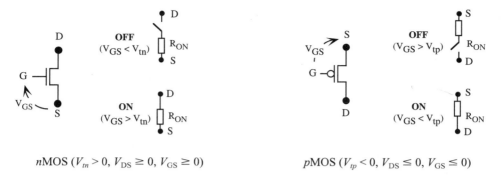

Figure 3.2. Transistor symbols and their equivalent ideal switches.

vice does not conduct, and it is modeled as an open switch. When $V_{GS} > V_{tn}$, the device is in the on-state and modeled as a closed switch in series with a resistor R_{ON}. The model in Figure 3.2 represents this on resistance as constant, whereas in a real transistor the drain–source current (and therefore its equivalent resistance) depends on the operating state of the device, and must be determined from the V_{GS} and V_{DS} relations.

The pMOS transistor model is equivalent, but the signals have an opposite terminal polarity. The source is the reference terminal, which is always at the highest voltage, so that $V_{DS} \leq 0$, and $V_{GS} \leq 0$. V_{tp} has negative voltages in pMOS transistors. The off-state is defined when $V_{GS} > V_{tp}$ and the on-state for $V_{GS} < V_{tp}$. Since V_{tp} is always negative, the pMOS transistor turns on when $V_{GS} < V_{tp}$, where both are negative numbers. This polarity confusion will become clear when we address pMOS transistor operation. For now, accept that the pMOS transistor has polarity control signals opposite to those of the nMOS transistor.

The ideal device characteristics for this simple model are

$$I_{DS} = 0, \text{(off-state) when} \begin{cases} V_{GS} < V_{tn} & n\text{MOS} \\ V_{GS} > V_{tp} & p\text{MOS} \end{cases}$$

$$I_{DS} = \frac{V_{DS}}{R_{ON}}, \text{(on-state) when} \begin{cases} V_{GS} \geq V_{tn} & n\text{MOS} \\ V_{GS} \leq V_{tp} & p\text{MOS} \end{cases} \qquad (3.1)$$

We will next consider the simple switch model and deepen our understanding of MOSFET structure, operating modes, and behavior models.

3.1.2 Physical Structure of MOSFETs

This section will describe transistors as they are today, and subsequent material will develop why they are so. MOSFET transistors are made from a crystalline semiconductor that forms the host structure called the substrate or bulk of the device. Substrates for nMOS are constructed from p-type silicon, whereas the pMOS substrates use n-type silicon. The thin oxide of the transistor electrically isolates the gate from the semiconductor crystalline structures underneath. The gate oxide is made of oxidized silicon, forming a noncrystalline, amorphous SiO_2. The gate oxide thickness is typically from near 15 Å to

100 Å (1 Å = 1 angstrom = 10^{-10} m). SiO_2 molecules are about 3.5 Å in diameter, so this vital dimension is a few molecular layers thick. A thinner gate oxide provides more gate terminal control over the device state.

Drain and source regions are made from crystalline silicon by implanting a dopant with polarity opposite to that of the substrate. The region between the drain and source is called the channel. The distance from the drain to the source is a geometrical parameter called the *channel length* (L) of the device, as shown in Figure 3.1(a). Another geometrical parameter of the device is the transistor *channel width* (W) [Figure 3.1(a)]. Transistor length and width are geometrical parameters set by the circuit designer. Other parameters, such as the transistor oxide thickness, threshold voltage, and doping levels, depend on the fabrication process, and cannot be changed by design; they are technology parameters.

The gate is the control terminal, and the source provides electron or hole carriers that are collected by the drain. Often, the bulk terminals of all transistors are connected to the ground or power rail that is often the source and, therefore, not explicitly drawn in most schematics.

Figure 3.3 shows *n*MOS and *p*MOS transistor structures. The *n*MOS transistor has a *p*-type silicon substrate with opposite doping for the drain and source. *p*MOS transistors have a complementary structure with an *n*-type silicon bulk and *p*-type doped drain and source regions. The gate region in both transistors is constructed with polysilicon, and is isolated from the drain and source by the thin oxide. The region between the drain and source under the gate oxide is called the channel, and is where conduction takes place.

The gate is electrically isolated from the drain, source, and channel by the gate oxide insulator. Since drain and source dopants are opposite in polarity to the substrate (bulk), they form *pn* junction diodes (Figure 3.3) that in normal operation are reverse-biased. CMOS logic circuits typically match one *n*MOS transistor to one *p*MOS transistor.

3.1.3 Understanding MOS Transistor Operation: A Descriptive Approach

Transistor terminals must have proper voltage polarity to operate correctly (Figure 3.4). The bulk or substrate of *n*MOS (*p*MOS) transistors must always be connected to the lower (higher) voltage that is the reference terminal. We will assume that the bulk and source terminals are connected, to simplify the description. The positive convention current in an *n*MOS (*p*MOS) device is from the drain (source) to the source (drain), and is referred to as I_{DS} or just I_D, since drain and source current are equal. When a positive (negative) voltage is applied to the drain terminal, the drain current depends on the voltage applied to the gate control terminal. Note that for *p*MOS transistors, V_{GS}, V_{DS}, and I_{DS} are negative.

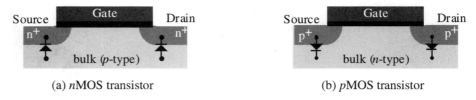

(a) *n*MOS transistor   (b) *p*MOS transistor

Figure 3.3. Relative doping and equivalent electrical connections between device terminals for (a) *n*MOS and (b) *p*MOS transistors.

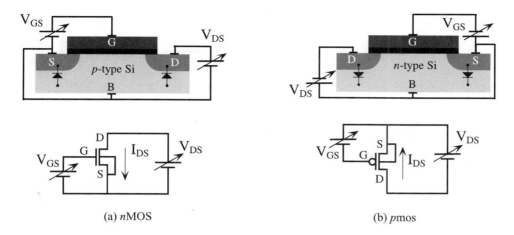

Figure 3.4. Normal transistor biasing (a) *n*MOS and (b) *p*MOS.

If V_{GS} is zero, then an applied drain voltage reverse-biases the drain–bulk diode (Figure 3.5), and there are no free charges between the drain and source. As a result, there is no current when $V_{GS} = 0$ for *n*MOS devices (the same hold for *p*MOS devices). This is the off, or nonconducting, state of the transistor.

We will first analyze transistor operation when the source and substrate are at the same voltage. When the gate terminal voltage of an *n*MOS (*p*MOS) transistor is slightly increased (decreased), a vertical electric field exists between the gate and the substrate across the oxide. In *n*MOS (*p*MOS) transistors, the holes (electrons) of the *p*-type (*n*-type) substrate close to the silicon–oxide interface initially "feel" this electrical field, and move away from the interface. As a result, a depletion region forms beneath the oxide interface for this small gate voltage (Figure 3.6). The depletion region contains no mobile carriers, so the application of a drain voltage provides no drain current, since free carriers still do not exist in the channel.

If the gate voltage of the *n*MOS (*p*MOS) device is further increased (decreased), then the vertical electric field is strong enough to attract minority carriers (electrons in the *n*MOS device and holes in the *p*MOS device) from the bulk toward the gate. These minority carriers are attracted to the gate, but the silicon dioxide insulator stops them, and the

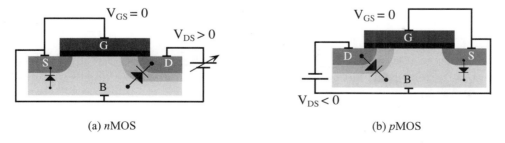

Figure 3.5. When the gate–source voltage is zero, the drain–source voltage reverses the built-in drain–bulk diode, preventing current from flowing from the drain to the source.

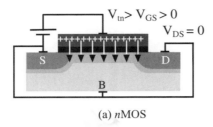 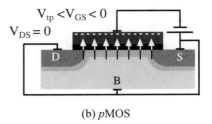

(a) nMOS (b) pMOS

Figure 3.6. (a) Depleting the nMOS channel of holes with small positive values of gate–source voltage, and (b) depleting the pMOS channel of electrons with small negative values of gate–source voltage.

electrons (holes) accumulate at the silicon–oxide interface. They form a conducting plate of minority mobile carriers (electrons in the p-type bulk of the nMOS device, and holes in the n-type bulk of the p-MOS device). These carriers form the inversion region or conducting channel, which can be viewed as a "short circuit" to the drain/source-bulk diodes. This connection is shown in Figure 3.7.

Since the drain and source are at the same voltage, the channel carrier distribution is uniform along the device. The gate voltage for which the conducting channels respond is an intrinsic parameter of the transistor called the *threshold voltage*, referred to as V_t. As a first approximation, V_t can be considered constant for a given technology. The threshold voltage of a nMOS transistor is positive, while for a pMOS transistor it is negative. Since nMOS and pMOS transistors have a different threshold voltages, V_{tn} refers to the nMOS transistor threshold voltage, and V_{tp} to the pMOS transistor.

An nMOS (pMOS) transistor has a conducting channel when the gate–source voltage is greater than (less than) the threshold voltage, i.e., $V_{GS} > V_{tn}$ ($V_{GS} < V_{tp}$).

When the channel forms in the nMOS (pMOS) transistor, a positive (negative) drain voltage with respect to the source creates a horizontal electric field, moving the channel carriers toward the drain and forming a positive (negative) drain current. If the horizontal electric field is of the same order or smaller than the vertical one, the inversion channel remains almost uniform along the device length. This happens when

$$V_{DS} < (V_{GS} - V_{tn}) \quad \text{nMOS transistor}$$
$$V_{DS} > (V_{GS} - V_{tp}) \quad \text{pMOS transistor} \qquad (3.2)$$

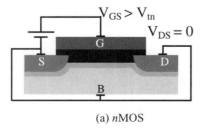 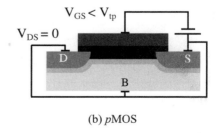

(a) nMOS (b) pMOS

Figure 3.7. Creating the conducting channel for (a) nMOS and (b) pMOS transistors.

3.1 PRINCIPLES OF OPERATION: LONG-CHANNEL TRANSISTORS

This condition will be explained below. It states that the vertical electric field dominates the horizontal one. The transistor is in its linear region, also called the ohmic or nonsaturated region.

If the drain voltage increases beyond the limit of Equation (3.2), the horizontal electric field becomes stronger than the vertical field at the drain end, creating an asymmetry of the channel carrier inversion distribution. The drain electric field is strong enough so that carrier inversion is not supported in this local drain region. The conducting channel retracts from the drain, and no longer "touches" this terminal. When this happens, the inversion channel is said to be "pinched off" and the device is in the *saturation region*. The pinch-off point is the location that separates the channel inversion region from the drain depletion region. It varies with changes in bias voltages. The channel distribution in this bias is shown in Figure 3.8.

Although there are no inversion charges at the drain end of the channel, the drain region is still electrically active. Carriers depart from the source and move under the effect of the horizontal field. Once they arrive at the pinch-off point of the channel, they travel from that point to the drain, driven by the high electric field of the depletion region.

CMOS ICs use all three states described here: off-state, saturated state, and the linear state. We will next look at real curves of MOS parameters, and learn how to use the analytical equations that predict and analyze transistor behavior in normal and defective environments. It is important to work through all examples and exercises. The examples will analyze MOS long- and short-channel transistor circuits.

3.1.4 MOSFET Input Characteristics

MOS transistors cannot be described with a single current–voltage curve, as can diodes and resistors, since they have four terminals. Transistors require that two sets of current–voltage curves be characterized: the input characteristic and the output characteristic. The input characteristic is a single curve that relates drain current to the input gate–source driving voltage. Since the gate terminal is electrically isolated from the remaining terminals (drain, source, and bulk), the current through the gate is essentially zero, so that gate current is not part of the device characteristics. The input characteristic curve can locate the voltage on the control terminal (gate) at which the transistor leaves the off-state.

Figure 3.9 shows measured input characteristics for an nMOS and pMOS transistor with a small, 0.1 V potential across their drain-to-source terminals. As V_{GS} increases for

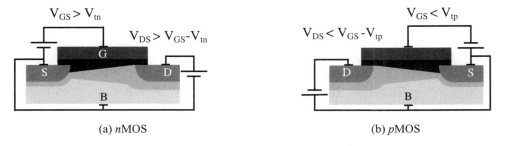

(a) nMOS (b) pMOS

Figure 3.8. Channel pinch-off for (a) nMOS and (b) pMOS transistor devices.

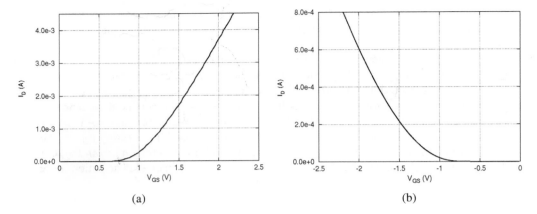

Figure 3.9. Measured input characteristics (I_D vs. V_{GS}) for (a) an nMOS, and (b) a pMOS transistor.

the nMOS transistor in Figure 3.9(a), a voltage is reached at which drain current begins. When the nMOS device conducts, the drain current is positive, since the current enters the drain. For V_{GS} between 0 V and 0.7 V, the drain current is nearly zero, indicating that the equivalent resistance between the drain and source terminals is extremely high. Once the gate–source voltage reaches 0.7 V, the current increases rapidly with V_{GS}, indicating that the equivalent resistance at the drain decreases with increasing gate–source voltage. Therefore, the threshold voltage of this device is about $V_{tn} \approx 0.7$ V. When a transistor turns on and current moves through a load, then voltage changes occur that translate into logic levels.

The pMOS transistor input characteristic in Figure 3.9(b) is analogous to the nMOS transistor except that the I_{DS} and V_{GS} polarities are reversed. V_{DS} is negative ($V_{DS} \approx -0.1$ V) and the drain current in a pMOS transistor is negative, indicating that it exits the drain terminal. Additionally, the gate is at a voltage lower than the source terminal voltage to attract holes to the channel surface. The threshold voltage of the pMOS device in Figure 3.9(b) can be seen as approximately $V_{tp} \approx -0.8$ V.

3.1.5 nMOS Transistor Output Characteristics

MOS transistor output characteristics plot I_D versus V_{DS} for several values of V_{GS}. Figure 3.10 shows such a measurement of an nMOS transistor. When the transistor is in the off-state ($V_{GS} < V_{tn}$), then I_D is near zero for any V_{DS} value. When the device is in the on-state, transistor conduction properties vary with the bias state of the device. Two states, described as the ohmic (or nonsaturated) state and the saturated state, are distinguished when the device is in the on-state. The boundary of the two bias states is seen for each curve in Figure 3.10 as the intersection of the straight line of the saturated region with the curving line of the ohmic region. This intersection point occurs at V_{Dsat}. In the ohmic state, the drain current initially increases almost linearly with the drain voltage before bending and flattening out. The drain current in saturation is virtually independent of V_{DS} and the transistor acts as a current source. A near-constant current is driven from the transistor no matter what the drain-to-source voltage is.

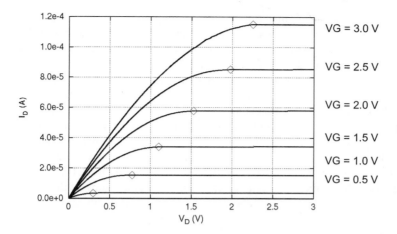

Figure 3.10. *n*MOS transistor output characteristics as a family of curves. The diamond symbol marks the pinch-off voltage, V_{DSAT}.

This family of curves is rapidly measured with a digital curve tracer called a parameter analyzer. Many useful parameters can be measured from data in the family of curves, and deviations in the curves are also a good indicator of a damaged transistor. Later, we will deepen our understanding of transistor operation expressed by Figure 3.10. *p*MOS transistor I_D versus V_{DS} curves have shapes similar to those in Figure 3.10, but the voltage and current polarities are negative to account for hole inversion and drain current that enters the transistor (*p*MOS device curves are shown later).

We next develop skills with the equations that predict voltages and currents in a transistor for any point in the family of curves in Figure 3.10. This capability is needed to analyze electronic behavior of CMOS circuits with bridge, open circuit, or parametric defects.

MOS equations can be derived by calculating the amount of charge in the channel at each point, and integrating such an expression from the drain to the source. This procedure is found in several books [3, 6, 7], and leads to expressions for the drain current in the linear and saturated states. Equations (3.3) and (3.4) are these equations for the *n*MOS transistor in the saturated and ohmic states.

$$I_D = \frac{\mu \varepsilon_{ox}}{2 T_{ox}} \frac{W}{L} (V_{GS} - V_{tn})^2 \quad \text{(saturated state)} \tag{3.3}$$

$$I_D = \frac{\mu \varepsilon_{ox}}{2 T_{ox}} \frac{W}{L} [2(V_{GS} - V_{tn})V_{DS} - V_{DS}^2] \quad \text{(ohmic state)} \tag{3.4}$$

where μ is the electron mobility, ε_{ox} is thin oxide (SiO$_2$) dielectric constant, T_{ox} is the transistor oxide thickness, and W and L are transistor effective gate width and length. A constant, K, is introduced to indicate the drive strength of the transistor as

$$K = \frac{\mu \varepsilon_{ox}}{2 T_{ox}} \tag{3.5}$$

If these constants are known, then Equation (3.3) can predict I_D for any value of V_{GS} in the saturated region, and Equation (3.4) can predict any I_D in the ohmic region if V_{GS} and V_{DS} are specified. Equation (3.3) is a square-law relation between I_D and V_{GS} that is independent of V_{DS}. Equation (3.3) is a flat line for a given V_{GS}, whereas Equation (3.4) is a parabola. For any V_{GS}, the two equations have an intersect point that is seen in Figure 3.10. The intersection point occurs at a parameter called V_{Dsat}, for which either equation describes the current and voltage relations.

We can solve for this important bias condition at which the saturated and ohmic states intersect (V_{Dsat}), and this knowledge is essential for solving problems that follow. Figure 3.11 plots three parabolas of Equation (3.4) at V_{GS} = 2.0 V, 1.6 V, and 1.2 V. Only the left-hand sides of the parabolas are used to predict the curves in Figure 3.10, but the parabolas also have a right-hand side. The dotted lines on the right-hand side of the curves are part of the continuous solution to the parabolas, but are electronically invalid, as examples will show.

The midpoint at zero slope defines the useful upper region of Equation (3.4), and also defines the boundary between the saturated and ohmic bias states. We can define the boundary bias condition by differentiating Equation (3.4) with respect to V_{DS}, setting the expression to zero, and then solving for the conditions. Equation (3.6) shows the derivative of Equation (3.4) set to zero:

$$\frac{dI_D}{dV_{DS}} = \frac{\mu\varepsilon_{ox}}{2T_{ox}} \frac{W}{L}[2(V_{GS} - V_{tn}) - 2V_{DS}] = 0 \tag{3.6}$$

Terms cancel, giving the bias condition at the transition between saturation and nonsaturation states as

$$V_{GS} = V_{DS} + V_{tn} \tag{3.7}$$

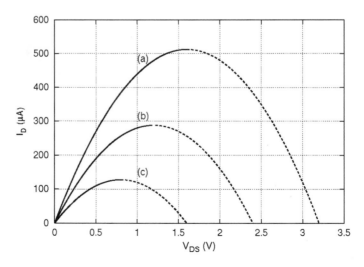

Figure 3.11. Plot of parabola of the nonsaturation state equation [Equation (3.4)]. K_n = 100 μA/V², V_{tn} = 0.4 V, and W/L = 2. Solid lines indicate valid regions, but dotted lines do not. (a) V_{GS} = 2.0 V, (b) V_{GS} = 1.6 V, (c) V_{GS} = 1.2 V.

This equation holds for each of the intersection points in Figure 3.11 denoted at the peak of each curve. Equation (3.7) can be extended to define the nMOS saturated bias condition:

$$V_{DS} > V_{GS} - V_t \quad \text{or} \quad V_{GS} < V_{DS} + V_t \qquad (3.8)$$

and the nMOS ohmic condition:

$$V_{DS} < V_{GS} - V_t \quad \text{or} \quad V_{GS} > V_{DS} + V_t \qquad (3.9)$$

We use these relations to analyze the effect of defects on CMOS circuits. A series of examples and exercises will illustrate their use. We emphasize that drill imparts the intuition that experienced failure analysts and test, reliability, and product engineers use in CMOS IC manufacturing.

■ **EXAMPLE 3.1**

Determine the bias state for the three conditions in Figure 3.12 if $V_{tn} = 0.4$ V.

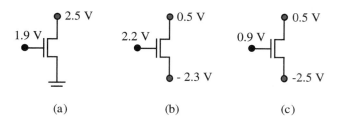

Figure 3.12. Transistor bias-state examples.

(a) $V_{GS} = 1.9$ V, $V_{DS} = 2.5$ V, and $V_{tn} = 0.4$ V, therefore $V_{GS} = 1.9$ V < 2.5 V $+ 0.4$ V $= 2.9$ V. Equation (3.8) is satisfied, and the transistor is in the saturated state described by Equation (3.3).
 (b) $V_{GS} = V_G - V_S = 2.2$ V $- (-2.3$ V$) = 4.6$ V. $V_{DS} = V_D - V_S = 0.5$ V $- (-2.3) = 2.8$ V. Therefore, $V_{GS} = 4.6$ V > 2.2 V $+ 0.4$ V $= 2.6$ V. Equation (3.9) is satisfied, and the transistor is in the nonsaturated state.
 (c) $V_{GS} = V_G - V_S = 0.9$ V $- (-2.5$ V$) = 3.4$ V. $V_{DS} = V_D - V_S = 0.5$ V $- (-2.5$ V$) = 3$ V. Therefore, $V_{GS} = 3.4$ V $= V_{DS} + V_t = 3$ V $+ 0.4$ V $= 3.4$ V, and the transistor is at the boundary of the saturated and ohmic regions. Either Equation (3.3) or (3.4) can be used to calculate I_D. ■

Self-Exercise 3.1

Determine the bias state for the three conditions in Figure 3.13 if $V_{tn} = 0.4$ V.

After solving bias Example 3.1 and Self-Exercise 3.1 with the proper bias-state equations, you may check your work by referring to the nMOS transistor family of curves in Figure

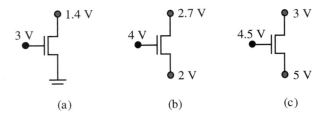

Figure 3.13. Bias-state exercises.

3.10. Find the coordinates in the example and exercise, and verify that the bias state is correct. A series of examples and exercises with the *n*MOS transistor will reinforce these important relations.

■ EXAMPLE 3.2

Calculate I_D and V_{DS} if $K_n = 100$ μA/V^2, $V_{tn} = 0.6$ V, and $W/L = 3$ for transistor M1 in the circuit in Figure 3.14.

The bias state of M1 is not known, so we must initially assume one of the two states, solve for bias voltages, and then check for consistency against that transistor's bias condition. Initially, assume that the transistor is in the saturated state so that

$$I_D = \frac{\mu \varepsilon_{ox}}{2T_{ox}} \frac{W}{L}(V_{GS} - V_{tn})^2 = K_n \frac{W}{L}(V_{GS} - V_{tn})^2$$
$$= (100 \; \mu A)(3)(1.5 - 0.6)^2$$
$$= 243 \; \mu A$$

Using Kirchhoff's voltage law (KVL):

$$V_{DS} = V_{DD} - I_D R$$
$$= 5 - (243 \; \mu A)(15 \; k\Omega)$$
$$= 1.355 \; V$$

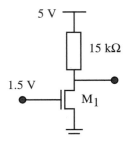

Figure 3.14.

We assumed that the transistor was in saturation, so we must check the result to see if that is true. For saturation,

$$V_{GS} < V_{DS} + V_{tn}$$
$$1.5\ V < 1.355\ V + 0.6\ V$$

so the transistor is in saturation, and our assumption and answers are correct. ■

■ EXAMPLE 3.3

Repeat Example 3.2, finding I_D and V_{DS} if $V_G = 1.8$ V.
Assume a transistor saturated state and

$$I_D = (100\ \mu A)(3)(1.8 - 0.6)^2$$
$$= 432\ \mu A$$
$$V_{DS} = 5 - (432\ \mu A)(15\ k\Omega)$$
$$= -1.48\ V$$

This value for V_{DS} is clearly not reasonable since there are no negative potentials in the circuit. Also, the bias check gives

$$V_{GS} > V_{DS} + V_{tn}$$
$$1.8V > -1.48\ V + 0.6\ V$$

The initial saturated state assumption was wrong, so we repeat the analysis using the ohmic state assumption:

$$I_D = \frac{\mu \varepsilon_{ox}}{2T_{ox}} \frac{W}{L} [2(V_{GS} - V_{tn})V_{DS} - V_{DS}^2]$$

$$I_D = K_n \frac{W}{L} [2(V_{GS} - V_{tn})V_{DS} - V_{DS}^2]$$

$$= (100\ \mu A)\ (3)\ [2(1.8 - 0.6)\ V_{DS} - V_{DS}^2]$$
$$= 300\ \mu A\ [2.4\ V_{DS} - V_{DS}^2]$$

This equation has two unknowns, so another equation must be found. We will use the KVL statement,

$$V_{DD} = I_D R + V_{DS}$$
$$I_D = (V_{DD} - V_{DS})/R$$
$$= (5 - V_{DS})/15\ k\Omega$$

The two equations can be equated to their I_D solution, giving

$$\frac{(5 - V_{DS})}{15\ k\Omega} = 300\ \mu A(2.4\ V_{DS} - V_{DS}^2)$$

After some algebra, this reduces to

$$V_{DS}^2 - 2.622 V_{DS} + 1.11 = 0$$

The two quadratic solutions are

$$V_{DS} = 0.531 \text{ V, } 2.09 \text{ V}$$

The valid solution is $V_{DS} = 0.531$ V, since this satisfies the nonsaturation condition that was used in its solution:

$$V_{GS} > V_{DS} + V_{tn}$$

$$1.8 \text{ V} > 0.531 \text{ V} + 0.6 \text{ V}$$

and

$$\begin{aligned} I_D &= (V_{DD} - V_{DS})/15 \text{ k}\Omega \\ &= (5 \text{ V} - 0.531 \text{ V})/15 \text{ k}\Omega \\ &= 298 \text{ }\mu\text{A} \end{aligned}$$

■

■ EXAMPLE 3.4

What value of R_d will drive transistor M1 in Figure 3.15 just into nonsaturation if $K_n = 50$ μA/V², $V_{tn} = 0.4$ V, and $W/L = 10$?

Since the bias state is at the boundary, either Equation (3.3) or (3.4) can be used. Equation (3.3) is simpler so

$$I_D = \frac{\mu \varepsilon_{ox}}{2 T_{ox}} \frac{W}{L} (V_{GS} - V_{tn})^2 = K_n \frac{W}{L} (V_{GS} - V_{tn})^2$$

$$= (50 \text{ }\mu\text{A})(10)(1.0 - 0.4)^2$$

$$= 180 \text{ }\mu\text{A}$$

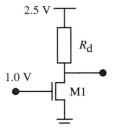

Figure 3.15.

The bias boundary condition

$$V_{GS} = V_{DS} + V_{tn}$$

becomes

$$V_G - V_S = V_D - V_S + V_{tn}$$
$$V_D = V_G - V_{tn}$$
$$V_D = 1.0 \text{ V} - 0.4 \text{ V} = 0.6 \text{ V}$$

Then

$$R_d = \frac{V_{DD} - V_D}{I_D}$$
$$= \frac{2.5 - 0.6}{180 \text{ μA}}$$
$$= 10.56 \text{ kΩ}$$

■ **EXAMPLE 3.5**

Transistors emit light from the drain depletion region when they are in the saturated bias state.
 (a) Show whether this useful failure analysis technique will work for the circuit in Figure 3.16. $V_{tn} = 0.6$ V, $K_n = 75$ μA/V², and $W/L = 2$.
 (b) Find I_D, V_{GS}, and V_{DS}.
 The saturated bias condition is

$$V_{GS} < V_{DS} + V_{tn}$$

or

$$V_G < V_D + V_{tn}$$
$$1.2 \text{ V} < 3.3 \text{ V} + 0.6 \text{ V}$$

so, transistor M1 is saturated and emitting visible light from its drain-channel region.

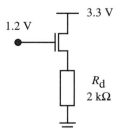

Figure 3.16.

Since M1 is in saturation,

$$I_D = \frac{\mu \varepsilon_{ox}}{2T_{ox}} \frac{W}{L}(V_{GS} - V_{tn})^2 = K_n \frac{W}{L}(V_{GS} - V_{tn})^2$$
$$= (75\ \mu\text{A})(2)[(V_G - V_S) - 0.6]^2$$
$$= (150\ \mu\text{A})[(1.2 - V_S) - 0.6]^2$$
$$= (150\ \mu\text{A})(0.6 - V_S)^2$$

Also,

$$I_D = \frac{V_S}{R_d} = \frac{V_S}{2\ \text{k}\Omega}$$

Then,

$$\frac{V_S}{2\ \text{k}\Omega} = (150\ \mu\text{A})(0.6\ \text{V} - V_S)^2$$

This reduces to

$$V_S^2 - 4.533\ V_S + 0.36 = 0$$

whose two solutions are $V_S = 80.85$ mV and 4.452 V. The valid solution is

$$V_S = 80.85\ \text{mV}$$

and

$$I_D = I_S = V_S/R_d = 80.85\ \text{mV}/2\ \text{k}\Omega = 40.43\ \mu\text{A}$$
$$V_{GS} = 1.2 - 0.08085 = 1.119\ \text{V}$$
$$V_{DS} = 3.2 - 0.08085 = 3.119\ \text{V}$$

∎

Self-Exercise 3.2

Find I_D and V_D in Figure 3.17. Verify the bias state consistency of your choice of MOS drain-current model for $V_{tn} = 1.0$ V, $K_n = 25\ \mu\text{A/V}^2$, and $W/L = 2$.

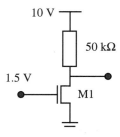

Figure 3.17.

Self-Exercise 3.3

Repeat Self-Exercise 3.2 (Figure 3.17) if $V_G = 3.0$ V and $W/L = 3$.

Self-Exercise 3.4

Calculate V_{GS} and give the correct bias state for transistor M1 in Figure 3.18. $V_{tn} = 0.5$ V.

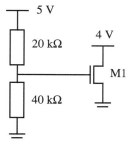

Figure 3.18.

Self-Exercise 3.5

Adjust R_1 in Figure 3.19 so that M1 is on the saturated/ohmic border where $V_{tn} = 0.5$ V.

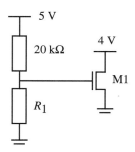

Figure 3.19.

Self-Exercise 3.6

Calculate R_0 in Figure 3.20 so that $V_0 = 2.5$ V. Given: $K_n = 300$ μA/V², $V_{tn} = 0.7$ V, and $W/L = 2$.

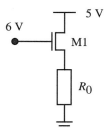

Figure 3.20.

3.1.6 *p*MOS Transistor Output Characteristics

*p*MOS transistor analysis is similar to that for the *n*MOS transistor with a major exception: care must be taken with the polarities of the current and voltages. The *p*MOS transistor major carrier is the hole that emanates from the source, enters the channel, and exits the drain terminal as a negative current. The gate-to-source threshold voltage V_{tp} needed to invert an *n*-substrate is negative to attract holes to the channel surface. The equations to model the *p*MOS transistor in saturation and nonsaturation conditions have a form similar to those for the *n*MOS device, but modified for these polarity considerations. We will choose a *p*MOS transistor equation form that is close to the *n*MOS transistor equations. Equations (3.10) and (3.11) describe the terminal behavior of a *p*MOS device. Remember that V_{GS}, V_{DS}, V_{tp}, and I_D are negative.

$$I_D = -\frac{\mu \varepsilon_{ox}}{2T_{ox}} \frac{W}{L} (V_{GS} - V_{tp})^2 \quad \text{(saturated state)} \quad (3.10)$$

$$I_D = -\frac{\mu \varepsilon_{ox}}{2T_{ox}} \frac{W}{L} [2(V_{GS} - V_{tp})V_{DS} - V_{DS}^2] \quad \text{(ohmic state)} \quad (3.11)$$

Figure 3.21 shows a measured *p*MOS transistor family of curves with all voltages given with respect to the source. The plot is shown in quadrant I, even though the drain current and voltage are negative. This is an author's choice, made to to retain similarity to the *n*MOS transistor family of curves.

The boundary of the bias states can again be found by differentiating Equation (3.11), setting the result to zero, and solving for the conditions to get

$$V_{DS} = V_{GS} - V_{tp} \quad (3.12)$$

The condition for transistor saturation is

$$V_{DS} < V_{GS} - V_{tp} \quad \text{or} \quad V_{GS} > V_{DS} + V_{tp} \quad (3.13)$$

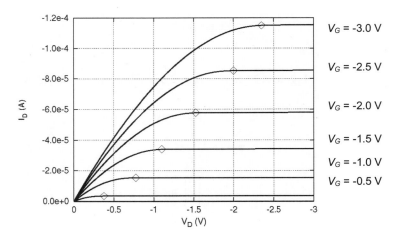

Figure 3.21. *p*MOS transistor family of curves.

and the condition for transistor nonsaturation is

$$V_{DS} > V_{GS} - V_{tp} \quad \text{or} \quad V_{GS} < V_{DS} + V_{tp} \quad (3.14)$$

An example is given using these equations.

■ **EXAMPLE 3.6**

Determine the bias state for the *p*MOS transistors in Figure 3.22, where $V_{tp} = -0.4$ V. The gate terminal has its most negative voltage with respect to the source terminal.

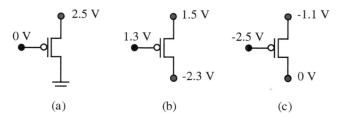

Figure 3.22. Transistors for Example 3.6, with $V_{tp} = -1.2$ V.

(a) $V_{GS} = -2.5$ V and $V_{DS} = -2.5$ V, therefore $V_{GS} > V_{DS} + V_{tp}$, or -2.5 V $> -2.5 + (-0.4)$ V, so the transistor is in saturation.

(b) The gate voltage is not sufficiently more negative than either the drain or source terminal so that the transistor is in the off-state.

(c) $V_{GS} = -2.5 - (-1.1) = -1.4$ V and $V_{DS} = 0 - (-1.1) = 1.1$ V. What is wrong? The gate voltage is sufficiently negative to turn on the transistor, but the source-to-drain voltage is negative. Holes must leave the source and flow to the drain, but they can't under this condition. The answer is that the drain terminal is on the top and the source on the bottom so that $V_{GS} = -2.5 - 0 = -2.5$ V and $V_{DS} = -1.1 - 0 = -1.1$ V. Therefore $V_{GS} < V_{DS} + V_{tp}$, or -2.5 V $< -1.1 + (-1.2)$ V, so the transistor is in nonsaturation. The source terminal always has a higher or equal voltage than the drain terminal in a *p*MOS transistor. ■

Self-Exercise 3.7

Give the correct bias state for the three *p*MOS's shown in Figure 3.23, where $V_{tp} = -0.4$ V.

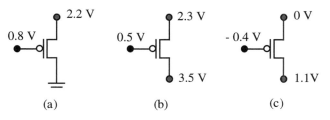

Figure 3.23.

After solving Example 3.6 and Self-Exercise 3.7 with the proper bias state equations, you may check your work by referring to the pMOS transistor family of curves in Figure 3.21. Find the coordinates in the example and exercise, and verify that the bias state is correct. A series of examples and exercises with the pMOS transistor will reinforce these important relations.

■ **EXAMPLE 3.7**

Calculate I_D and V_{DS} for circuit in Figure 3.24. $V_{tp} = -1.0$ V, $K_p = 100$ μA/V², and $W/L = 4$.

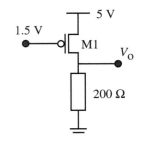

Figure 3.24.

Assume a saturated bias state:

$$I_D = -\frac{\mu\varepsilon_{ox}}{2T_{ox}}\frac{W}{L}(V_{GS} - V_{tp})^2 = -100 \text{ μA}(4)[-3.5 - (-1)]^2$$

$$= -2.5 \text{ mA}$$

$$V_0 = -I_D(200 \text{ Ω}) = (2.5 \text{ mA})(200 \text{ Ω}) = 0.5 \text{ V}$$

then

$$V_{DS} = 0.5 \text{ V} - 5 \text{ V} = -4.5 \text{ V}$$

The bias state consistency is

$$V_{GS} > V_{DS} + V_{tp}$$
$$-3.5 > -4.5 + (-1.0)$$

so the transistor is in the saturated bias state and the solutions are correct. ■

■ **EXAMPLE 3.8**

Calculate I_D and V_{DS} in Figure 3.25. $V_{tp} = -0.6$ V, $K_p = 80$ μA/V², and $W/L = 10$.
Assume a saturated bias state:

3.1 PRINCIPLES OF OPERATION: LONG-CHANNEL TRANSISTORS 73

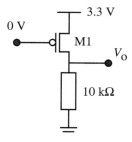

Figure 3.25.

$$I_D = -\frac{\mu\varepsilon_{ox}}{2T_{ox}}\frac{W}{L}(V_{GS} - V_{tp})^2 = -80\ \mu A(10)[-3.3 - (-0.6)]^2$$
$$= -5.832\ \text{mA}$$

Then

$$V_O = -I_D(10\ k\Omega) = -(-5.832\ \text{mA})(10\ k\Omega)$$
$$= 58.32\ V$$

This voltage is beyond the power supply value, and is not possible. The saturated state assumption was wrong, so we must start again using the ohmic state equation:

$$I_D = -\frac{\mu\varepsilon_{ox}}{2T_{ox}}\frac{W}{L}[2(V_{GS} - V_{tp})V_{DS} - V_{DS}^2] = -80\ \mu A(10)[2(-3.3 + 0.6)V_{DS} - V_{DS}^2]$$

Another equation is required to solve the problem, so using the KVL (Ohm's law, here)

$$-I_D = \frac{V_D}{R_d} = \frac{V_{DD} + V_{DS}}{10\ k\Omega}$$
$$= \frac{3.3 + V_{DS}}{10\ k\Omega} = 80\ \mu A(10)[2(-3.3 + 0.6)V_{DS} - V_{DS}^2]$$

The two quadratic solutions are: $V_{DS} = -75.70$ mV and -5.450 V. The correct solution is $V_{DS} = -75.70$ mV. Therefore

$$V_O = 3.3\ V + (-75.70\ \text{mV}) = 3.224\ V$$

■ **EXAMPLE 3.9**

Calculate I_D and V_{SD}, and verify the assumed bias state of transistor M1 for the circuit in Figure 3.26. $V_{tp} = -0.4$ V, $K_p = 60\ \mu A/V^2$, and $W/L = 2$.

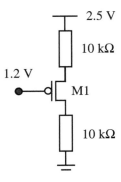

Figure 3.26.

Assume a saturated bias state and

$$I_D = -\frac{\mu\varepsilon_{ox}}{2T_{ox}}\frac{W}{L}(V_{GS} - V_{tp})^2$$

Since V_{GS} is not known, we must search for another expression to supplement this equation. We can use the KVL statement:

$$V_{GS} = 1.2 - [V_{DD} - (-I_D R_S)]$$
$$= 1.2 - 2.5 + I_D R_S$$
$$= -1.3 + (10 \text{ k}\Omega)I_D$$

We equate this expression to the saturated current expression to get

$$I_D = -60 \text{ μA}(2)[-1.3 + (10 \text{ k}\Omega)I_D + 0.4]^2$$
$$= -120 \text{ μA}[-0.9 + (10 \text{ k}\Omega)I_D]^2$$

This quadratic equation in I_D gives solutions

$$I_D = -35.56 \text{ μA} \quad \text{and} \quad -227.8 \text{ μA}$$

The valid solution is $I_D = -35.56$ μA, since the other solution for I_D, when multiplied by the sum of the two resistors, gives a voltage greater than the power supply. V_{SD} is then

$$V_{SD} = V_{DD} - I_D(20 \text{ k}\Omega)$$
$$V_{SD} = I_D(20 \text{ k}\Omega) - V_{DD}$$
$$= 2.5 \text{ V} - (35.56 \text{ μA})(20 \text{ k}\Omega)$$
$$= 1.789 \text{ V}$$

and

$$V_S = V_{DD} - I_D R_S = 2.5 + (-35.56\ \mu A)(10\ k\Omega)$$
$$= 2.144\ V$$

so that

$$V_{GS} = V_G - V_S = 1.2 - 2.144 = -0.944\ V$$

Transistor M1 is in saturation since

$$V_{GS} > V_{DS} + V_{tp}$$
$$-0.944\ V > -1.789\ V - 0.4\ V$$

■

■ EXAMPLE 3.10

What value of R_d in Figure 3.27 will raise V_0 to half of the power supply voltage (i.e., $V_0 = 0.5\ V_{DD}$). $V_{tp} = -0.7\ V$, $K_p = 80\ \mu A/V^2$, and $W/L = 5$.

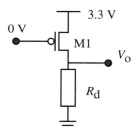

Figure 3.27.

Check for bias state consistency:

$$V_{GS} < V_{DS} + V_{tp}$$
$$-3.3\ V < -1.65\ V - 0.7\ V$$

So M1 is in ohmic bias state. Therefore, we use the ohmic state equation, where $V_{DS} = -1.65\ V$:

$$I_D = -\frac{\mu \varepsilon_{ox}}{2T_{ox}} \frac{W}{L} [2(V_{GS} - V_{tp})V_{DS} - V_{DS}^2]$$
$$= -80\ \mu A(5)\{2[-3.3 - (-0.7)](-1.65) - 1.65^2\}$$
$$= -2.343\ mA$$

Then

$$R_d = \frac{V_0}{-I_D} = \frac{1.65}{2.343 \text{ mA}} = 704.2 \text{ }\Omega$$

∎

Self-Exercise 3.8

Calculate I_D and V_0 for circuit in Figure 3.28. $V_{tp} = -0.8$ V, $K_p = 30$ µA/V², and $W/L = 2$.

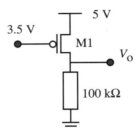

Figure 3.28.

Self-Exercise 3.9

Repeat Self-Exercise 3.8, but let $V_G = 1.5$ V.

Self-Exercise 3.10

Find I_D and V_0 for the circuit in Figure 3.29. $V_{tp} = -0.6$ V, $K_p = 20$ µA/V², and $W/L = 3$.

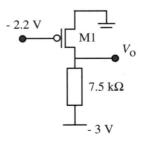

Figure 3.29.

Self-Exercise 3.11

The voltage drop across each of the two identical resistors and V_{DS} are equal for the circuit in Figure 3.30. $V_{tp} = -0.5$ V, $K_p = 100$ μA/V², and $W/L = 2$. Find the value of the resistors.

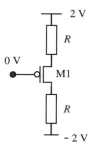

Figure 3.30.

Self-Exercise 3.12

For the circuit in Figure 3.31, $V_{tp} = -0.8$ V and $K_p = 100$ μA/V². What is the required W/L ratio if M1 is to pass 0.5 A and keep $V_{SD} < 0.1$ V.

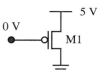

Figure 3.31.

These many examples and exercises with MOS transistors have a purpose. These problems, when combined with transistor family of curves plots, should now allow you to think in terms of a transistor's reaction to its voltage environment. This is basic to electronics engineering instruction. It should allow you to quickly anticipate and recognize aberrations caused by defective circuits, and later to predict what category of defect exists. The techniques needed to solve these problems should become reflexive.

3.2 THRESHOLD VOLTAGE IN MOS TRANSISTORS

Until now, we assumed that the transistor source and substrate terminals were connected to the same voltage. This is valid for isolated transistors, but when transistors are connected in CMOS circuits, this condition may not hold for all devices.

Figure 3.32 is a circuit cross section of two *n*MOS and one *p*MOS transistors fabricated in a CMOS process. All devices are constructed on the same *p*-type silicon substrate. Since *p*MOS transistors are formed on *n*-type substrates, there must be a region of the circuit that is oppositely doped to the initial bulk, forming what is called a *well*.

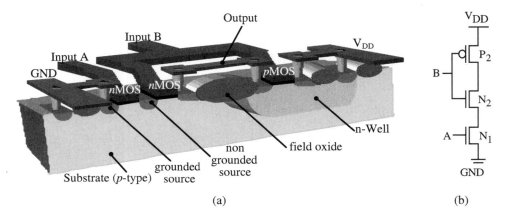

Figure 3.32. (a) Structure for two series-connected *n*MOS transistors and one *p*MOS transistor in a CMOS technology. (b) Circuit schematic.

The *p*-type substrate for *n*MOS transistors is connected to zero (or ground, GND), whereas the *n*-type well is connected to V_{DD}, since it forms the bulk of the *p*MOS transistors. The source of the *n*MOS device N_1 is connected to ground, so that previous equations are valid for this device. The transistor N_2 source is connected to the drain of N_1 to make a series connection of both devices, required to implement the operation of the logic gate (a detailed analysis of transistor interconnection to form gates is given in Chapter 5). As a result, the source of transistor N_2 is not grounded, and it can acquire voltages close to V_{DD}, whereas its substrate is connected to ground through the polarizing contact. Therefore, the condition $V_{SB} = 0$ will not hold in some bias cases for transistor N_2.

When the source and substrate voltages differ, the gate–source voltage is not fully related to the vertical electric field responsible for creating the channel. The effect of the higher source voltage above (below) the substrate for an *n*MOS (*p*MOS) transistor is to lower the electric field induced from the gate to attract carriers to channel. The result is an effective raising of the transistor threshold voltage. The threshold voltage can be estimated as [8]

$$V_t = V_{t0} \pm \gamma \sqrt{V_{SB}} \qquad (3.15)$$

where V_{t0} is the threshold voltage when the source and the substrate are at the same voltage, and γ is a parameter dependent on the technology. The parameter γ is called the body effect constant. The positive sign is used for *n*MOS transistors, and the negative sign for *p*MOS transistors. When the source and substrate are tied together, $V_{SB} = 0$, and the threshold voltage is constant.

The significance of the threshold body effect lies with certain circuit configurations whose transistor thresholds will be altered, generally being higher than expected. This can lead to conduction states and changes in transistor delay time. We will return to this topic when we discuss pass transistor properties, particularly in memories, and circuits such as that in Figure 3.32.

3.3 PARASITIC CAPACITORS IN MOS TRANSISTORS

We have learned the equations that describe the static operation of the transistor, i.e., the current into the device when voltage nodes remain stable with time, but the dynamic operation requires knowledge of other aspects of the devices.

One limitation of high-speed digital ICs is the time required to switch a transistor between the on- and off-states. This delay mechanism is primarily due to transistor parasitic capacitors that fall into two types: voltage-dependent and non-voltage-dependent capacitors. Non-voltage-dependent capacitors are characterized by physical overlap of the gate terminal with the drain and source areas. The voltage-dependent capacitors are the reverse-biased drain–substrate and source–substrate diodes, plus those characterized by the creation of depletion regions and conducting channels [3]. Another significant cause of delay in modern ICs is the capacitance of the interconnect wires between transistors.

3.3.1 Non-Voltage-Dependent Internal Capacitors

Non-voltage-dependent capacitor values are calculated from device dimensions using a parallel plate model. A parallel plate capacitor has two conductors of area (A) separated by a distance (d). The space between the conductors can be empty or filled with an insulator to increase the capacitance value. The capacitance is

$$C_p = \left(\frac{\varepsilon_0 \varepsilon_m}{d}\right) A \tag{3.16}$$

where ε_0 is the permittivity of free space, and ε_m is the relative permittivity of the material filling the capacitor. The non-voltage-dependent capacitors in a MOSFET device are C_{GDov} and C_{GSov} shown in Figure 3.33. Their value is found by applying Equation (3.16) to the overlap region between the gate and the drain.

$$C_{ov} = \left(\frac{\varepsilon_0 \varepsilon_{Si}}{T_{ox}}\right) W \cdot L_D \tag{3.17}$$

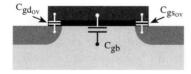

(a) No bias applied (all terminals grounded)

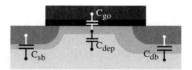

(b) Depletion or weak inversion

(c) Nonsaturation

(d) Saturation

Figure 3.33. Parasitic capacitors of MOS transistors for four operating regions.

where ε_{Si} is the relative permittivity of the silicon dioxide, W is the transistor width, L_D is the overlap distance between the gate and the drain or source, and T_{ox} is the gate oxide thickness.

3.3.2 Voltage-Dependent Internal Capacitors

Once the capacitor is biased, the depletion and inversion regions change the effective internal parasitic capacitors. These capacitors are voltage-dependent since their value is related to charge redistribution within the device. We will provide approximate expressions for such capacitors.

In all operation regions, consider the following three terminal capacitors:

C_{gb} Gate–substrate (gate–bulk) capacitor. It has a strong dependence on the transistor biasing.

C_{gs}, C_{gd} Gate–channel capacitors. They are divided into gate–drain and gate–source capacitors, since the channel distribution is not uniform along the device in saturation.

C_{sb}, C_{db} Source and drain-to-bulk capacitors. Their capacitance is due to the reverse-bias, built-in diodes.

The total voltage-dependent gate capacitance of a MOS is found by summing the capacitors:

$$C_g = C_{gb} + C_{gs} + C_{gd} \quad (3.18)$$

The drain/source-to-bulk capacitors do not impact the gate voltage. These capacitors can be calculated from the reverse diode capacitor expression in Equation 2.12, so that

$$C_{xb} = \frac{C_{xb0}}{\left(1 - \dfrac{V_{xb}}{V_{bi}}\right)^{1/2}} \quad (3.19)$$

where x must be replaced by d or s to refer to the drain and source terminals respectively. The value of C_g must be computed for each transistor operation region.

No Biasing (All Terminals at the Same Voltage). The gate–substrate capacitance is due to the MOS structure, and is calculated as a parallel plate capacitor filled by the gate oxide. This "physical" capacitor is referred to as C_{g0} and has the expression

$$C_{gb} = C_{g0} = \frac{\varepsilon_0 \varepsilon_{Si}}{T_{ox}} W \cdot L_{eff} \quad (3.20)$$

where L_{eff} is the transistor effective length. It differs from the physical length L drawn in the design because the drain and source regions diffuse under the gate making $L_{eff} = -2L_{overlap}$. The only difference between Equations (3.17) and (3.20) is the capacitor area. In Equation (3.20) the area is that of the whole device, whereas in Equation (3.17) the area was only the overlap region between the gate and the drain or source.

Depletion or Weak Inversion. Although conceptually different, depletion and weak inversion will be treated as equivalent. In these regions, the gate voltage is not sufficient to create a channel, and the electric field from the gate induces a depletion region within the bulk (Figure 3.33(b)). The gate capacitor is the series connection of C_{g0} and the depletion capacitor C_{dep}. The computation of the depletion capacitance is complicated, and beyond the scope of this book.

Nonsaturation. When the transistor is in nonsaturation, the conducting channel "touches" the drain and source terminals (Figure 3.33(c)). Thus, C_{gs} and G_{gd} dominate and the gate–bulk capacitor is now negligible, since the conducting layer "disconnects" the gate from the bulk. As a result, the gate–bulk capacitor C_{g0} is shared between C_{gd} and C_{gs} so

$$C_{gs} = C_{gd} = \frac{\varepsilon_0 \varepsilon_{Si}}{2T_{ox}} W \cdot L_{eff} \quad (3.21)$$

Saturation. Once in saturation, the conduction channel no longer touches the drain end. The gate–drain channel capacitance is now negligible, and all gate capacitances are connected to the source terminal. The gate–source capacitance can be approximated as [8]

$$C_{gs} = \frac{2\varepsilon_0 \varepsilon_{Si}}{3T_{ox}} W \cdot L_{eff} \quad (3.22)$$

The simplifications made in the computation of the gate capacitance are summarized in Table 3.1.

Several forms of transistor capacitances have been described. Although we did not stress numerical work, you should assimilate the locations and different properties of each. Dynamic performance of an IC depends upon these capacitances, and also upon parasitic resistances of the transistors and interconnect wires. The latter elements are discussed in Chapter 9. We will next explore the influence on the basic transistor properties described thus far of the extreme scaling of dimensions of modern ICs.

3.4 DEVICE SCALING: SHORT-CHANNEL MOS TRANSISTORS

Device scaling is relentless in the microelectronics industry since transistor miniaturization allows faster device operation, yield improvement, and cost savings resulting from

Table 3.1. Simplified Intrinsic MOS Capacitor Expressions for Each Operating Region

	C_{gb}	C_{gs}	C_{gd}
Cutoff	$\frac{\varepsilon_0 \varepsilon_{Si}}{T_{ox}} W \cdot L_{eff}$	0	0
Ohmic	0	$\frac{\varepsilon_0 \varepsilon_{Si}}{2T_{ox}} W \cdot L_{eff}$	$\frac{\varepsilon_0 \varepsilon_{Si}}{2T_{ox}} W \cdot L_{eff}$
Saturation	0	$\frac{2\varepsilon_0 \varepsilon_{Si}}{3T_{ox}} W \cdot L_{eff}$	0

82 CHAPTER 3 MOSFET TRANSISTORS

more dies per wafer. There is an important barrier encountered when minimum dimensions go below about 0.5 μm channel lengths since new device physical properties appear. Devices with dimensions more than 0.5 μm are usually called long-channel devices, whereas smaller transistors are called deep submicron or short-channel devices. The division at 0.5 μm is slightly arbitrary, but gives an approximate dimension for entering the short channel and deep submicron regions of electronics.

Figure 3.34 compares the output current characteristics for a long-channel transistor (1.2 μm technology) and a submicron technology device (0.25 μm technology). The major differences are

- The spacing between equally incremented drain–current versus gate–voltage curves is constant in short-channel transistors in the saturation region, whereas the long-channel transistor spacing between these curves increases nonlinearly for increasing gate voltages.
- Once the device reaches saturation, the current remains nearly flat for the long-channel device (it has small dependence on V_{DS} for this region), whereas for the short-channel transistor the saturated current shows more slope.
- The third difference relates to the total amount of current. If two transistors have equal gate size and are at the same gate and drain voltage, then the short-channel transistor passes more drain current than the long-channel one. This is largely due to the thinner oxides used in short-channel transistors.

We will identify the mechanisms causing these differences, and present equations describing these behaviors for short-channel transistors. Short-channel transistor analysis has more complexity than long-channel analysis. It requires detailed concentration on your part, but we hope that you will obtain an understanding of the difficulty of manual analysis of the sort similar to that for long-channel transistors. We acknowledge that we build on the impressive work of others, such as Foty [2], Tsividis [7], and Weste and Eshraghian [8].

There are many submicron effects, but we will focus on channel length modulation, velocity saturation, subthreshold current, DIBL, and hot-carrier effects, since they may have

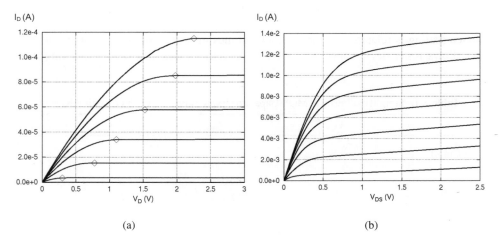

Figure 3.34. Current characteristics for (a) long-channel, and (b) short-channel transistors.

a significant impact on IC testing and reliability. Our goal is to understand short-channel effects, and develop tools that allow manual analysis as we did for long-channel transistors.

3.4.1 Channel Length Modulation

Channel length modulation appears when the device is in saturation and is related to the pinch-off effect described earlier. When the device reaches saturation, the channel no longer "touches" the drain and acquires an asymmetric shape that is thinner at the drain end (Figure 3.33(d)). As the drain–source voltage increases, the channel saturation or depletion region moves further away from the drain end because the drain electric field "pushes" it back. The reverse-bias depletion region widens, and the effective channel length decreases by an amount ΔL (Figure 3.35) for increasing V_{DS}. In large devices, the relative change of the effective channel length ΔL with respect to the total channel length with V_{DS} is negligible, but for shorter devices $\Delta L/L$ becomes important.

The effective length of the device varies with V_{DS} once the device is in saturation, and, as a result, the curves are no longer flat in this region compared to long-channel devices (Figure 3.34(b)).

The small drain current dependence on the drain–source voltage in the saturation region can be modeled by multiplying the saturation current by a slope factor that depends on this voltage. The resulting equation is

$$I_D = I_{Dsat}(1 + \lambda V_{DS}) \qquad (3.23)$$

where λ is a parameter that can be measured for each particular technology.

3.4.2 Velocity Saturation

When charged carriers move in a solid under the force of an electric field, they acquire a velocity proportional to the magnitude of this electric field. The applied electric field ($\mathscr{E}$) and the carrier velocity (v) are related through the mobility parameter μ, introduced in Chapter 2 as

$$v = \mu \mathscr{E} \qquad (3.24)$$

For small electric fields, μ is constant and independent of the applied electric field.

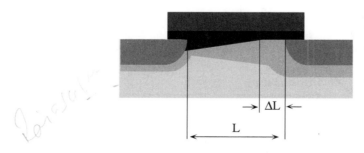

Figure 3.35. The channel length modulation effect.

$$\mu = \mu_0 \text{ (constant)} \tag{3.25}$$

As a result when carrier velocity (v) is plotted versus the applied electric field ($\mathscr{E}$), the result is a straight line (low electric field region of Figure 3.36).

The reason for this linear dependence between velocity and small electric fields is that electrons moving in semiconductors collide with silicon atoms, an effect known as scattering. Electron scattering is linear with small electric fields. If the electric field further increases, the carrier velocity enters a region in which it is said to move at *velocity saturation*. As device dimensions scale down, the electric fields within the transistor increase, making the velocity saturation more important. Velocity saturation due to mobility reduction is important in submicron devices.

If Equation (3.24) holds for carriers moving in small electric fields and carriers moving at the velocity saturation, then the mobility μ must change with the electric field (Figure 3.36). Two effects are combined in transistors to account for this mobility: reduction due to the horizontal electric field and mobility reduction due to the vertical electric field.

Horizontal Electric Field Mobility Reduction. Carriers in short-channel devices reach velocity saturation at lower values of V_{DS} than for long-channel transistors. Figure 3.37(a) illustrates the effect of V_{Dsat} moving toward small values, and shows that saturation current is smaller in short-channel transistors. This effect is due to channel length reduction that implies higher horizontal electric fields for equivalent drain–source voltages. The horizontal electric field within the channel is due to the voltage applied to the drain terminal. Horizontal mobility (μ_H) reduction can be related to the drain voltage by

$$\mu_H = \frac{\mu_0}{1 + \dfrac{V_{DS}}{L_{eff}\mathscr{E}_{crit}}} = \frac{\mu_0}{1 + \theta_2 V_{DS}} \tag{3.26}$$

$1/(L_{eff}\mathscr{E}_{crit})$ is a parameter called *drain bias mobility reduction* that in some texts is referred to as θ_2. $\mathscr{E}_{crit}$ is the electric field shown in Figure 3.36, and depends on the technology. For large transistors, $\theta_2 V_{DS}$ is smaller than 1, and, therefore, the mobility is constant ($\mu_H \approx \mu_0$), giving the linear relation between velocity and electric field for this region. When L_{eff} decreases ($L_{eff} \rightarrow 0$), θ_2 increases, and $\theta_2 V_{DS}$ becomes important, lowering the mobility below μ_0. Even for short-channel transistors, when V_{DS} is small, the denominator

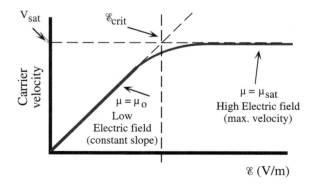

Figure 3.36. Electric field to carrier velocity dependence in a solid.

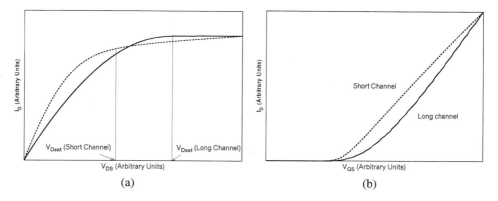

Figure 3.37. Two effects of velocity saturation on submicron devices. (a) The saturation voltage V_{Dsat} moves to lower values, and (b) the I_D versus V_{GS} relationship becomes linear.

of Equation (3.26) is almost equal to 1, and the mobility is constant. When V_{DS} increases, the denominator of Equation (3.26) becomes much larger than one, and the mobility is reduced. Equation (3.26) is a simple model of mobility reduction in submicron transistors caused by the lateral (horizontal) electric field.

Vertical Electric Field Mobility Reduction. There is also a vertical electric field due to the gate voltage that creates the conduction channel. When carriers move within the channel under the effect of the horizontal field, they "feel" the effect of this gate–substrate-induced electrical vertical field, pushing carriers toward the gate oxide. This provokes carrier collisions with the oxide–channel interface, reducing their mobility. The interface of Si and SiO_2 is rough and imperfect, so that carriers move with more difficulty.

The mobility reduction from this effect can be described in a similar way to that in the horizontal field mobility reduction [Equation (3.26)]. Now, the expression of the vertical mobility (μ_V) reduction will contain the gate–source voltage instead of the drain–source voltage:

$$\mu_V = \frac{\mu_0}{1 + \theta_1(V_{GS} - V_t)} \quad (3.27)$$

Similar to Equation (3.26), θ_1 is a parameter called *gate bias mobility reduction* that depends on the technology.

3.4.3 Putting it All Together: A Physically Based Model

We discussed two mechanisms that impact transistor behavior when technology scales down. Although other effects appear in submicron technologies, channel length modulation and mobility reduction are the two main effects used in the device model equations in CAD simulators. We now introduce model equations to describe the drain current of submicron devices. The objective is to give the reader basic and accurate expressions to account for the most relevant submicron effects. Device modeling is an active research area. For deeper discussions refer to [2] and [7].

The long-channel transistor drain-current square law dependence on the gate–source voltage for ohmic and saturation conditions are repeated from Equations (3.3) and (3.4):

$$I_D = \mu_0 C_{ox} \left[\frac{W}{L}(V_{GS} - V_{tn})V_{DS} - \frac{V_{DS}^2}{2} \right] \quad \text{(ohmic)}$$

$$I_D = \mu_0 C_{ox} \frac{W}{L}(V_{GS} - V_{tn})^2 \quad \text{(saturation)}$$

(3.28)

where $C_{ox} = \varepsilon_{ox}/T_{ox}$. These expressions can be modified to incorporate the effects described previously for short-channel transistors.

Ohmic State. The ohmic region expression for short-channel devices is similar to Equation (3.28), including the velocity saturation effects:

$$I_D = \mu_0 C_{ox} \frac{W}{L_{eff}} \frac{(V_{GS} - V_t)V_{DS} - \frac{V_{DS}^2}{2}(1 + \delta)}{[1 + \theta_1(V_{GS} - V_t)]\left(1 + \frac{V_{DS}}{L_{eff}\mathscr{E}_{crit}}\right)} \quad (3.29)$$

Equation (3.29) combines the horizontal and vertical mobility reduction [the two expressions in brackets of the denominator taken from Equations (3.26) and (3.27)]. The parameter δ relates the charge at the inversion channel to the surface potential and the oxide capacitance, and will not be described in detail.

In addition to the drain current expression, we need an expression for the saturation voltage V_{Dsat} that describes carrier saturation at smaller drain voltages (Figure 3.37(a)). This saturation voltage depends on the gate voltage, and is calculated by differentiating the drain current in the ohmic region [Equation (3.29) in this case] and setting the expression equal to zero. This leads to

$$V_{Dsat} = \frac{2(V_{GS} - V_t)}{(1 + \delta)\left(1 + \sqrt{1 + \frac{2(V_{GS} - V_t)}{L_{eff}\mathscr{E}_{crit}(1 + \delta)}}\right)} \quad (3.30)$$

The short-channel drain saturation voltage depends on the square root of the gate voltage, whereas for long-channel devices the term in the square root tends to 1, and the saturation voltage depends on $(V_{GS} - V_t)$.

Saturated State ($V_{DS} > V_{Dsat}$). The short-channel drain current expression in saturation is obtained from Equation (3.29), substituting V_{DS} by V_{Dsat} and including the channel modulation effect from Equation (3.23):

$$I_D = \mu_0 C_{ox} \frac{W}{L_{eff}} \frac{(V_{GS} - V_t)V_{Dsat} - \frac{V_{Dsat}^2}{2}(1 + \delta)}{[1 + \theta_1(V_{GS} - V_t)]\left(1 + \frac{V_{Dsat}}{L_{ef}\mathscr{E}_{crit}}\right)} [1 + \lambda(V_{DS} - V_{Dsat})] \quad (3.31)$$

Equations (3.29) and (3.31) describe the drain current of submicron MOS transistors for the ohmic and saturated states, respectively. The transition of the drain current between states described by these equations occurs sharply and can lead to computation errors from derivative discontinuities. Equation (3.32) [7] describes the drain current for a MOS transistor using only one expression, which is valid for ohmic and saturated states. It is taken from Equation (3.31), where V_{Dsat} is substituted by V_{DS1}. V_{DS1} is an internal drain voltage that makes a smooth transition from V_{DS} in Equation (3.29) to V_{Dsat} in Equation (3.31) when changing from the ohmic state to the saturated one, as shown in Figure 3.38:

$$I_D = \mu_0 C_{ox} \frac{W}{L_{eff}} \frac{(V_{GS} - V_t)V_{DS1} - \frac{V_{DS1}^2}{2}(1+\delta)}{[1 + \theta_1(V_{GS} - V_t)]\left(1 + \frac{V}{L_{eff}\mathcal{E}_{crit}}\right)}[1 + \lambda(V_{DS} - V_{DS1})] \quad (3.32)$$

Figure 3.38 shows that the internal voltage V_{DS1} is mainly the drain–source voltage for small applied V_{DS}. Once the device enters saturation, the internal drain–source voltage becomes V_{Dsat}.

Figure 3.39 compares measured drain current for nMOS and pMOS transistors versus the drain current equation [Equation (3.32)], showing that the model fits the measured transistor curves.

3.4.4 An Empirical Short-Channel Model for Manual Calculations

Equations (3.29) and (3.31) have a physical basis. They provide understanding of how these physical effects impact the device characteristics, but lack simplicity for hand calculations. Another approach derives mathematical expressions that fit transistor curves over both bias ranges. These expressions use empirical parameters to specifically match measured device characteristics. These parameters typically do not have a physical relation with the underlying device mechanism, and are called fitting parameters (they fit a curve with a mathematical expression).

Empirically based models afford the possibility of deriving a mathematical expression for hand calculations. This section introduces an empirical model for short-channel tran-

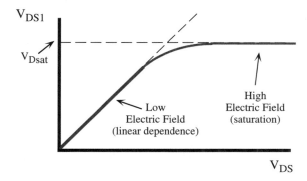

Figure 3.38. Internal drain–source voltage versus applied drain source voltage [V_{DS1} in Equation (3.29)].

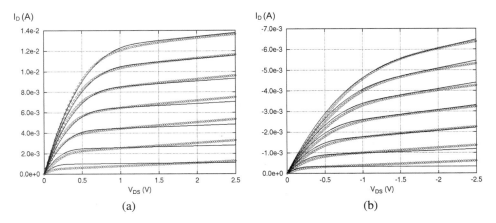

Figure 3.39. Comparison of experimental data (diamonds) with the drain current model of Equation (3.32) (lines) for submicron (a) *n*-type and (b) *p*-type transistors.

sistors applicable for manual calculations. These equations can compute voltage and current in simple circuits with just a few transistors. This is valuable when computing electrical parameters in small circuits in which fabrication defects are present. This empirical model is based on the Sakurai model [5].

We begin with the transistor in saturation. The model introduces a maximum current parameter I_{D0} that is the current when $V_{DS} = V_{GS} = V_{DD}$ (V_{DD} is the maximum voltage in the circuit). The drain current in saturation has a small linear dependence with the drain voltage. This dependency is described with a second parameter, λ, similar to Equation (3.23).

Finally, we know that the drain current has a quadratic dependence with the gate voltage for long-channel devices and a linear dependence for submicron devices [Figure 3.37(b)]. To account for this variation, we use a third fitting parameter called α (taken from Sakurai [5]). This parameter is equal to 1 for short-channel devices and equal to 2 for long-channel transistors, and can be somewhat related to carrier velocity saturation. A mathematical expression for the drain current in saturation is

$$I_D = I_{D0}\left(\frac{V_{GS} - V_{th}}{V_{DD} - V_{th}}\right)^\alpha [1 + \lambda(V_{DS} - V_{DD})] \quad \text{for} \quad V_{DS} > V_{Dsat} \quad (3.33)$$

We "constructed" this equation to fulfill these conditions:

- The drain current is equal to I_{D0} when $V_{GS} = V_{DD}$ and $V_{DS} = V_{DD}$.
- The drain current dependence on the drain–source voltage is linear with slope λ.
- When $\alpha = 1$ (submicron transistors), the drain current dependence with the gate voltage is linear, and for large devices the dependence is quadratic ($\alpha = 2$).

Equation (3.33) was constructed to fit experimental results. None of the parameters introduced are founded on physical phenomena in the transistor, although we could provide a physical meaning for some of them.

Equation (3.33) holds when the drain voltage is beyond the saturation voltage V_{Dsat} that is empirically defined as

$$V_{\text{Dsat}} = V_{\text{D0}} \left(\frac{V_{\text{GS}} - V_{th}}{V_{\text{DD}} - V_{th}} \right)^{\alpha/2} \qquad (3.34)$$

where V_{D0} is the saturation voltage when $V_{\text{DS}} = V_{\text{GS}} = V_{\text{DD}}$, and the exponential dependence with $\alpha/2$ was observed experimentally.

We next need an expression for the drain current in nonsaturation. In this region, drain current has a quadratic dependence on the drain voltage, and when $V_{\text{DS}} = V_{\text{Dsat}}$ its value must match the current value for saturation from Equations (3.33) and (3.34). This gives

$$I_{\text{DS}} = \left(2 - \frac{V_{\text{DS}}}{V_{\text{Dsat}}}\right) \frac{V_{\text{DS}}}{V_{\text{Dsat}}} I_{\text{D0}} \left(\frac{V_{\text{GS}} - V_{th}}{V_{\text{DD}} - V_{th}} \right)^{\alpha} [1 + \lambda(V_{\text{Dsat}} - V_{\text{DD}})] \quad \text{for} \quad V_{\text{DS}} < V_{\text{Dsat}} \qquad (3.35)$$

We introduced no new parameters for this region. The reader can easily verify that Equations (3.35) and (3.33) have the same expression for $V_{\text{DS}} = V_{\text{Dsat}}$. It is also easy to verify that for $\lambda \neq 0$ there is a slope discontinuity at this point. A discontinuity is not desirable for models used in CAD tools or simulators because convergence problems must be avoided. In our case, the aim of this model is hand calculations in which this problem does not arise.

Combining both expressions and neglecting subthreshold leakage:

$$I_D = \begin{cases} 0 & \text{when } V_{\text{GS}} < V_{th} \\ \left(2 - \frac{V_{\text{DS}}}{V_{\text{Dsat}}}\right) \frac{V_{\text{DS}}}{V_{\text{Dsat}}} I_{\text{D0}} \left(\frac{V_{\text{GS}} - V_{th}}{V_{\text{DD}} - V_{th}} \right)^{\alpha} [1 + \lambda(V_{\text{Dsat}} - V_{\text{DD}})] & \text{when } V_{\text{DS}} < V_{\text{Dsat}} \\ I_{\text{D0}} \left(\frac{V_{\text{GS}} - V_{th}}{V_{\text{DD}} - V_{th}} \right)^{\alpha} [1 + \lambda(V_{\text{DS}} - V_{\text{DD}})] & \text{when } V_{\text{DS}} \geq V_{\text{Dsat}} \end{cases} \qquad (3.36)$$

with V_{Dsat} given by Equation (3.34).

Figure 3.40 compares experimental data for a submicron process (0.25 μm) and the

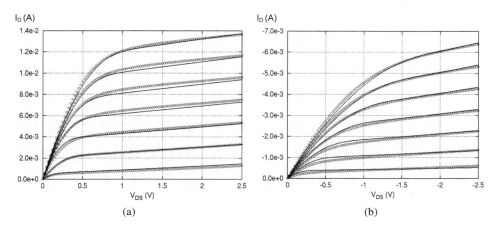

Figure 3.40. Comparison of experimental data for (a) *n*MOS, and (b) *p*MOS short-channel transistors with the empirical model of Equation (3.36).

empirical model. The fitting parameters are $\alpha = 1.12$, $I_{D0} = 13.46$ mA, $V_{D0} = 1$ V, $\lambda = 0.08$, $V_{DD} = 2.25$ V, and $V_{th} = 0.64$ V.

■ EXAMPLE 3.11

Calculate I_D and V_{DS} for the circuit in Figure 3.41 using the empirical model with the parameters $\alpha = 1.12$, $I_{D0} = 13.46$ mA, $V_{D0} = 1$ V, $\lambda = 0.08$, and $V_{th} = 0.64$ V.

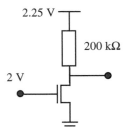

Figure 3.41.

First, find the region of operation for the device. We initially assume that the device is in saturation. From Equation (3.33), the drain current would be

$$I_D = I_{D0}\left(\frac{V_{GS} - V_{th}}{V_{DD} - V_{th}}\right)^\alpha [1 + \lambda(V_{DS} - V_{DD})]$$

$$= 13.46 \text{ mA}\left(\frac{2 - 0.64}{2.25 - 0.64}\right)^{1.12}[1 + 0.08(V_{DS} - 2.25)]$$

$$= (9.136 + 0.8914 V_{DS}) \text{ mA}$$

Applying KVL to the circuit,

$$V_{DD} = V_{DS} + I_D R$$
$$V_{DS} = V_{DD} - I_D R$$
$$V_{DS} = 2 \text{ V} - I_D \, 200 \, \Omega$$

Solving both equations:

$$V_{DS} = 0.147 \text{ V}$$

We must verify that the device is in saturation, so we calculate V_{Dsat} and make sure that $V_{DS} > V_{Dsat}$. Using Equation (1.34),

$$V_{Dsat} = V_{D0}\left(\frac{V_{GS} - V_{th}}{V_{DD} - V_{th}}\right)^{\alpha/2}$$

$$= 1\left(\frac{2 - 0.65}{2.25 - 0.65}\right)^{0.56}$$

$$= 0.91 \text{ V}$$

since $V_{DS} < V_{Dsat}$, the device is in the ohmic region. We must recalculate the drain current using Equation (3.35):

$$I_D = \left(2 - \frac{V_{DS}}{V_{DSsat}}\right)\frac{V_{DS}}{V_{DSsat}} I_{D0}\left(\frac{V_{GS} - V_{th}}{V_{DD} - V_{th}}\right)^\alpha [1 + \lambda(V_{DS} - V_{DD})]$$

$$\text{Combining with } I_D = \frac{2.25 - V_{DS}}{200}$$

$$12.01 V_{DS}^2 - 26.86 V_{DS} + 11.5 = 0$$

Solving with the KVL equation derived above we get

$$V_{DS} = 0.56 \text{ V} \quad \text{and} \quad V_{DS} = 1.68 \text{ V}$$

This solution is valid for the device to be in ohmic state. Therefore, the solution of the problem is

$$I_D = 8.47 \text{ mA}$$
$$V_{DS} = 0.56 \text{ V}$$

∎

Self-Exercise 3.13

Repeat the problem of Example 3.11 with $V_{GS} = 1.5$ V.

Self-Exercise 3.14

Repeat the problem of Example 3.11 with $V_{GS} = 0.75$ V.

Short-channel transistors require specific constants for each technology, and they are not intuitive. As a result, many engineers still use long-channel equations for the "back of the envelope" estimations, acknowledging the increased error. Computers use complicated models to obtain accurate results [2], but do not give a feel for the underlying electronic physics. This conflict of rapid, more inaccurate hand calculations versus accurate computer calculation is unavoidable. We need an accurate approach, and we need an approach that rapidly gives us insight into physical behavior.

3.4.5 Other Submicron Effects

MOSFET devices show other effects when scaled down. We will not give analytical descriptions for these effects, but will briefly introduce them. The effects are: *subthreshold current, drain-induced barrier lowering (DIBL),* and *hot carrier effects.*

Subthreshold Current. Subthreshold current is the drain–source current when the gate–source voltage is below the transistor threshold voltage. The threshold voltage distinguishes the conduction from the nonconduction states of a MOS transistor. This gives the threshold voltage a vague definition. The transition from the conducting to the nonconducting state is not sharp, but continuous. This means that when the gate–substrate voltage increases, the charge in the channel is not created abruptly, but appears gradually with V_{GS}. There is a range of gate voltages lower than V_t for which there are carriers in the inversion layer that contribute to the drain current. In this region, the number of carriers that constitute the channel varies exponentially with the gate voltage.

We want no current when the transistor is in the off-state, i.e., when the gate voltage is below the threshold voltage. If V_t is large so that $|V_{GS}| < |V_t|$, then the number of carriers into the channel approaches zero. However, a high V_t increases the time required to switch between the on and off conducting states, resulting in slower devices.

Smaller transistors require lower operating voltages to restrict the internal electric field within reasonable margins. This in turn requires lowering the threshold voltage to maintain the operating speed of the device. This tendency is used in today's processes to maintain circuit performance at the cost of power increase. Modern devices show considerable current leakage even at $V_{GS} = 0$.

Threshold voltage reduction increases transistor leakage since an appreciable subthreshold current occurs during the off-state of the transistor. The subthreshold current can be easily observed from the device characteristics by plotting the logarithm of I_D versus V_{GS} for the subthreshold region (Figure 3.42). The subthreshold slope (S_t) is the amount of gate voltage required to increase the drain current one decade and is measured from Figure 3.42.

Subthreshold current has impact at the circuit level, since it is a fixed current contribution from all devices in the off-state. A subthreshold current of 10 nA at $V_{GS} = 0$ is insignificant for a single device, but in a 100 million transistor circuit the impact on the overall power consumption can be significant. Some technologies have MOSFETs with two different V_t's to reduce this problem. This separates the design into high-speed and low-speed transistors. High speed devices with lower V_t contribute higher leakage, and are

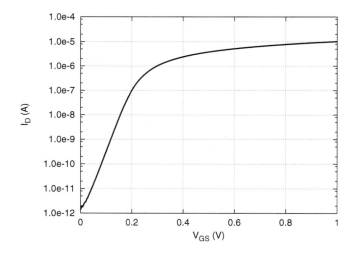

Figure 3.42. Effect of the subthreshold current in submicron devices.

used in critical delay paths or circuit blocks that must operate at high speed. Circuit blocks that do not require high speed are designed with high V_t transistors, and contribute less to the overall leakage.

Drain-Induced Barrier Lowering (DIBL). The population of channel carriers in long-channel devices is controlled by the gate voltage through the vertical electric field, whereas the horizontal field controls the current between the drain and the source. In large-channel devices, the horizontal and vertical electrical fields can be treated as having separate effects on the device characteristics. When the device is scaled down, the drain region moves closer to the source, and its electric field influences the whole channel. In this situation, the drain-induced electric field also plays a role in attracting carriers to the channel without control from the gate terminal. This effect is called *drain-induced barrier lowering* (DIBL), since the drain lowers the potential barrier for the source carriers to form the channel. The threshold voltage "feels" the impact of this effect on the transistor curves. Since DIBL attracts carriers into the channel with a loss of gate control, DIBL lowers the threshold voltage and increases off-state leakage.

An expression of the subthreshold leakage of a MOSFET including DIBL is given by Chandrakasan et al. [1]:

$$I_{\text{subth}} = A \cdot e^{(1/nV_T)V_{GS} - V_{t0} - \gamma \cdot V_S + \eta \cdot V_{DS}} \cdot (1 - e^{-V_{DS}/V_T}) \tag{3.37}$$

with

$$A = \mu_0 \cdot C_{\text{ox}} \frac{W}{L} (V_T)^2 e^{1.8} e^{-\Delta V_{TH}/\eta V_T} \tag{3.38}$$

where μ_0 is the zero bias mobility, $V_T = kT/q$ is the thermal voltage, γ is the linearized body effect coefficient, η is the DIBL coefficient, and n is the subthreshold swing coefficient. The term ΔV_{TH} is introduced to account for transistor-to-transistor leakage variations.

Hot Carrier Effects. In saturation, carriers crossing from the inverted channel pinch-off point to the drain travel at their maximum saturated speed, and so gain their maximum kinetic energy. These carriers collide with atoms of the bulk, causing a weak avalanche effect that creates electron–hole pairs. These carriers have high energy, and are called hot carriers.

Some carriers interact with the bulk, giving rise to a substrate current. Hot carriers can significantly affect reliability since some of these carriers generated by impact ionization have enough energy to enter the gate oxide and cause damage (traps) in the SiO_2 region. The accumulation of such traps can gradually degrade device performance, change the device V_t, and can increase conduction through the oxide, giving rise to oxide wearout and breakdown. This phenomena and its effects are detailed in Chapter 6.

Very Short Channel Devices. When the channel length is drastically reduced, the horizontal electric field increases and the effects of velocity saturation become much stronger. Horizontal mobility reduction becomes more important than vertical mobility reduction, and the drain saturation voltage is reduced. In these cases, Equation (3.32) can be rewritten neglecting the quadratic term in the drain internal voltage (V_{DS1}) since it is very small, and considering only the horizontal mobility effect (since it becomes predominant), leading to

$$I_D \approx \mu_0 C_{ox} \frac{W}{L_{eff}} \frac{(V_{GS}-V_t)V_{DS1}}{1+\dfrac{V_{DS1}}{L_{eff}\mathscr{E}_{crit}}} \qquad (3.39)$$

and for $V_{DS1}/(L_{eff}\mathscr{E}_{crit}) \gg 1$ gives

$$I_D \approx WC_{ox}(V_{GS}-V_t)\mu\mathscr{E}_{crit} \qquad (3.40)$$

This assumption and the result states that for very short channel devices, carriers are always velocity saturated, and the drain current does not depend on the transistor length. Equations (3.32) and (3.40) apply to transistors saturated and ohmic states under the assumption that $V_{DS1}/(L_{eff}\mathscr{E}_{crit}) \gg 1$.

3.5 SUMMARY

This chapter examined MOSFET transistors using the physical description of semiconductors and diodes in Chapter 2. Transistor operation was explained with figures showing the interaction of gate, drain, source, and bulk regions with external bias, minority carrier inversion, and diodes. Abundant examples with *n*MOS and *p*MOS transistors model equations emphasized reflexive approaches to analyze circuits with transistors and resistors. The chapter closed with descriptions of the body effect and short-channel transistors. A modeling approach was given for short-channel transistors that stretches the outer limit of manual calculations for transistors.

REFERENCES

1. A. Chandrakasan, W. Bowhill, and F. Fox (Eds.), *Design of High-Performance Microprocessor Circuits,* IEEE Press, 2001.
2. D. Foty, *MOSFET Modeling with SPICE,* Prentice-Hall, 1997.
3. R. F. Pierret, *Semiconductor Device Fundamentals,* Addison-Wesley, 1996.
4. J. Rosselló and J. Segura, "Charge-based analytical model for the evaluation of power consumption in submircrometer CMOS buffers," *IEEE Transactions on Computer Aided Design, 21,* 4, 433–448, April 2002.
5. T. Sakurai and R. Newton, "Delay analsysis of series-connected MOSFET circuits," *IEEE Journal of Solid-State Circuits, 26,* 122–131, February 1991.
6. Y. Taur and T. Ning, *Fundamentals of Modern VLSI Devices,* Cambridge University Press, 1998.
7. Y. Tsividis, *Operation and Modeling of The MOS Transistor,* 2nd ed., McGraw-Hill, 1999.
8. N. Weste and K. Eshraghian, *Principles of CMOS VLSI Design,* Addison-Wesley, 1993.

EXERCISES

3.1. For the three circuits in Figure 3.43, (a) give the transistor bias state, (b) write the appropriate model equation, and (c) calculate I_D, where $V_{tp} = -0.4$ V and $K_p = 100$ μA/V².

3.2. Repeat the same steps as in the previous exercise if $V_{tn} = 0.4$ V and $K_n = 200$ μA/V² (Figure 3.44).

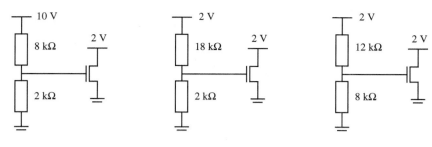

Figure 3.43.

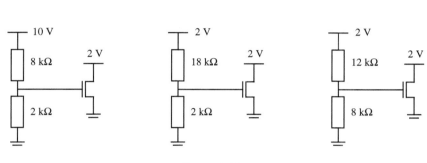

Figure 3.44.

3.3. Given $V_{tp} = -0.6$ V and $K_p = 75$ µA/V^2, and $W/L = 5$ (Figure 3.45), (a) solve for source voltage V_S, (b) solve for drain voltage.

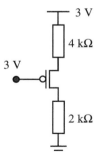

Figure 3.45.

3.4. Given the circuit in Figure 3.46 and the transistor parameters of Problem 3.2, (a) find the value of R_D to satisfy $V_0 = 1.2$ V. (b) As V_{DD} drops, find the value of V_D at the transition point where the transistor enters saturation and the new value of V_{DD}.

3.5. The transistor parameters in the circuit in Figure 3.47 are: $V_{tp} = -0.5$ V, $K_p = 75$ µA/V^2, and $W/L = 4$. If $V_0 = 1.2$ V, what is V_{IN}?

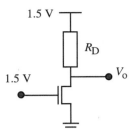

Figure 3.46.

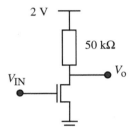

Figure 3.47.

3.6. Calculate V_G in Figure 3.48 so that $I_D = 200$ μA, given that $V_{tn} = 0.8$ V and $K_n = 100$ μA/V², and $W/L = 4$.

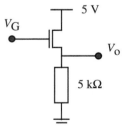

Figure 3.48.

3.7. Given that $R_1 = R_2$ and $K_n = 200$ μA/V², determine the resistance values in Figure 3.49 so that $V_D = 1$ V and $V_S = -1$ V.

3.8. Given $V_{tp} = -0.6$ V, $K_p = 50$ μA/V², $W/L = 3$, and $V_D = 0.8$ V. If $V_0 = 1.2$ V, what is R in Figure 3.50?

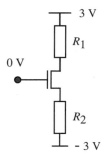
Figure 3.49.

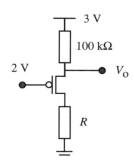
Figure 3.50.

3.9. In the circuit in Figure 3.51 $V_{tp} = -0.6$ V, $K_p = 75$ μA/V^2, and $W/L = 2$. (a) What value of R will place the transistor on the boundary between saturation and ohmic? (b) If R doubles its value, what are I_D and V_o?

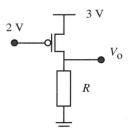
Figure 3.51.

3.10. The pMOSFET in Figure 3.52 has $V_{tp} = -0.5$ V, $K_p = 150$ μA/V^2, and $W/L = 3$, and a body effect constant $\gamma = 0.1$. The bulk voltage is at -0.3 V. (a) Calculate I_D. (b) If $V_{IN} = -1$ V, find I_D.

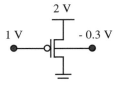
Figure 3.52.

3.11. Repeat Exercise 3.7 using the empirical model of Equation (3.36) with $\alpha = 1.14$, $I_{D0} = 14$ mA, $V_{D0} = 1$ V, $\lambda = 0.08$, and $V_{th} = 0.64$ V.

3.12. MOS transistors have two forms of capacitance. Describe each type and where you find them in the device.

3.13. What are the major differences between a short-channel transistor and a long-channel transistor?

3.14. Determine V_0 in the circuit of Figure 3.53, assuming that the topmost device is long channel, while the bottom one is short channel and needs the empirical model of Equation (3.36). Use: $V_{IN} = 1.5$ V, $V_{tn(up)} = 0.4$ V, $V_{tn(dwn)} = 0.3$ V, $L_{up} = 2$ μm, $W_{up} = 30$ μm, $K_{up} = 0.9$ mA/V^2, $I_{D0} = 5$ mA, $\alpha = 1.2$, $\lambda = 0.09$, and $V_{D0} = 1$V.

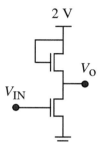

Figure 3.53.

3.15. Observe the log I_D versus V_{GS} curve in Fig. 3.22.
   (a) Define subthreshold current for short channel transistors
   (b) Why is subthreshold current in short channel transistors a problem in advanced CMOS technologies

3.16. Describe how DIBL lowers V_t in short channel transistors.

CHAPTER 4

CMOS BASIC GATES

4.1 INTRODUCTION

This chapter describes the electronics of basic logic gates, starting with the CMOS inverter whose simple appearance hides its complexity. Knowledge of inverter properties leads to knowledge of larger gates, such as NAND and NOR gates and their complicated properties. We will relate CMOS digital circuits to logic behavior and to CMOS failure mechanisms that typically involve small defects that alter normal inverter properties.

CMOS logic gates are digital cells, meaning that they perform Boolean algebra and their input and output voltages take one of the two possible logic states (high/low or 1/0). The output logic states have terminal voltages that respond to a range of input voltages but map into one of the two logic states. For example, a 1 V power supply technology has nominal output logic levels of 1 V (high) and 0 V (low). However, the input may range from 0 V to 0.3 V and the output still remains at a logic high of about 1 V.

This mapping of an input voltage range to a logic state gives noise immunity to digital circuits that is a major difference between analog and digital circuits. A small voltage fluctuation in an analog circuit node can cause significant error in the output signal. The same fluctuation in digital ICs is tolerated if it remains within the assigned range, and no error occurs.

There is a third range of digital voltage levels that is not mapped to any logic state. These voltages are between the logic levels, and they occur during an input/output voltage transition. None of the circuit nodes take these voltages in a normal or quiescent operation state since they have no logic meaning.

4.2 THE CMOS INVERTER

An inverter circuit converts a logic high-input voltage, such as 1 V, to a low logic voltage of 0 V (or 0 V to 1 V). The electronic symbol and truth table are shown in Figures 4.1(a) and (b). The Boolean statement is $V_{out} = V'_{in}$. The nMOS and pMOS transistors of the CMOS inverter (Figure 4.1(c)) act as complimentary switches. A logic high-input voltage turns on the nMOS transistor, driving the output node to ground, and also turns off the pMOS transistor. A low input voltage turns on the pMOS transistor and the nMOS off driving the output node to a high voltage.

Boolean values are read in the quiescent state that occurs when all signal nodes settle to their steady-state values. Only one inverter transistor is on connecting the output terminal V_{out} to one of the power rails, and there is no current in the circuit since the other transistor is off, thus eliminating a DC path between the rails. A capacitor load C_L is shown in Figure 4.1(c) as it is unavoidable in any circuit. The capacitance is from transistor node and wiring capacitances and does not affect static properties, but hinders the speed of logic transitions. We will analyze the static and dynamic operation.

4.2.1 Inverter Static Operation

Voltage Characteristic. The static voltage characteristic measures the logic gate input and output voltage over the whole voltage range. This curve defines the voltage levels mapped to each logic state. Figure 4.2 shows an inverter static voltage transfer curve (V_{out} versus V_{in}). Noise margin refers to the amount of input signal variation allowed before the output voltage shows significant change. Noise margins are often simplistically defined at the points of the curve where the slope is –1. There are five bias state regions corresponding to the transistor operating regions.

These five regions are:

1. *Region I. nMOS off, pMOS ohmic.* This voltage range exists for $V_{in} < V_{tn}$. The nMOS transistor is off, and the pMOS transistor is driven into nonsaturation since $V_{GS} \approx -V_{DD} < V_{DS} + V_{tp}$ (see Equation 3.14). The pMOS drain node at V_{out} is pulled up to a logic high V_{DD} through the low impedance of the pMOS channel.
2. *Region II. nMOS saturated, pMOS ohmic.* V_{in} goes just above the nMOS threshold voltage ($V_{in} > V_{tn}$), and the nMOS transistor is barely turned on and in saturation

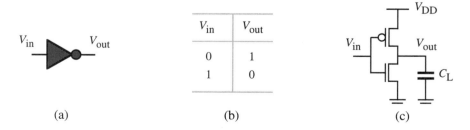

Figure 4.1. Inverter (a) symbol, (b) truth table, and (c) schematic.

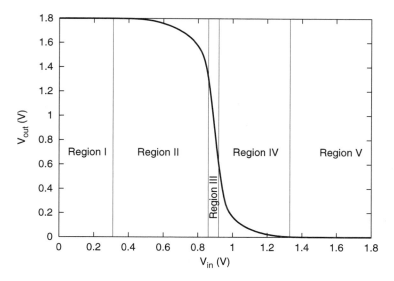

Figure 4.2. Inverter V_{in} versus V_{out} current transfer curve with five bias states.

($V_{DS} = V_{out} > V_{in} - V_{tn}$). Current now passes through both transistors and V_{out} drops as V_{in} is increased. The pMOS transistor remains in the ohmic state, but with decreasing gate drive.

3. *Region III. nMOS saturated, pMOS saturated.* When the output voltage goes below $V_{in} - V_{tp}$ and remains above $V_{in} - V_{tn}$, the nMOS and pMOS transistors are both in saturation, and the region has a straight line. Since V_{out} and V_{in} are linearly related, analog amplification occurs here. The drain voltage is a faithful replica of small changes in the input waveform, but amplified by a value equal to the slope of the straight line. MOS analog circuit designs use this property. It is also good for digital circuits that demand rapid V_{out} change during logic transitions of V_{in}. A digital goal is to get through the transition region as quickly as possible, and what better way than to have the circuit behave as an amplifier.

4. *Region IV. nMOS ohmic, pMOS saturated.* As V_{in} further increases, it approaches a value such that the difference between V_{in} and V_{DD} is close to the pMOS transistor threshold voltage. This is similar to Region II, but the roles of the transistors are reversed. The pMOS transistor is in saturation and the nMOS enters nonsaturation.

5. *Region V. nMOS ohmic, pMOS off.* When V_{in} goes to a logic high voltage, then $V_{in} \gg V_{out} + V_{tn}$. The pMOS transistor is turned off, and the nMOS transistor is in its ohmic state pulling the drain voltage V_{out} down to the source ground.

Inverter logic threshold voltage (V_{thr}) is the point at which $V_{in} = V_{out}$. V_{thr} is typically near $V_{DD}/2$. This is a unique condition since $V_{in} = V_{out}$ occurs only once in the inverter voltage range. The logic state changes as V_{in} moves through V_{thr}. V_{thr} is important when analyzing defect properties in CMOS circuits, since many defects cause intermediate voltages at some circuit nodes. Whether these defects cause a logic malfunction depends on the logic threshold voltage and the input voltage. Voltages slightly less than the logic high voltages and slightly more than logic low voltages are called weak logic voltages. Weak logic states are read correctly, but noise margins and gate driving voltages are compromised.

Except for Region I and Region V, the point at which transistors change from one zone to another depends on the inverter input and output voltages (Regions I and V depend only on the input). The input voltage at which these changes occur depends on the relative sizing of the devices, since the transistor width-to-length dimension (W/L) determines the current for a given gate–source voltage and, therefore, the effective equivalent resistance between drain and source.

In practice, input and output voltage ranges differ partly due to design and partly due to electrical noise. Figure 4.3 shows the voltage levels for the logic high and low values at a gate input and output. The following terms are defined

V_{IL} = input low voltage: maximum input voltage recognized as a logic low.

V_{IH} = input high voltage: minimum input voltage recognized as a logic high.

V_{OL} = output low voltage: maximum voltage at a gate output for a logic low for a specified load current.

V_{OH} = output high voltage: minimum voltage at a gate output for a logic high for a specified load current.

These voltage-based logic levels define the noise margin or immunity needed when connecting logic gates. *Noise Margin* (*NM*) is a parameter obtained from these levels and is defined for each logic value. NM_H and NM_L for the high and low logic values are

$$NM_H = V_{OH} - V_{IH}$$
$$NM_L = V_{IL} - V_{OL}$$

Noise margins must be positive for proper logic operation and the higher these values, the better the circuit noise immunity. These parameters are an essential measurement during production testing of ICs. Board designers must know that ICs that connect to each other are within specification and interface properly.

The logic threshold voltage of a CMOS inverter is set by the *p*MOS and *n*MOS transistor width and length ratios in the aspect ratio $(W_p/L_p)/(W_n/L_n)$ [see Figure 3.1(a)]. Usually, both transistors have the same channel length set by the minimum value of the technology.

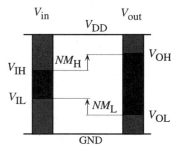

Figure 4.3. Voltage ranges mapped to logic Boolean values.

Therefore the W_p/W_n ratio determines the inverter logic threshold voltage. Inverters are often designed for a symmetric static transfer characteristic, so that the V_{out} versus V_{in} curve intersects the unity–gain line $V_{in} = V_{out}$ at about $V_{DD}/2$. A symmetrical transfer curve denotes a circuit with equal pullup and pulldown current drive strength. The aspect ratio giving a symmetric transfer characteristic for constant gate oxide thickness is

$$\frac{W_p}{W_n} = \frac{\mu_n}{\mu_p}\left(\frac{1 - \frac{2V_{tn}}{V_{DD}}}{1 - \frac{2|V_{tp}|}{V_{DD}}}\right)^2 \tag{4.1}$$

Equation (4.1) is found by equating the saturation current for nMOS and pMOS transistors and setting the input voltage equal to $V_{DD}/2$. If it is possible for the technology to have $\mu_n = \mu_p$, and $V_{tn} = |V_{tp}|$, then a symmetric inverter requires $W_n = W_p$. In practice, electron and hole mobility are never equal, and the threshold voltages of nMOS and pMOS transistors have slightly different absolute values. Symmetrical CMOS inverter designs often have $(W_p/W_n) \approx 2$ to compensate for the smaller hole mobility in pMOS transistors.

> *Self-Exercise 4.1*
>
> Compute the ratio between nMOS and pMOS transistor width to obtain a symmetric inverter for a 0.18 μm technology in which $\mu_n = 360$ cm²/V · s, $\mu_p = 109$ cm²/V · s, $V_{tn} = 0.35$ V, and $V_{tp} = -0.36$ V with $V_{DD} = 1.8$ V.

The inverter logic threshold voltage can be made smaller or larger within the range $V_{tn} < V_{thr} < V_{DD}+V_{tp}$ by setting the ratio W_p/W_n above or below the value of Equation (4.1). $V_{thr} = 0.5V_{DD}$ when the pull-up and pull-down transistor current drive strength are equal. If $V_{thr} < 0.5V_{DD}$ then the nMOS pull-down transistor is stronger than the pull-up. If $V_{thr} > 0.5V_{DD}$ then the pMOS pull-up is stronger.

Current Characteristic. The DC power supply current transfer curve is equally important. Figure 4.4 shows the I_{DD} versus V_{in} characteristic. At $V_{in} < V_{tn}$ in Region I, the nMOS transistor is off and no current passes through the circuit. When $V_{in} = V_{DD}$ in Region V, the pMOS transistor is off and, again, no current passes from the power supply to ground. Typical inverter current at these quiescent logic levels is in the low pA's and is mostly drain–substrate reverse-bias saturation current or subthreshold current (Chapters 2–3). Virtually no power is dissipated in the quiescent logic states, giving CMOS ICs their traditional technology advantage. The peak current near $V_{DD}/2$ depends upon transistor drive strength (the size of the width-to-length ratio). Both transistors are in the saturated state and the current peaks. When an inverter changes logic state, the transient current is wasted power. Peak currents in large microprocessor designs are many Amperes, as this is the sum of the transient currents of millions of logic gates.

Transistor off-state leakage in deep submicron technologies is a concern and the major cause is the intentional threshold voltage reduction needed to maintain circuit performance. Deep submicron ICs must reduce V_{DD} to contain the internal electric fields and power dissipation, but V_t is kept at about 15%–25% of V_{DD} to ensure strong gate overdrive voltage. The design tradeoff is stronger gate overdrive (faster IC) with lower V_t versus significantly higher off-state leakage (higher off-state power). Total IC off-

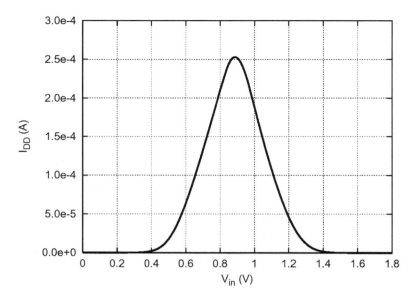

Figure 4.4. Inverter power supply current transfer curve.

state leakage can approach Ampere levels when high-speed performance is the dominant issue. Subthreshold current rises rapidly when V_t is lowered. One solution reduces off-state leakage by using transistors with different threshold voltages in the same circuit. This technique uses low V_t transistors in the speed-critical paths and high V_t devices in the noncritical paths. Another approach uses high V_t transistors to disconnect logic gates (designed with low V_t devices) from the power supply when they are inactive. Low V_t transistors have higher performance and higher leakage, whereas high V_t devices are slower with smaller leakage current.

Graphical Analysis of Bias Regions. Figure 4.5 distinguishes the bias state regions with the voltage transfer function and two 45° lines. The lower unity–slope line separates the bias condition, mutually satisfying saturation and nonsaturation for an *n*MOS:

$$V_{GS} = V_{DS} + V_{tn} \tag{4.2}$$

or

$$V_{in} = V_{out} + V_{tn} \Rightarrow V_{out} = V_{in} - V_{tn} \tag{4.3}$$

Equation (4.3) is a straight line superimposed on the voltage transfer curve in Figure 4.5 and labeled as (a). All points on the transfer curve lying above line (a) represent the *n*MOS transistor in saturation or the off-state. All points below line (a) represent the *n*MOS transistor in the ohmic state. A similar derivation leads to the *p*MOS transistor bias boundary line labeled (b) in Figure 4.5. The *p*MOS transistor is saturated or off for all points on the curve below line (b) and in the ohmic state above line (b). Both transistors are saturated in the region between the two lines.

4.2 THE CMOS INVERTER   105

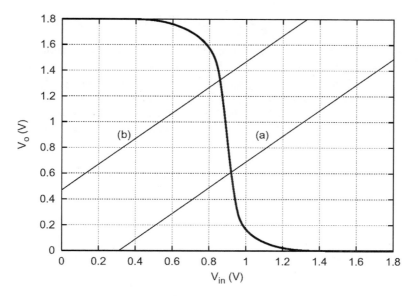

Figure 4.5. Inverter transfer curves and transistor state.

■ EXAMPLE 4.1

Figure 4.6 is an inverter transfer curve. Estimate V_{tn} and V_{tp} using bias line concepts.

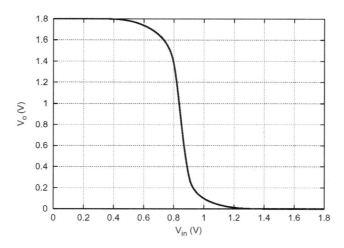

Figure 4.6. CMOS inverter voltage transfer curve.

Put a small mark on the ends of the linear region in Figure 4.6 and draw 45° lines. The threshold values are the intercepts. $V_{tn} = 0.42$ V and $V_{tp} = -0.44$ V. ■

Graphical analysis allows visualization of transistor states during logic transitions. The maximum gain region is where both transistors are saturated, as seen between the two dotted bias lines (a, b) in Figure 4.5. An example emphasizes this thinking.

EXAMPLE 4.2

When V_{out} switches in an inverter from V_{DD} to 0 V, estimate the fraction of this voltage range (V_{DD}) for which the nMOS transistor is in saturation. Let $V_{tn} = 0.2V_{DD}$, $V_{tp} = -0.2V_{DD}$, and $K'_n = K'_p$, where $K'_n = K_n (W/L)_n$, and $K'_p = K_p (W/L)_p$.

We know $I_{Dn} = -I_{Dp}$ for all the points in the static curve and also the point on line (a) in Figure 4.5 where the nMOS transistor leaves saturation. At this point, both transistors can be treated as in the saturation state. This is the transition between Regions III and IV in Figure 4.2. So,

$$K'_n(V_{GS} - V_{tn})^2 = K'_p(V_{GS} - V_{tp})^2$$
$$K'_n(V_{in} - V_{tn})^2 = K'_p(V_{DD} - V_{in} - V_{tp})^2$$

Solve for

$$V_{in} = \frac{V_{DD} + V_{tn}\sqrt{\frac{K'_n}{K'_p}} - V_{tp}}{1 + \sqrt{\frac{K'_n}{K'_p}}}$$

Substituting

$$V_{in} = V_{out} + V_{tn}$$

into the above V_{in} equation:

$$V_{out} = \frac{V_{DD} - V_{tn} - V_{tp}}{1 + \sqrt{\frac{K'_n}{K'_p}}}$$

Substituting

$$K'_n = K'_p, \quad V_{tn} = 0.2V_{DD}, \quad \text{and } V_{tp} = -0.2V_{DD}$$

then

$$V_{out} = 0.5V_{DD}$$

at the transition point.
The fraction is

$$\frac{V_{DD} - V_{out}}{V_{DD}} = 0.7 = 70\%$$

The point is made that for these conditions, the nMOS and pMOS transistors are individually in saturation for about 70% of the transition and jointly in saturation for about 40%. This has strong implications for failure analysis and design debug tools. ∎

CMOS inverter theory underlies failure analysis, test, and reliability electronics. For example, Chapter 3 described drain region light (photon) emission when a transistor was saturated. The high electric field of the drain depletion region accelerates channel charge, causing impact ionization and subsequent photon emission. Figure 4.5 shows where and when light is expected in the inverter. During normal logic transitions, light is emitted from the nMOS transistor in the saturated state region above line (a). The pMOS transistor emits light in the region below line (b). Both transistors emit light in the region between lines (a) and (b). The IBM PICA timing path analyzer with 15 ps timing resolution is based on these principles [7]. Failure analysts must know when defects force a transistor into an inadvertent saturated bias state, causing light emission [4]. Visible light from a defect location is an efficient failure analysis tool that, in addition to locating a defect region, says that a transistor is in its saturated state (or a diode is breaking down) with a weak gate and drain voltage. Failure analysts must describe defect properties consistent with all electrical clues.

> **Self-Exercise 4.2**
>
> A CMOS inverter has transistor parameters: $K_n (W/L)_n = 265$ μA/V^2, $K_p (W/L)_p = 200$ μA/V^2, $V_{tn} = 0.55$ V, $V_{tp} = 0.63$ V, and $V_{DD} = 2.5$ V. (a) For what fraction of the total output voltage swing will the nMOS transistor be in saturation. (b) Same question for the pMOS transistor. (c) What fraction will both be in saturation at the same time.

> **Self-Exercise 4.3**
>
> Estimate the voltage analog gain for the inverter whose transfer curve is in Figure 4.5

4.2.2 Dynamic Operation

The transfer curve of Figure 4.2 gives essential inverter information, but does not represent the circuit behavior during its rapid transition. The switching time for modern inverters can be a few tens of picoseconds, and parasitic capacitance of the transistors and the external wiring load alter the phase relation between V_{in} and V_{out}. An inverter and its transfer curve phase relation are drawn in Figure 4.7 for different input speed transitions, showing that for rapid transitions, the drain voltage lags input gate voltage.

The circuit model for the inverter dynamic analysis in Figure 4.7(a) shows two parasitic capacitors that are important during the transition. The input–output capacitor (called the coupling capacitor C_{coup}) comes from the gate–channel device capacitance that is strongly bias-dependent and the overlapping gate–drain capacitors from both the nMOS and pMOS devices. For high-speed transitions, the coupling capacitor tries to maintain its initial voltage difference between the input and the output ($-V_{DD}$ for a low–high input transition and V_{DD} for a high–low one). This temporarily drives the output voltage beyond V_{DD} (overshoot) for an input rising transition and below ground for a falling edge (undershoot). This induces noise in the supply (ground) node while increasing transition delay since the output voltage swing is higher than V_{DD}.

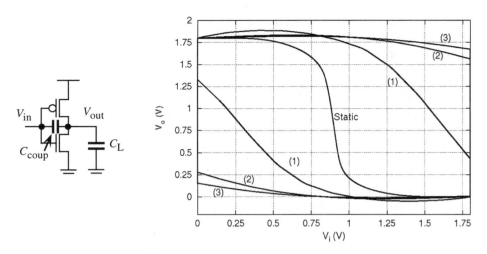

Figure 4.7. (a) Dynamic CMOS inverter circuit model, and (b) transfer curves for high-to-low output and low-to-high output for different input ramp speeds.

Curves (2) and (3) in Figure 4.7(b) correspond to very high speed input transistors and show that the output drain node remains at a relatively high voltage when the gate input has almost completed its transition. The same phenomena holds when the gate input switches rapidly from high to low. The drain remains in a low-voltage state until the input has almost completed its transition. The circuit parasitic capacitances cause this phase relation. The slow transition response of the static curve allows time for the drain nodes to exactly follow the input gate voltage in time. Curve (1) is an intermediate case for a medium speed input transition, in which the output is beyond $V_{DD}/2$ when the input reaches its final value, although it is far from the almost-static transfer curve.

This effect is more important at the beginning of the transition since for a rising (falling) input swing the *n*MOS (*p*MOS) device is off until $V_{in} = V_{tn}$ ($V_{in} = V_{DD} + V_{tp}$). During this period, the output voltage is disconnected from ground (supply) and there is no pull-down (pull-up) element to compensate for the charge injected from the input. This effect is shown in Figure 4.8 for the input and output voltage evolution for high–low and low–high input transitions.

Three parameters determine the duration and magnitude of overshoot (undershoot):

1. *The value of the input–output coupling capacitance.* The larger the capacitance, the more charge is transferred to the output during the input rising (falling) transition and the higher (lower) the output voltage overshoot (undershoot).
2. *The input transition time.* The charge injection from the input to the output through the coupling capacitance depends on the time derivative of the input voltage ($i = C\, dV/dt$). The shorter the input transition times, the higher the overshoot or undershoot.
3. *The width of the nMOS (pMOS) transistor.* Overshoot (undershoot) occurs at the beginning of the input transition, since the *n*MOS (*p*MOS) device is off and does not pull down (pull up) the output voltage. Once the input voltage goes beyond (below) V_{tn} ($V_{DD} + V_{tp}$) the *n*MOS (*p*MOS) device turns on and pulls the output voltage

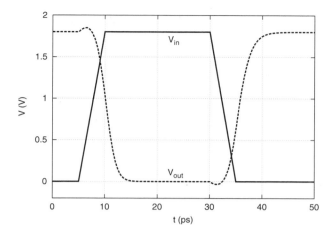

Figure 4.8. Time evolution of the input and output voltage of an inverter for an input high–low and low–high transitions. The output voltage exceeds the power and ground levels during the transition.

down (up). The larger the W/L, the larger its current drive and the smaller the time required to pull down (pull up) the output.

The inverter output capacitance is the sum of the drain diffusion capacitance, interconnect wiring, and the input capacitance of the load gates. The transition time is sensitive to the load capacitance since the output must be charged–discharged during the transitions. Figure 4.8 also illustrates the definition of logic gate propagation delay time T_{PD}, which is the time between the input and output waveform measured at the 50% amplitude points.

4.2.3 Inverter Speed Property

The dynamic relation between the input and the output voltages defines the different operating regions, but in the dynamic case the input voltage transition time does not uniquely determine the output voltage transition time. Sometimes, the input reaches its final value while the output voltage is still close to its initial value because the transistor that must sink or source current is too small compared to the output load. Different situations appear depending on transistor-to-load capacitance sizing.

Figure 4.9 plots the output voltage responses for different values of the output capacitance when the input low–high transition time is fixed. Curve V_{03} corresponds to a small output capacitor, and the output voltage is almost zero when the input reaches V_{DD}. This case is similar to the static transfer curve. In curve V_{02}, the output voltage is equal to $V_{in} - V_{tn}$ when the input voltage reaches V_{DD}, whereas in curve V_{01} the output is still above that value when the input gets to its steady state. In all cases, the output voltage initially goes beyond V_{DD} due to overshoot caused by the charge injected from the input through the coupling capacitor. During this period, there is current from the output node through the pMOS transistor back to the supply terminal. In none of the three cases does the output voltage start to decay until the input voltage goes beyond V_{tn} at time $t = t_n$. Fast and slow input ramps can distinguish the operating regions of the nMOS transistor when the input voltage reaches V_{DD}. In Curve V_{03}, the nMOS device is in its linear region when the input

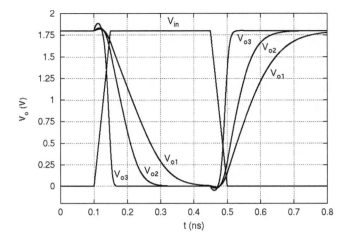

Figure 4.9. Different cases for a dynamic inverter transition.

transition is finished, whereas in curve V_{o1} this transistor is still saturated when V_{tp} reaches V_{DD}.

Curve V_{o1} is interesting because the pMOS transistor never passes a positive current. When overshoot ceases and the output voltage goes below V_{DD}, the input is below V_{tp} and the pMOS is off. This condition results in a slight reduction in power consumption.

An exact calculation of the propagation delay of an inverter requires a complex differential equation. We derive a simple formulation, assuming that the device is an ideal current source (i.e., the transistor is always in saturation).

The speed with which an inverter switches logic states depends on V_{tn}, V_{tp}, V_{DD}, W/L, temperature, and the coupling and load capacitances. A simple model shows these parameters that affect inverter rise and fall time. The current source I_0 in Figure 4.10(a) represents the MOS transistor in saturation since that bias state dominates the transition time and C_L is the load capacitance. The Shockley MOSFET model is

$$I_0 = I_D = \frac{\mu_n C_{ox}}{2} \frac{W}{L} (V_{GS} - V_{tn})^2 \quad (4.4)$$

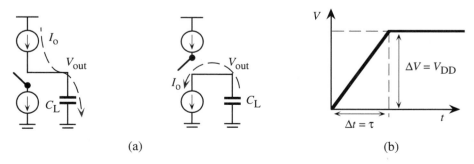

Figure 4.10. (a) Circuit model to estimate rise and fall delays in a CMOS inverter, and (b) voltage response of capacitor to constant current.

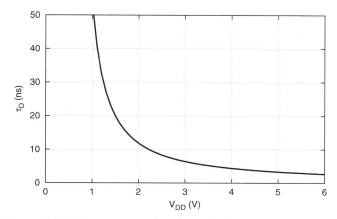

Figure 4.11. Delay versus supply voltage for the model in Equation (4.8).

The capacitor expression for current, voltage, and time from Chapter 1 is

$$i(t) = C_L \frac{dV(t)}{dt} \quad (4.5)$$

If $i(t) = I_0$ (a constant current) and we approximate $dv(t)/dt = \Delta V(t)/\Delta t$, then

$$\Delta V(t) = \frac{I_0}{C_L} \Delta t \quad (4.6)$$

If Δt is the delay time τ_D in Figure 4.10(b) for the signal to rise to $\Delta V(t) = V_{DD}$, then Equation (4.6) is rewritten as

$$\tau_D = \frac{C_L V_{DD}}{I_0} \quad (4.7)$$

Equation (4.7) adequately matches experimental data [2]. Substituting Equation (4.4) into Equation (4.7) with $V_{GS} = V_{DD}$ gives

$$\tau_D = C_L V_{DD} \frac{2L}{W \mu_n C_{ox}} \frac{1}{(V_{DD} - V_{tn})^2} \quad (4.8)$$

Equation (4.8) shows that time delay is geometrically related to the difference in V_{DD} and V_t. Figure 4.11 plots the time delay versus V_{DD} for $C_L = 20$ fF, $\mu_n C_{ox} = 150$ µA/V^2, $V_t = 0.5$ V, and $W/L = 2$. Time delay asymptotically approaches infinity as V_{DD} approaches V_t.

> *Self-Exercise 4.4*
>
> If a pMOS transistor has $\mu_n C_{ox} = 56$ µA/V^2, $V_{tp} = -0.6$ V, and $W/L = 6$, what is the expected additional time delay if the gate V_{DD} is reduced from a normal $V_{DD} = 2.5$ V to $V_{DD} = 1.8$ V with $C_L = 25$ fF?

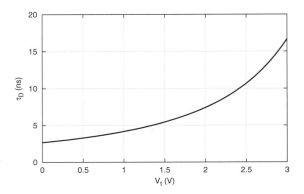

Figure 4.12. Delay versus threshold voltage for the model in Equation (4.8).

The result is similar if V_t varies for a fixed V_{DD}. Figure 4.12 is a plot similar to Figure 4.11, but with time delay plotted against V_t. These figures illustrate why deep submicron technologies strive for low V_t. This is a complicated trade-off for logic speed against the increase in off-state leakage current when V_t is lowered. Chapter 3 discussed these mechanisms.

■ EXAMPLE 4.3

It is given that $C_L = 10$ fF, $\mu_n C_{ox} = 118$ μA/V^2, $W/L = 6$, and $V_{DD} = 2.3$ V. Initially, $V_t = 0.6$ V. If V_t is reduced to 0.2 V, what is the percent decrease in speed and what is the percent increase in I_{OFF} if subthreshold slope of I_D versus V_G is 83 mV/decade?

You can substitute the values into Equation (4.8) and take the ratio or divide Equation (4.8) by itself, substituting $V_t = 0.6$ V and $V_t = 0.2$ V. You get

$$\frac{\tau_D(V_t = 0.6 \text{ V})}{\tau_D(V_t = 0.2 \text{ V})} = \frac{(2.3 - 0.2)^2}{(2.3 - 0.6)^2} = 1.526$$

The threshold setting of $V_t = 0.6$ V slows the transistor by about 53%

The reduction is V_t affects the off-state leakage current by shifting the log (I_D) versus V_G curve to the right by (0.6 V – 0.2 V) = 400 mV (see Chapter 3). The increase in off-state leakage current is

$$\frac{400 \text{ mV}}{83 \text{ (mV/decade)}} = 4819 \text{ decades} \Rightarrow 65.96 \times 10^3$$

The off-state leakage is increased over four orders of magnitude. A battery-operated circuit, such as a watch or pocket calculator, would choose the higher threshold voltage setting as speed is not a concern and power reduction is. ■

Equation (4.8) is valid for long-channel devices, since the current expression in Equation (4.4) has a quadratic dependence on the gate voltage. An analysis for deep-submicron devices substitutes any of their equations for the drain current into Equation (4.7). In this case, the delay is approximately [1]

$$\tau_D = K \frac{C_L V_{DD}}{(V_{DD} - V_{tn})} \qquad (4.9)$$

K is a constant that depends on device size and technology parameters. The main difference is the inverse dependence on V_{DD} instead of the inverse square in Equation (4.9). Although deep-submicron delay appears larger than for long-channel transistors, the transistor thin oxide (T_{OX}) is much smaller, reflecting a larger value of K in Equation (4.8).

4.2.4 CMOS Inverter Power Consumption

The energy dissipated by an inverter has static and dynamic components. Dynamic dissipation is due to the charge–discharge of the gate output load capacitance (transient component) and to the short-circuit current from the supply to ground created during the transition. Static dissipation for long-channel transistors is due mainly to reverse bias drain–substrate (–well) *pn* junction leakage current from transistors in the off-state. The deep-submicron off-state current is mostly subthreshold leakage, and it dominates with technology scaling, since the threshold voltage V_{th} is reduced to maintain circuit performance [6]. Gate oxide tunneling current in ultrathin oxides also contributes significantly to IC off-state leakage. The dynamic power calculation requires computation of transient and short-circuit components.

Transient Component. The dynamic power (P_d) to charge and discharge a capacitor C_L for a period T of frequency f is

$$P_d = \frac{1}{T} \int_0^T i_L(t) v_0(t) dt \qquad (4.10)$$

In one period interval, the output voltage changes from 0 to V_{DD} and vice-versa. Equation (4.5) relates the current and voltage of the output capacitor, so Equation (4.10) is rewritten as

$$P_d = \frac{1}{T} \left[\int_0^{V_{DD}} C_L v_0 dv_0 + \int_{V_{DD}}^0 C_L (V_{DD} - v_0) d(V_{DD} - v_0) \right] \qquad (4.11)$$

giving

$$P_d = \frac{C_L V_{DD}^2}{T} = C_L V_{DD}^2 f \qquad (4.12)$$

Equation (4.12) shows that transient power can be lowered by reducing the output capacitance, the supply voltage, or the operating frequency. Since the power dependence on the supply voltage is quadratic, lowering V_{DD} is more efficient for reducing power dissipation than the other two parameters.

Short-Circuit Component. When the input is changing and its voltage is between V_{tn} and $V_{DD} - |V_{tp}|$, then both transistors are simultaneously conducting creating a current path from V_{DD} to ground. This power component (P_{sc}) depends on device size, input transition time, and the output and coupling capacitors that consume about 10%–20% of the overall power. It was recently shown that the ratio of the short-circuit to the dynamic current remains constant for submicron technologies if the ratio V_{th}/V_{DD} is constant. The exact computation is complex, so we present an approximation.

Consider a symmetric inverter (i.e., $K'_n = K'_p$, and $V_{tn} = -V_{tp}$) with no output load and an input voltage transition having equal rise and fall times. The time interval when both transistors simultaneously conduct is from t_1 to t_3 in Figure 4.13. During the interval $t_1 - t_2$, the short-circuit current increases from zero to its maximum value I_{max}. Since the nMOS transistor is saturated during this period, its drain current is Equation (4.4)

$$I = K'_n(V_{in} - V_{tn})^2 \quad \text{for } 0 < I < I_{max}$$

where

$$K'_n = \frac{\mu_n C_{ox}}{2} \frac{W}{L} \quad (4.13)$$

Since the inverter was assumed to be symmetric with no load, the maximum current occurs at $V_{in} = V_{DD}/2$ and its shape is symmetric along the vertical axis at $t = t_2$. We compute a mean current by integrating from $t = 0$ to $t = T$ and dividing by the period T. There are four equal area current segments to integrate in Figure 4.13 over the whole period T:

$$I_{mean} = \frac{1}{T}\int_0^T I(t)dt = \frac{4}{T}\int_{t_1}^{t_2} K'_n(V_{in}(t) - V_t)^2 dt \quad (4.14)$$

If the input voltage is a linear ramp of duration τ,

$$V_{in} = \frac{V_{DD}}{\tau} t \quad (4.15)$$

t_1 and t_2 are given by

$$t_1 = \frac{V_t}{V_{DD}}\tau \quad \text{and} \quad t_2 = \frac{\tau}{2} \quad (4.16)$$

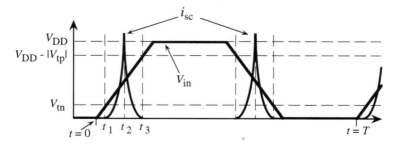

Figure 4.13. A simplified view of the short-circuit current contribution.

Substituting (4.15) and (4.16) into (4.14)

$$I_{\text{mean}} = K_n' \int_{\left(\frac{V_t}{V_{\text{DD}}}\right)\tau}^{\tau/2} \left(\frac{V_{\text{DD}}}{\tau}t - V_t\right)^2 dt \tag{4.17}$$

This integral is of the type $\int x\,dx$ with $x = (V_{\text{DD}}/\tau)t - V_t$, so the result is

$$I_{\text{mean}} = \frac{1}{6}\frac{K_n'}{V_{\text{DD}}}(V_{\text{DD}} - V_T)^3 \frac{\tau}{T} \tag{4.18}$$

Finally, the power contribution is given by

$$P_{\text{sc}} = V_{\text{DD}} I_{\text{mean}}$$

Power Supply Scaling. The ratio of V_t to V_{DD} impacts several inverter properties. Equation (4.4) is repeated:

$$I_D = \frac{\mu_n \varepsilon}{2T_{\text{ox}}} \frac{W}{L}(V_{\text{GS}} - V_{tn})^2 \tag{4.19}$$

$V_{\text{GS}} = V_{\text{DD}}$ for logic circuits, so that Equation (4.19) becomes

$$I_D = \frac{\mu_n \varepsilon}{2T_{\text{ox}}} \frac{W}{L}(V_{\text{DD}} - V_{tn})^2 \tag{4.20}$$

When V_{DD} drops, several things happen:

- The voltage difference in the parentheses (the gate overdrive) is smaller, so the current drive is lower.
- When $V_{\text{DD}} < (V_{tn} - V_{tp}) \approx |2V_t|$ the transition still occurs, but only one transistor is on at a time. There is essentially no transient current spike. Figure 4.14 shows a transfer curve at $V_{\text{DD}} = 1$ V and $V_{\text{DD}} = 0.5$ V for transistors with thresholds on the order of 0.35 V. There is no current spike for the $V_{\text{DD}} = 0.5$ V measurement since $V_{\text{DD}} < 2 \times |V_t|$. V_{in} was swept from 0.5 to 0 V. The power reduction for this condition is large. Low-power, battery-operated products such as electronic watches and medical implants use this technique.
- The fraction of the V_{in} sweep in which both transistors are simultaneously on also drops, reducing the peak I_{DD} current.

There is a trade-off between power savings and delay increase when reducing the supply voltage. Technology scaling includes thinner gate oxides, forcing lower voltages to contain the gate oxide and drain–substrate electric fields to subcritical values. We present some approaches to power supply voltage scaling that focus on reliability, speed, or energy–delay.

Reliability Driven Supply Scaling. The most general voltage scaling trades off long-term reliability, operating speed, and energy. Circuit speed increases with higher V_{DD} but the higher device electric fields increase carrier velocity, creating more hot carriers. Hot carriers contribute to oxide degradation and can limit circuit lifetime. It is possible to de-

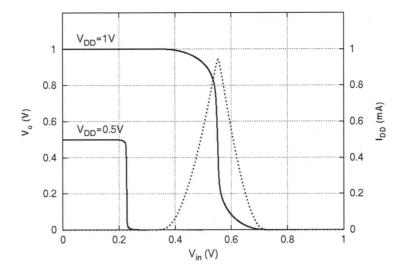

Figure 4.14. Inverter transfer curves at two V_{DD} values. Notice the absence of a short-circuit current spike for $V_{DD} = 0.5$ V, where V_{in} was swept from 0.5 to 0 V.

velop a model in which circuit delay and hot carrier effects are included and from which an optimum power supply can be determined [2].

Technology Driven Supply Scaling. Transistor current drive in the saturated state for submicron technologies is not quadratic but linear. It is dominated by carrier velocity saturation (v_{max}), as explained in Chapter 3 through Equation (3.40). Since $v_{max} = \mu \mathcal{E}_{crit}$, Equation (3.40) is rewritten as

$$I = WC_{ox}(V_{DD} - V_t)v_{max} \qquad (4.21)$$

The first-order delay model of Equation (4.7) with this current expression gives a delay almost independent of the supply voltage, provided that carriers move at the velocity saturation v_{max} i.e., at high electric fields. Mathematically, this implies that $V_{DD} - V_t \approx V_{DD}$ and V_{DD} vanishes in the delay equation. A "technology" based criterion chooses the power supply voltage based on the desired speed–power performance for a given deep-submicon technology [5]. The relative independence of delay on supply voltage at high electric fields allows voltage reduction for a velocity-saturated device with little penalty in speed performance. This concept of operating above a certain voltage was formalized by Kakamu and Kingawa [5], where the concept of a "critical voltage" was developed.

Energy–Delay Minimum Supply Scaling. Another approach reduces the voltage supply by minimizing the energy–delay product [2]. For a fixed technology, there is a supply voltage that trades off the quadratic dependence of energy and the increased circuit delay.

4.2.5 Sizing and Inverter Buffers

A speed design problem exists when a signal passes through a series of logic gates. A fast charge–discharge of a load capacitor requires large W/L of the driving transistors. However, the large *W/L* is defeating since its large gate area increases the load capacitance for its own driving logic gate. Working backward, that would cause all preceding logic gates to have ever larger *W/L* ratios. A better solution exists.

A better approach for driving large loads at high speed uses successively larger channel widths in a cascade of inverters to sufficiently increase the current drive of the last stage. A circuit driving a large load is commonly known as a buffer, and a circuit designed with successive inverters is known as a tapered buffer (Figure 4.15). When the area of each stage increases by the same factor, the circuit is called a fixed-taper buffer, and if this ratio is not constant it is called a variable-taper buffer.

The fixed-taper buffer structure was proposed by Linholm in 1975 [8]. He used a simple capacitance model, making the output load of a stage proportional to the size of the input capacitance of the next stage, while the area of each inverter was proportional to the channel width of the transistors. The overall buffer delay was optimized by minimizing the delay of each stage. A better result is obtained if the system delay is considered instead of individual stages.

Let each succeeding stage in the buffer in Figure 4.15 have transistor widths larger than the previous one by a factor α. The first inverter is the smallest, with an input capacitance C_{in}, whereas the ith stage has an input capacitance given by

$$C_i = \alpha^{i-1} C_{in} \qquad i = 1, 2, \ldots, n \tag{4.22}$$

The number of stages n is computed from

$$C_L = \alpha^n C_n \tag{4.23}$$

then

$$\alpha^n = \frac{C_L}{C_{in}} \tag{4.24}$$

and

$$n = \frac{\ln(C_L/C_{in})}{\ln \alpha} \tag{4.25}$$

α is computed by optimizing the delay. Assuming that the delay of the first stage driving an identical one is τ_0, the delay of the ith stage is

$$t_{di} = \alpha \tau_0 \qquad i = 1, 2, \ldots, n \tag{4.26}$$

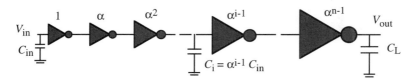

Figure 4.15. Tapered buffer structure.

The global delay of the n stages is

$$t_d = \sum_{i=1}^{n} t_{di} = n\alpha\tau_0 \qquad (4.27)$$

giving

$$t_d = \ln\left(\frac{C_L}{C_{in}}\right)\frac{\alpha}{\ln \alpha}\tau_0 \qquad (4.28)$$

Differentiating (4.28) with respect to α and equating to zero, the optimum α_{opt} is

$$\alpha_{opt} = e \approx 2.7 \qquad (4.29)$$

and the optimum number of stages n_{opt} is

$$n_{opt} = \ln(C_L/C_{in}) \qquad (4.30)$$

Other buffer designs addressed power dissipation, circuit area, and system reliability, many of them leading to results different than shown here. Additionally, more accurate models for capacitance and delay estimation exist. Methods exist that consider interrelated issues of circuit speed, power dissipation, physical area, and system reliability [3]. The point of this section is to impress upon the reader that care must be taken in a design when a logic gate of one size and capacitance drives a logic gate of larger size and capacitance.

4.3 NAND GATES

CMOS technology implements negated functions. This means that the output signals, although controlled by many input lines, are inverted with respect to one or more controlling inputs. Simple examples are the inverter, the NAND gate, and the NOR gate.

A 2NAND gate symbol and truth table are shown in Figures 4.16(a) and (b). There are two properties to note. The first is that any logic 0 to the inputs of a NAND gate causes logic 1 output. The other property is more subtle, but vital to logic design and testing ICs. Certain input levels are called *noncontrolling states*. When A = 1 in rows 3 and 4, then the

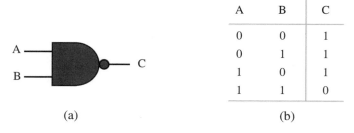

Figure 4.16. (a) 2NAND gate symbol and (b) truth table.

output C is the negation or complement of B (C = B′). The output C depends only on the value of B if A = 1. If B = 1, C is the complement of A. IC testing requires that signal information from specific nodes be read at an output pin without interference from the other input lines of that gate. The property is essential for passing specific logic values deep in the logic blocks to an observable circuit node such as an output pin.

When a NAND gate input is set to logic 1, then the effects of that signal node are neutralized with respect to signals on the other input lines. For example, if a fault were suspected on the input of B, then a test pattern would drive a signal to B and measure C, but ensure that node A = 1. The noncontrolling logic state for an AND gate is also logic 1.

■ EXAMPLE 4.4

If in Figure 4.17 you want to examine the signal at (a) node B, what should node I_1 be set to? (b) To examine node A, what should I_2 and I_3 be set to? (c) To examine node I_3, what should I_1 and I_2 be set to?

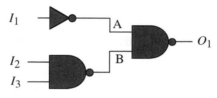

Figure 4.17.

(a) To pass a signal from node B to the output O_1 requires that node A is set at logic 1. Therefore I_1 must be logic 0.
(b) To pass a signal from node A to the output O_1 requires that node B is set at logic 1. Therefore $I_2 I_3$ must be 00, 01, or 10.
(c) To pass a signal from node I_3 to the output O_1 requires that node I_2 be set at logic 1 and I_1 is set at logic 0. ■

Figure 4.18 shows the 2NAND gate transistor schematic. The electronic operation follows the truth table in Figure 4.16(b). A logic 0 on any input line turns off an nMOS pull-down transistor and closes the path from the output V_C to ground. A logic 0 ensures that a pMOS is turned on. Therefore, for any logic 0 on the inputs, the output is at logic 1, or $V_C = V_{DD}$. If both logic inputs are logic 1 ($V_A = V_B = V_{DD}$), then both nMOS transistors turn on, both pMOS transistors are off, and $V_C = 0$ V. The noncontrolling logic state for a NAND (and AND gate) input node is logic 1. AND gates can be made by adding an inverter to the output of a NAND gate.

The NAND gate has most of the inverter properties developed in Sections 4.1.2–4.1.5. If we set input B in Figure 4.18 to its noncontrolling state $V_B = V_{DD}$, then a voltage sweep at node A produces static and dynamic transistor curves similar to those measured for the inverter and shown in Figures 4.2 and 4.7.

Inverter speed and power properties apply to the NAND gate with minor exceptions. The goal of matching inverter rise and fall times led to design of $K'_n = K'_p$. This was done

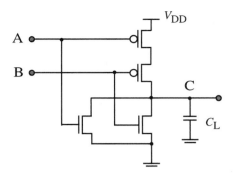

Figure 4.18. Transistor level structure of a static CMOS NAND gate.

by making $(W/L)_p > (W/L)_n$ by a factor of about 2–3.3 depending upon the technology. This ratio compensates for the lower *p*MOS transistor carrier mobility. The NAND gate has more input signal possibilities to deal with if equal rise and fall times are desired. The pull-up current drive strength depends upon the number of *p*MOS transistors that are activated. Two parallel *p*MOS transistors have twice the pull-up strength of a single *p*MOS. Also, when the pull-down path is activated, two series *n*MOS transistors have about half the current drive strength of just a single *n*MOS. Compromises are made with NAND gate *p*MOS transistors and, typically, the $(W/L)_p$ ratios are not as large as the *n*MOS transistor $(W/L)_n$ ratios of inverters.

4.4 NOR GATES

A NOR gate symbol and truth table are shown in Figures 4.19(a) and (b). There are again two properties to note. The first is that any logic 1 to the inputs of a NOR gate causes a logic 0 output. The noncontrolling states are different than the NAND and AND gates. When A = 0 in rows 1 and 2, then the output C is the complement of B (C = B′). A similar property is seen in rows 1 and 3, where the output C = A′ when B = 0. When an input to a NOR gate is set to logic 0, then the effects of that signal node are removed with respect to signals on the other input lines. The noncontrolling logic state for a NOR

(a) (b)

Figure 4.19. (a) 2NOR gate symbol and (b) truth table.

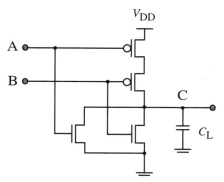

Figure 4.20. 2NOR transistor-level schematic.

gate (and OR gate) is logic 0. OR gates can be made by adding an inverter to the output of a NOR gate.

The NOR gate schematic in Figure 4.20 shows that any high logic input of V_{DD} turns on one of the nMOS transistors, forcing the output node C to 0 V. If both inputs are driven high with V_{DD}, then the pull-down strength is large since the nMOS transistor mobility is higher than the series pMOS transistors and the nMOS transistors are in parallel. Two pMOS transistors in series are potentially very slow, so the W/L adjustments for equal rise and fall times favor larger pMOS transistors. The NOR gate also has inverter static and dynamic properties. These are seen by setting one of the inputs to its noncontrolling logic state and measuring a transfer curve at the other terminal.

Self-Exercise 4.5

(a) Specify the input signals that allow node I_3 contents to be measured at O_1 (Figure 4.21). (b) Repeat for reading node I_2.

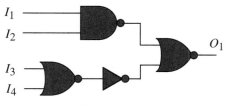

Figure 4.21.

Self-Exercise 4.6

(a) Specify the input signals that allow node I_5 contents to be measured at O_1 (Figure 4.22). (b) Repeat for reading node I_3. (c) Repeat for reading node I_1.

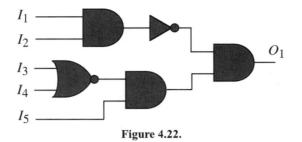

Figure 4.22.

4.5 CMOS TRANSMISSION GATES

A CMOS transmission gate, or T-gate, is a switch with many useful functions. Figure 4.23 shows an early T-gate design with symbol, truth table, and schematic. Signal transmission is controlled by the gating or control signal G in Figure 4.23. When G = 1, both transistors turn on and the signal passes to the output node B. When G = 0, no signal can pass since both transistors are off. The T-gate shown in Figure 4.23 can cause circuit problems in combinational logic since it has a high impedance state (also called floating, hi-Z, or tri-state) when the control signals turn off. Floating nodes allow voltage drift on transistor gates that can upset logic states. T-gates typically appear in CMOS flip-flop designs and to control tri-state levels in IC output buffers. In both applications, the floating node does not cause core logic instability.

Single transistors acting as T-gates are called pass transistors, but have a weakness. Assume that only the *n*-channel transistor in Figure 4.23(c) drives C_L and the *p*MOS transistor is removed. When the input A is high, the T-gate opens and charge passes through to C_L. When node B rises to a voltage one threshold drop below V_G, then the *n*-channel transistor turns off, so node B cannot rise above $V_G - V_{tn}$. An *n*-channel transistor passes a weak logic high. Similarly, a *p*-channel transistor will pass a weak logic zero that is one threshold drop above ground.

The cure is to put the *n*-channel and *p*-channel transistors in parallel. The *n*MOS transistor passes low logic levels with no voltage degradation, while the *p*MOS transistor passes the high logic levels with no V_t degradation. Ideally, a switch should have a constant resistance once turned on for any voltage that is transferred from node A to node B

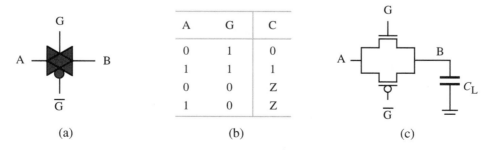

Figure 4.23. (a) Transmission gate symbol, (b) truth table, and (c) transistor-level representation.

(Figure 4.23), but since MOS transistors are nonlinear elements, the on resistance is not constant.

4.6 SUMMARY

This chapter examined detailed electronic properties of the inverter. Much of design and failure analysis uses this information. The inverter properties also align with NAND, NOR, and other logic gates. Static and dynamic transfer curves explain much of the speed and power behavior of integrated circuits. The important design technique of tapered buffers is commonly used in design to match small logic gate drives to larger high-input capacitance load gates. It is essential to understand the operation of NAND, NOR, and transmission gates at the transistor schematic level. The next chapter expands these concepts to show how design of higher functions is achieved.

BIBLIOGRAPHY

1. A. Bellaouar and M. Elmasry, *Low-Power Digital VLSI Design; Circuits and Systems,* Kluwer Academic Publishers, 1995.
2. A. Chandrakasan and R. Brodersen, *Low Power Digital CMOS Design,* Kluwer Academic Publishers, 1995.
3. B. Cherkauer and E. Friedman, "A unified design methodology for CMOS tapered buffers," *IEEE Transactions on Very Large Scale Integration (VLSI) Systems, 3,* 1, March 1995.
4. C. Hawkins, J. Soden, E. Cole, and E. Snyder, "The use of light emission in failure analysis of CMOS ICs," in *International Symposium on Test and Failure Analysis (ISTFA),* pp. 55–67, 1990.
5. M. Kakamu and M. Kingawa, "Power supply voltage impact on circuit performance for half and lower submicrometer CMOS LSI," *IEEE Transactions on Electron Devices, 37,* 8, 1902–1908, Aug. 1990.
6. A. Keshavarzi, K. Roy, and C. Hawkins, "Intrinsic leakage in low power deep submicron CMOS ICs," in *IEEE International Test Conference (ITC),* pp. 146–155, Oct. 1997.
7. T. Tsang, J. Kash, and D. Vallett, "Time-resolved optical characterization of electrical activity in integrated circuits," *Proceeding IEEE,* 1440–1459, Nov. 2000.
8. H. C. Lin and L. Lindholm, "An optimized output stage for MOS integrated circuits," *IEEE Journal of Solid State Circuits, SC-10,* 2, 106–109, April 1975.

EXERCISES

4.1. A logic gate noise margin parameters are: $V_{IH} = 1.6$ V, $V_{IL} = 0.3$ V, $V_{OH} = 1.7$ V, and $V_{OL} = 0.2$ V.
   (a) Calculate NM_H.
   (b) Calculate NM_L.
   (c) The input voltage is down to 1.7 V and a negative 50 mV noise spike appears. What happens to the circuit fidelity?
   (d) The input voltage is down to 1.7 V and a negative 150 mV noise spike appears. What happens to the circuit fidelity?

4.2. Given an inverter A whose voltage transfer curve has a logic threshold at $V_{DD} = 3$ V of $V_{TL} = 1.5$ V, and a second inverter B with a logic threshold of $V_{TL} = 1.2$ V for the same V_{DD}:
   (a) When the *n*-channel transistor goes into nonsaturation, is V_{in} more or less for B than for A?
   (b) When the *p*-channel transistor in B goes into nonsaturation, is V_{in} more or less that that of the *p*-channel transistor in A when it goes into nonsaturation?
   (c) What is the design difference between the transistors in inverters A and B?

4.3. Graphically determine the change in logic threshold of the CMOS inverter transfer curve in Figure 4.24 if the curve shifts 0.2 V to the right in the mid-region.

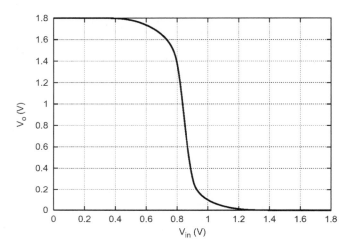

Figure 4.24.

4.4. Figure 4.7 shows how a CMOS inverter transfer curve changes for fast input transitions taking into account the load capacitance.
   (a) Do these curves affect logic threshold?
   (b) The hot carrier injection phenomenon that was described in Chapter 3 (Section 3.4.5) is aggravated by the electric field of the drain depletion region. Will hot carrier injection damage differ on the rising and falling input signal?

4.5. Calculate the power dissipated by a cardiac pacemaker circuit if $f_{clk} = 32.6$ kHz, $V_{DD} = 1.5$ V, C_L (per gate) = 300 fF, and the number of logic gates = 10 k.

4.6. Figure 4.9 shows that logic gate spikes can go above V_{DD} and below 0 V (GND). Why does this happen? Where does the charge come from, and where does it go?

4.7. Use the transition time delay model of Figure 4.10 if $C_L = 30$ fF, $V_{DD} = 1.5$ V, $K_p = 200$ µA/V^2, $V_{tp} = -0.35$ V, $V_{tn} = 0.35$ V, and $K_n = 125$ µA/V^2. What is the difference between rise and fall time of the transition if defined between 0 V and 1.5 V?

4.8. Repeat previous problem using the short channel model of Equation (4.9).

4.9. Figure 4.14 show CMOS transfer curves as V_{DD} goes above and below $V_{tn} + |V_{tp}|$. If V_{DD} goes below a single threshold voltage (i.e., V_{tn}), then the inverter still exhibits a valid transfer curve, although the curve is not as sharp. From your knowledge of transistors in Chapter 3, how is operation for $V_{DD} < V_t$ possible?

4.10. An output buffer has an input capacitance of 95 fF and a load capacitance of 100 pF. How many inverters are required in a fixed-taper design to minimize the propagation delay?

4.11. Figure 4.25 shows a single transistor transmission gate. At $t = 0$ the gate voltage moves to $V_G = 2$ V. The load capacitance C_L is initially at zero volts. As the node capacitance (the source) charges V_{GS} becomes smaller until it is less than V_{tn}. The transistor cuts off and the node is less than the input drain voltage by $V_S = V_D - V_{tn}$. If the body coefficient is $\gamma = 0.2V^{1/2}$ and $V_{t0} = 0.4$ V, calculate V_{GS} and V_S when the capacitor fully charges before cutoff.

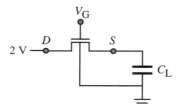

Figure 4.25.

4.12. Given an inverter with $V_{tn} = 0.4$ V and $V_{tp} = -0.4$ V, calculate the peak current during the transition if $W/L = 3$, $K_p = 50$ μA/V², and $K_n = 125$ μA/V².

4.13. Given the logic circuit in Figure 4.26:
   (a) What signal values must the input nodes be set to in order to read the contents of node C at output H?
   (b) What signal values must the input nodes be set to in order to read the contents of node G at output H?

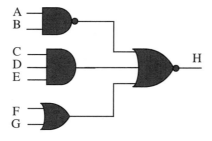

Figure 4.26.

CHAPTER 5

CMOS BASIC CIRCUITS

The previous chapter covered CMOS basic gate construction, emphasizing switching delay and power consumption characteristics. We now look at CMOS logic design styles, including static, dynamic, and pass-transistor logic. Input–output (I/O) circuitry and its protection problems are also discussed.

5.1 COMBINATIONAL LOGIC

Several design options exist for CMOS combinational gates. One reliable, lower-power design style uses complementary static gates, whereas high-performance circuits may use dynamic logic styles more suitable for high speed. Dynamic logic is more sensitive to noise and requires synchronization of signals (with a clock), even for combinational logic. Another logic design style uses pass-transistor or pass-gate elements as basic switches when fewer transistors are needed to implement a function. We want to understand these combinational logic design styles and their trade-offs.

5.1.1 CMOS Static Logic

Static, fully complementary CMOS gate designs using inverter, NAND, and NOR gates can build more complex functions. These CMOS gates have good noise margins and low static power dissipation at the cost of more transistors when compared with other CMOS logic designs. CMOS static complementary gates have two transistor nets (nMOS and pMOS) whose topologies are related. The pMOS transistor net is connected between the power supply and the logic gate output, whereas the nMOS transistor topology is connected between the output and ground (Figure 5.1). We saw this organization with the NAND and NOR gates, but we point out this topology to lead to a general technique to convert Boolean algebra statements to CMOS electronic circuits.

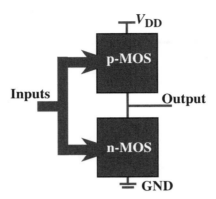

Figure 5.1. Standard configuration of a CMOS complementary gate.

The transistor network is related to the Boolean function with a straightforward design procedure:

1. Derive the *n*MOS transistor topology with the following rules:
   - Product terms in the Boolean function are implemented with series-connected *n*MOS transistors.
   - Sum terms are mapped to *n*MOS transistors connected in parallel.
2. The *p*MOS transistor network has a dual or complementary topology with respect to the *n*MOS net. This means that serial transistors in the *n*MOS net convert to parallel transistors in the *p*MOS net, and parallel connections within the *n*MOS block are translated to serial connections in the *p*MOS block.
3. Add an inverter to the output to complete the function if needed. Some functions are inherently negated, such as NAND and NOR gates, and do not need an inverter at the output state. An inverter added to a NAND or NOR function produces the AND and OR function. The examples below require an inverter to fulfill the function.

This procedure is illustrated with three examples.

■ EXAMPLE 5.1

Design a complementary static CMOS 2NAND gate at the transistor level.
   The Boolean function is simply A · B, therefore the *n*MOS net consists of two series-connected transistors, whereas the *p*MOS net will use the complementary topology, i.e., two transistors in parallel. The transistor structure was shown in Figure 4.18. ■

■ EXAMPLE 5.2

Design a complementary static CMOS XOR gate at the transistor level.
   The XOR gate Boolean expression F has four literals and is

$$F = x \oplus y = \bar{x}y + x\bar{y}$$

F is the sum of two product terms. The design steps are:

1. Derive the *n*MOS transistor topology with four transistors, one per literal in the Boolean expression. The transistors driven by $\bar{x}$ and y are connected in series, as well as the devices driven by x and $\bar{y}$. These transistor groups are connected in parallel, since they are additive in the Boolean function. The signals and their complements are generated using inverters (not shown). The *n*MOS transistor net is shown in Figure 5.2.

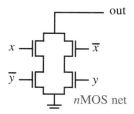

Figure 5.2.

2. Implement the *p*MOS net as a dual topology to the *n*MOS net. The *p*MOS transistors driven by $\bar{x}$ and y are connected in parallel, as are the devices driven by x and $\bar{y}$ (Figure 5.3). These transistor groups are connected in series, since they are parallel connected in the *n*MOS net. The *out* node now implements $\bar{F}$.

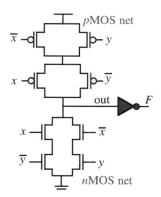

Figure 5.3.

3. Finally add an inverter to obtain the function F, so that $F = \overline{out}$

Steps 1–3 show that any Boolean function, regardless of its complexity, can be implemented with a CMOS complementary structure and an inverter. A more complicated example is developed below. ■

■ **EXAMPLE 5.3**

Design the *n*MOS transistor net for a Boolean function $F = x + \{\bar{y} \cdot [z + (t \cdot \bar{w})]\}$. We design this gate with a top-down approach. The *n*MOS transistor network

130 CHAPTER 5 CMOS BASIC CIRCUITS

is connected between the output and ground terminals, i.e., the lower box in Figure 5.1. The higher-level function F is a sum of two terms

$$F = x + \{operation\ A\}$$

where *operation A* stands for the logic within the brackets of F. The transistor version of this sum is shown in Figure 5.4.

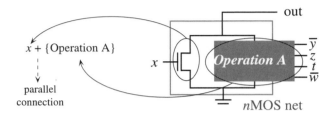

Figure 5.4.

Now we design the transistor topology that implements the block "*operation A*," whose higher level operation is an AND, i.e.:

$$operation\ A = \bar{y} \cdot \{operation\ B\}$$

Hence, the design topology is a transistor controlled by input $\bar{y}$ in series with a third box that will implement *operation B,* as shown in Figure 5.5.

We then design the topology of box B. This is a transistor controlled by input z, in parallel with two transistors connected in series; one controlled by input t, and the other by input $\bar{w}$. The complete *n*MOS network is shown in Figure 5.6.

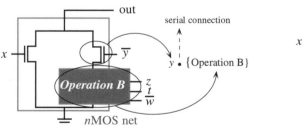

Figure 5.5.

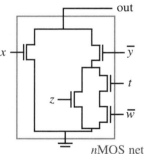

Figure 5.6.

Once the *n*MOS block is designed, we build the *p*MOS block with a dual topological structure and then connect an inverter to its output, as shown in Figure 5.7. ∎

Self-Exercise 5.1

Design the transistor level schematic of function $F = (x + y)[z + (wt)(\bar{z} + x)]$.

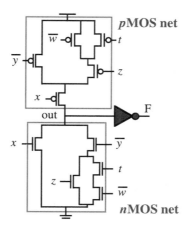

Figure 5.7.

5.1.2 Tri-State Gates

Many logic gates require a tri-state output—high, low, and high-impedance states. The high-impedance state is also called the high-Z state, and is useful when connecting many gate outputs to a single line, such as a data bus or address line. A potential conflict would exist if more than one gate output tried to simultaneously control the bus line. A controllable high-impedance-state circuit solves this problem.

There are two ways to provide high impedance to CMOS gates. One way provides tri-state output to a CMOS gate by connecting a transmission gate at its output (Figure 5.8). The control signal C sets the transmission gate conducting state that passes the non-tri-stated inverter output out' to the tri-stated gate output out. When the transmission gate is off ($C = 0$), then its gate output is in the high-impedance or floating state. When $C = 1$, the transmission gate is on and the output is driven by the inverter.

A transmission gate connected to the output provides tri-state capability, but also consumes unnecessary power. The design of Figure 5.8 contributes to dynamic power each time that the input and output (out') are switched, even when the gate is disabled in the tri-state mode. Parasitic capacitors are charged and discharged. Since the logic activity at the input does not contribute to the logic result while the output is in tri-state, the power consumption related to this switching is wasted.

This can be avoided by putting a transmission gate "inside" the inverter (Figure 5.9).

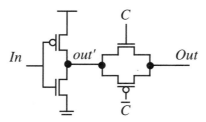

Figure 5.8. Inverter with a transmission gate to provide tri-state output.

132 CHAPTER 5 CMOS BASIC CIRCUITS

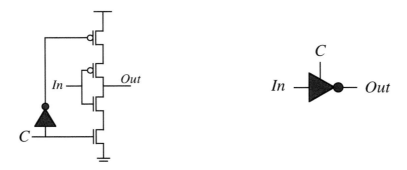

Figure 5.9. Schematic and symbol. The transmission gate "inside" the inverter provides tri-state output.

The *p*MOS and *n*MOS transistors of the transmission gate are in series within the conducting path between the power and ground rails and the inverter transistors. When the gate is in the tri-state mode, the inner transistor source nodes float, and the output is isolated from supply and ground. The activity at the inverter output signal node does not consume power as long as the gate is in the high-Z state ($C = 0$).

A tri-state capability adds delay independent of the configuration, due to the extra resistance and capacitance of the transistors driven by the tri-state control signal.

5.1.3 Pass Transistor Logic

There are many pass transistor (pass gate) logic subfamilies [3], and we will describe a few. Pass transistor logic uses transistors as switches to carry logic signals from node to node, instead of connecting output nodes directly to V_{DD} or ground. If a single transistor is a switch between two nodes, then a voltage degradation equal to V_t for the high or low logic level is obtained, depending on the *n*MOS or *p*MOS transistor type (Chapter 4). CMOS transmission gates avoid these weak logic voltages of single-pass transistors at the cost of an additional transistor per transmission gate.

Advantages are the low number of transistors and the reduction in associated interconnects. The drawbacks are the limited driving capability of these gates and the decreasing signal strength when cascading gates. These gates do not restore levels since their outputs are driven from the inputs, and not from V_{DD} or ground [6].

A typical CMOS design is the gate-level multiplexer (MUX) shown in Figure 5.10 for a 2-to-1 MUX. A MUX selects one from a set of logic inputs to connect with the output. In Figure 5.10, the logic signal c selects either *a* or *b* to activate the output (*out*). Figure 5.10(b) shows a MUX design with transmission gates. The complementary CMOS gates (Figure 5.10(a)) require 14 transistors (four transistors for each NAND and two transistors to complement the control signal), whereas the transmission gate design requires only six devices (more than 50% reduction). Each transmission gate has two transistors plus two more to invert the control signal.

Another pass gate design example is the XOR gate that produces a logic one output when only one of the inputs is logic high. If both inputs are logic one or logic zero, then the output is zero. Figure 5.11 shows an 8-transistor XOR gate having a tri-state buffer and transmission gate with their outputs connected. Both gates are controlled by the same input through a complementary inverter (A-input in this case).

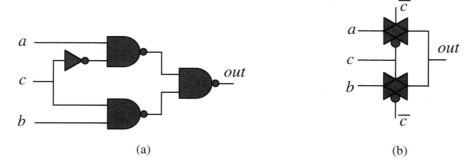

Figure 5.10. (a) Standard 2-to-1 MUX design. (b) Transmission (pass) gate-based version.

The XOR gate of Figure 5.11 is not a standard complementary static CMOS design since there is no *n*MOS transistor network between the output and ground, nor is there a *p*MOS transistor net between the output and the power rail. The XOR standard CMOS design built in Example 5.2 requires fourteen transistors, whereas the design in Figure 5.11 requires only eight.

5.1.4 Dynamic CMOS Logic

Previous sections showed conventional static CMOS circuit design techniques and designs based on tri-state gates and pass transistors. These designs are static, since they do not require a clock signal for combinational circuits. So, if circuit inputs are stopped (elapsed), then the circuits retain their output state (all circuit nodes remain at their valid quiescent logic values) as long as power is maintained. Dynamic CMOS logic families do not have this property, but do have the following advantages:

- They use fewer transistors and, therefore, less area.
- Fewer transistors result in smaller input capacitance, presenting a smaller load to previous gates, and therefore faster switching speed.
- Gates are designed and transistors sized for fast switching characteristics. High-performance circuits use these families.

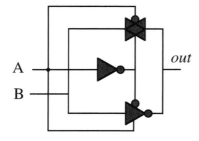

A	B	out
0	0	0
0	1	1
1	0	1
1	1	0

Figure 5.11. 8-transistor XOR gate and truth table.

- The logic transition voltages are smaller than in static circuits, requiring less time to switch between logic levels.

The disadvantages of dynamic CMOS circuits are

- Each gate needs a clock signal that must be routed through the whole circuit. This requires precise timing control.
- Clock circuitry runs continuously, drawing significant power.
- The circuit loses its state if the clock stops.
- Dynamic circuits are more sensitive to noise.
- Clock and data must be carefully synchronized to avoid erroneous states.

Dynamic CMOS Logic Basic Structure. A dynamic CMOS gate implements the logic with a block of transistors (usually *n*MOS). The output node is connected to ground through an *n*MOS transistor block and a single *n*MOS evaluation transistor. The output node is connected to the power supply through one precharge *p*MOS transistor (Figure 5.12). A global clock drives the precharge and evaluation transistors. The gate has two phases: evaluation and precharge. During precharge, the global clock goes low, turning the *p*MOS transistor on and the evaluation *n*MOS off. The gate output goes high (it is precharged) while the block of *n*MOS transistors float.

In the evaluation phase, the clock is driven high, turning the *p*MOS device off and the evaluation *n*MOS on. The input signals determine if there is a low or high impedance path from the output to ground since the global clock turns on the *n*MOS evaluation transistor. This design eliminates the speed degradation and power wasted by the short-circuit current of the *n*- and *p*-channel transistors during the transition of static complementary designs. If the logic state determined by the inputs is a logic one (V_{DD}) then the rise time is zero. The precharge and evaluation transistors are designed to never conduct simultaneously.

Dynamic circuits with an *n*-input gate use only $n + 2$ transistors instead of the $2n$ devices required for the complementary CMOS static gates. Dynamic CMOS gates have a drawback. If the global clock in Figure 5.12 is set high, then the output node could be in high-Z state with no electrical path to V_{DD} or ground. This exposes the node to noise fluctuations and charge sharing within the logic block, thus degrading its voltage. Also, the output load capacitor will slowly discharge due to transistor off-state leakage currents and lose its logic value. This limits the low-frequency operation of the circuit. The gate inputs

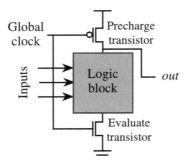

Figure 5.12. Basic structure of a dynamic CMOS gate.

can only change during precharge, since charge redistribution from the output capacitor to internal nodes of the nMOS logic block may drop the output voltage when it has a logic high.

Finally, dynamic gate cascading is challenging since differences in delay between logic gates may cause a slow gate to feed an erroneous logic high (not yet evaluated to zero because of the delay) to the next gate. This would cause the output of the second gate to be erroneously zero. Different clocking strategies can avoid this, as shown next.

Domino CMOS Logic. Domino CMOS was proposed in 1982 by Krambeck, et al., [4]. It has the same structure as dynamic logic gates, but adds a static buffering CMOS inverter to its output. In some cases, there is also a weak feedback transistor to latch the internal floating node high when the output is low (Figure 5.13). This logic is the most common form of dynamic gates, achieving a 20%–50% performance increase over static logic [3].

When the nMOS logic block discharges the *out'* node during evaluation (Figure 5.13), the inverter output out goes high, turning off the feedback pMOS. When *out'* is evaluated high (high impedance in the dynamic gate), then the inverter output goes low, turning on the feedback pMOS device and providing a low impedance path to V_{DD}. This prevents the *out'* node from floating, making it less sensitive to node voltage drift, noise, and current leakage.

Domino CMOS allows logic gate cascading since all inputs are set to zero during precharge, avoiding erroneous evaluation from different delays. This logic allows static operation from the feedback latching pMOS, but logic evaluation still needs two subcycles: precharge and evaluation. Domino logic uses only noninverting gates, making it an incomplete logic family. To achieve inverted logic, a separate inverting path running in parallel with the noninverted one must be designed.

Multiple output domino logic (MODL) is an extension of domino logic, taking internal nodes of the logic block as signal outputs, thus saving area, power, and performance. Compound domino logic is another design that limits the length of the evaluation logic to prevent charge sharing, and adds other complex gates as buffer elements (NAND, NOR, etc., instead of inverters) to obtain more area compaction. Self-resetting domino logic (SRCMOS) has each gate detect its own operating clock, thus reducing clock overhead and providing high performance. These and other dynamic logic designs are found in [3].

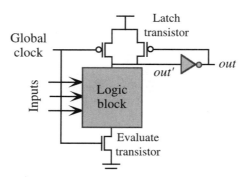

Figure 5.13. Domino CMOS logic gate with feedback transistor

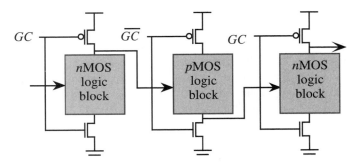

Figure 5.14. NORA CMOS cascaded gates.

NORA CMOS Logic. This design alternative to domino CMOS logic eliminates the output buffer without causing race problems between clock and data that arise when cascading dynamic gates. NORA CMOS (No-Race CMOS) avoids these race problems by cascading alternate nMOS and pMOS blocks for logic evaluation. The cost is routing two complemented clock signals. The cascaded NORA gate structure is shown in Figure 5.14. When the global clock (GC) is low ($\overline{GC}$ high), the nMOS logic block output nodes are precharged high, while outputs of gates with pMOS logic blocks are precharged low. When the clock changes, gates are in the evaluate state.

Other CMOS Logic Families. Dynamic circuits have a clock distribution problem, since all gates must be functionality synchronized. Self-timed circuits are an alternative to dynamic high-performance circuits, solving the clock distribution by not requiring a global clock. This simplifies clock routing and minimizes clock skew problems related to clock distribution. The global clock is replaced by a specific self-timed communication protocol between circuit blocks in a request–acknowledge scheme. Although more robust than dynamic circuits, self-timed logic requires a higher design effort than other families. These gates implement self-timing (i.e., derivation of a completion signal) by using a differential cascode voltage switch logic (known as DCVS) based on an extension of the domino logic.

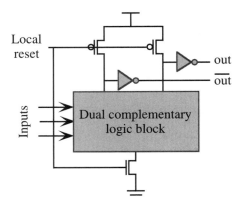

Figure 5.15. Basic DCVS logic gate.

The DCVS logic family (Figure 5.15) uses two complementary logic blocks, each similar to the domino structure. The gate inputs must be in the true and complementary form. Since output true and output negated are available, they can activate a completion signal when the output is evaluated. Since the gate itself signals when the output is available, DCVS can operate at the maximum speed of the technology, providing high-performance asynchronous circuits. The major drawbacks are design complexity and increased size.

5.2 SEQUENTIAL LOGIC

Nonstandard complementary CMOS designs are widely used in sequential logic to achieve compaction, high speed, and data storage. Latches, flip-flops, and registers are basic to many IC circuit designs. We present some of the better known static and dynamic memories.

5.2.1 Register Design

Registers are made with flip-flops that are in turn made with latches. Latches are memory elements whose transparent/memory states depend on the logic value (level) of a control signal. Flip-flops are constructed using latches to obtain a memory element that is transparent during the transition (edge) of the control signal for a better command of the time instant at which data are captured. We describe the basic latch, and then build the higher blocks.

CMOS Latch with Tri-State Inverters. Figure 5.16 shows the gate level and tri-state inverter design of a compact CMOS latch with two tri-state inverters and one regular inverter (Figure 5.16(b)). When $clk = C = 1$, the outputs of the first set of 2NOR gates are logic zero. This is the noncontrolling logic state feeding the D signal to the two output 2NOR gates. Therefore, the Q and $\overline{Q}$ signals feeding the inputs of the two output 2NOR gates set a stable logic condition. If $\overline{Q} = 1$, then the bottom output 2NOR gate is driven to $Q = 0$. The Q signal feeds a logic zero to the upper 2NOR gate, setting and holding $Q = 1$ (and $\overline{Q} = 0$). The latch holds its logic state indefinitely unless input signals change or the power is lost. When $C = 0$ (noncontrolling logic state to the input 2NOR gates), the Q outputs respond to the data input signal D. This is an example of a circuit that loads data on the low or negative portion of the clock signal.

In Figure 5.16(b), a level-sensitive clock controls the tri-state input of both inverters

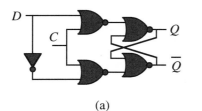 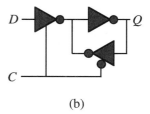

(a) (b)

Figure 5.16. (a) Basic gate-level CMOS latch design. (b) Tri-state inverter-level schematic.

such that when one is in tri-state, the other one is not. When the output of the first tri-state inverter stage is active (C = high), the feedback inverter is in tri-state (off), and the latch output is transparent. When C is low, the output of the first inverter floats, and the feedback tri-state inverter latches the value maintaining a feedback recovery configuration, holding the value. When $C = 0$, the latch is in its memory state. This is an example of a circuit that loads data on the positive portion of the clock.

> *Self-Exercise 5.2*
>
> Compare the number of transistors in the latch of Figure 5.16(a) with a D latch designed with tri-state inverters [Figure 5.16(b).]

CMOS Latch with Transmission Gates. Another transmission gate latch design further reduces transistor count. The circuit in Figure 5.17 uses two transistors less than that shown in Figure 5.16(b).

CMOS Flip-Flop with Tri-State Inverters. Flip-flops are edge-sensitive memory elements using latches in a "master–slave" (MS) configuration. This edge-sensitive circuit changes logic state not on the level of the clock, but on the leading or falling edge of the clock. This eliminates the transparency properties of the latch since the output signal never sees a direct path to the input. The output is sensitive to change on one of the clock edges, and insensitive to the clock level.

The clock drives the master latch with the slave latch clock signals inverted. The master and slave are coupled through a transmission gate. The master latch configuration captures data at one clock level (high or low), and the slave captures data on the opposite value. The transmission gate between the master and slave latches controls the timing for capture of output data Q.

Figure 5.18 shows a flip-flop design with unequal master and slave cells. The master cell (left portion of the circuit) is the latch design described earlier, and is connected to the slave (right portion of the circuit) through a transmission gate. When the clock is low, the master and slave are isolated, with the master active and the slave in memory. The action of the master tri-state circuit generates a logic value at the master inverter output that equals the input data D. When the clock goes high, the transmission gate connecting the master and slave opens, and data are transferred. Data are read directly to the Q output on the rising edge of the clock. The data could be transferred on the clock falling edge if the coupling transmission gate (and the other clocked signals) reversed their clock signal polarities. The MS design differs from a latch, since the MS output Q sees very little of the

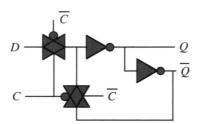

Figure 5.17. Alternate design of a latch cell with transmission gates.

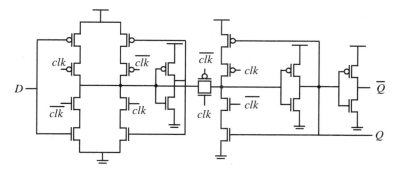

Figure 5.18. CMOS design of a flip-flop combining tri-state inverters and transmission gate design. The slave cell (right side) is only half of the master latch design to further reduce the number of transistors. Data are loaded into the first master latch on the negative clock edge, and data are read by the output Q on the rising clock edge. Then data are stored when the clock returns to logic zero.

input signal D directly. There is a small transient period when all transmission gates are in switching conduction states, and an electrical path may exist throughout the MS flip-flop. However, modern transition times are in the tens of picoseconds, and small clock timing skews make the overlap time very short.

5.2.2 Semiconductor Memories (RAMs)

Memories are a high-volume product in the IC market. The original phrase "random access memory" (RAM) refers to a memory in which all data have equal access procedures. There is no shifting of registers to capture a data bit. Test and reliability engineers also use memories to screen and verify emerging technologies, since they are relatively easy to test and failure analyze for process debugging. Their regularity and high density make them

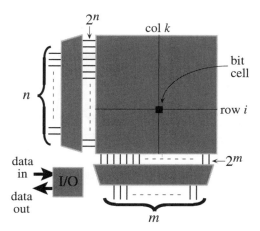

Figure 5.19. General architecture of a semiconductor memory.

good process monitors. Design regularity makes failure analysis easier than in random logic, since it is straightforward to map a logic failure to a physical location. High-density design provides good process monitoring, since transistors are designed for minimum dimensions of the technology, and conducting lines are kept as close as possible. These tight dimensions increase the probability of exposing process deficiencies.

The architecture of a static or dynamic semiconductor memory is shown in Figure 5.19. Memories have three major blocks: the memory array cells, the decoders, and the input–output circuitry. Memories can be bit- or word-oriented, accessing a single bit of the memory or the whole word (8, 16, 32, or 64 bits). In any case, the memory array is organized in rows and columns, with bits located at the intersection between a row and a column.

Each bit (or word) has a unique address that is mapped to physical locations with row and column decoders. The input–output circuitry performs the read or write data operations, i.e., store or retrieve the information in the memory.

Row and column decoders take an address of $n + m$ bits, and select one word line out of 2^n and one column out of 2^m for bit-oriented memories. In word-oriented memories, the column decoder selects as many columns as the number of bits per word.

Static and dynamic memories have different cell designs. Dynamic memories store information in a capacitor, retaining data for a limited time, after which the information is lost due to leakage. Information can be retained at the expense of additional external circuitry and dedicated working modes to allow memory refreshment. When the memory is being refreshed, it cannot be accessed, and is said to be in a latency period.

Static memories store information in feedback structures (two cross-coupled inverters). They are faster than dynamic memories since static RAMs do not have latency periods, whereas dynamic memory cost per bit is cheaper because fewer transistors per cell are required.

Static Memories (SRAMs). Static semiconductor memories use two inverters in a bi-stable feedback design (Figure 5.20(a)). Bi-stable operation is illustrated by plotting the output versus input voltages on the same axes for both inverters [Figure 5.20(b)]. The stable quiescent states of the circuit are at the intersections where $V_i = 0$ and $V_i = V_{DD}$, whereas the intersection voltage at $V_i = V_o$ is not a stable state (called a metastable state). The system is called bi-stable, since only two states are stable.

The inverter feedback circuit retains its state as long as the power supply is maintained.

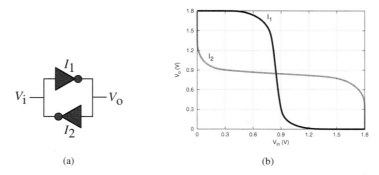

Figure 5.20. (a) Basic storage mechanism for static memories. (b) Input–output characteristics of the circuit.

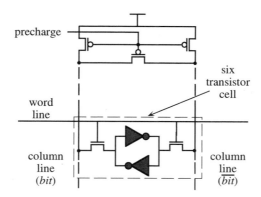

Figure 5.21. Six-transistor CMOS SRAM cell architecture.

Any "soft" voltage perturbation or possible current leakage in one node tending to switch the cell will be compensated for and overridden by the inverter output connected to this node. V_i in Figure 5.20(a) must have a stronger drive than the output of I_2. Memory cells typically set the memory state by driving V_i and V_0 simultaneously with opposite polarity signals.

The six-transistor cell architecture for a CMOS static RAM is given in Figure 5.21. All cell transistors and their interconnections are minimally sized to keep the array as small as possible. The word line controls the access transistors connecting the cell nodes to the column *bit* lines that run in pairs *bit* and $\overline{bit}$. When the word line is high, all cells in that row are connected to their corresponding *bit* and $\overline{bit}$ lines, and can, therefore, be accessed to read or write.

Memory read–write access time is reduced by precharging the *bit* and $\overline{bit}$ lines, i.e., forcing lines to the same voltage before any operation takes place. The precharge signal at the top of Figure 5.21 turns on all three *p*-channel transistors, forcing V_{DD} on both *bit* lines. When a write operation is performed, the *bit* and $\overline{bit}$ line drivers rapidly unbalance these lines, so that the correct value is stored in the memory. The precharge avoids the significant time for charging the highly capacitive *bit* lines when signals go from low to high. *n*MOS transistors pull down faster than equal sized *p*MOS pull-up transistors.

Memory cell inverters are minimally sized, but must drive long *bit* lines through a pass transistor during read operations. This potential delay can be improved using small analog circuits, called *sense amplifiers,* that are placed at each *bit* column output. Figure 5.22

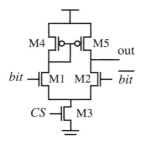

Figure 5.22. A differential sense amplifier used in SRAM memories.

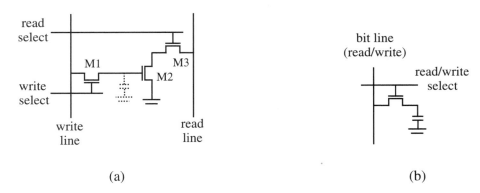

Figure 5.23. DRAM cells (a) Three-transistor cell. (b) One-transistor cell.

shows a typical differential sense amplifier used in CMOS SRAM designs. When the control signal CS is low, M_3 is off, and the sense amplifier output is floating. This corresponds to write operations. When CS is high, the circuit is activated. The sense amplifier reads *bit* and $\overline{bit}$ line voltages after precharge, and quickly transfers the cell value to the input–output circuitry, even before internal *bit* and $\overline{bit}$ lines reach steady voltages. If *bit* and CS are high, then M_1 drives current through M_4. The voltage drop across M_4 reduces the drain voltage at M_1. M_2 is off, and the *out* signal is pulled to V_{DD} through M_5. When *bit* is low and CS high, then M_2 turns on and *out* goes low. Sense amplifiers are only used during the read phase, and are disabled in other operations.

Dynamic Memories (DRAMs). Dynamic memory retains data as charge stored on a capacitor. This allows smaller memory cells, but since charge is not maintained by a feedback structure, stored values are lost with time and require refresh periods.

Two dynamic cell configurations are shown in Figure 5.23. Both cells use the parasitic gate capacitance of a MOS transistor to store the charge. The three-transistor cell (Figure 5.23(a)) has separate read and write select lines, giving a faster operation, but occupying more space. When the write select line is high, M_1 acts as a pass transistor, transferring the write line logic state to M_2 and putting M_2 in the off or conducting state. The drain $M_1 - M_2$ node capacitance holds that state. The read signal turns on M_3 and the data bit on the M_2 drain is passed through M_3 to the read line. This configuration allows for a nondestructive read operation, meaning that the cell does not lose its contents after a read is performed.

The single-transistor cell (Figure 5.23(b)) is popular since it has the smallest memory cell area. The charge stored in the cell storage capacitor is lost during the read operation because of charge sharing with the *bit* line parasitic capacitor, thus requiring a refresh operation during the same access cycle. The refresh operation uses circuitry that restores the original value in the cell once it is read.

5.3 INPUT–OUTPUT (I/O) CIRCUITRY

Input–output circuitry must link logic signals inside the IC to the outside world. The major I/O design problems are sufficient signal strength to drive large loads on printed circuit boards (PCBs) and IC internal circuitry protection from outside electrical assaults. Output current drive is typically achieved by using large output buffers that can have W/L

ratios in the range of 1000–4000. I/O design is challenging and very technology dependent.

5.3.1 Input Circuitry: Protecting ICs from the Outside Environment

CMOS circuits need protection from electrical assaults of the outside environment, especially for circuit inputs since they are connected to transistor gates. Input devices are often exposed to electrical overstress (EOS) and electrostatic discharge (ESD) phenomena that are responsible for gate oxide ruptures and interconnect damage [1]. A person walking on a carpeted floor can accumulate over 20 kV of static charge. Contact between a charged human body and an IC pin can cause a several nanosecond discharge, leading to Ampere current peaks and pin voltages up to 4,000 V or greater!

ESD is the rapid transient discharge from picoseconds to nanoseconds of static charges when two dissimilar bodies come in contact. The transistor thin silicon dioxide (SiO_2) film of less than 25 Å is easily damaged. The operating electric fields that gate oxides typically use are between 2–5 MV/cm, whereas breakdown occurs between 10–18 MV/cm. It takes only a small extra gate voltage to push the oxides into rupture. Some IC fabrication steps induce ESD on internal transistors of the circuit, so that the phenomena are not just related to those transistors physically driven by pin connected inputs. ESD protection structures are designed within the IC, and can protect the circuit if designed well.

EOS delivers a high voltage for a longer time than ESD. EOS times between microseconds to seconds cause more visible damage to the IC than ESD. ESD and EOS have different properties and root causes, but both destroy ICs. ESD typically occurs when a circuit contacts a charged machine or human, and EOS comes from aberrant longer pulses from power supplies, testers, lightning, or general misuse, such as mounting a package backward. EOS protection strategies often seek to eliminate the problem at the system level.

Different strategies are adopted to protect input structures against ESD. Elements are connected between the input PAD, the transistor gates, and the power rails to provide safe discharge paths when ESD occurs. These elements are inactive as long as the voltage levels of the node are within the normal operating conditions of the device.

When ESD assaults the IC, the protection circuit must drive the excess charge to the power or ground rails, steering that damaging energy away from the transistor gate oxide or metal interconnects. These protection devices are diodes and/or transistors working out of their normal operating ranges at high voltages and currents. Protection devices must sink large currents in nanosecond response times, suppressing heating effects and high electric fields. The protection circuits must survive the static energy assault to continue their protective function.

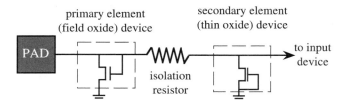

Figure 5.24. Example of input protection scheme against ESD.

ESD protection circuit design greatly depends on the technology, and is very layout sensitive. A common protection circuit with two protecting elements and a resistor is shown in Figure 5.24. The primary element takes most of the current during the ESD event, whereas the secondary element gives rapid initial protection to the logic gate input until the primary device turns on. The resistor provides a voltage drop to isolate both elements, allowing high voltage operation of the primary element while the voltage at the gate input can be maintained at a lower value. There are several ways to design the primary and secondary elements, so only the basics are described here (for more information refer to [1]).

The primary input protection circuitry in MOS technologies may use a field oxide transistor with a triggering voltage of about 30–40 V (this greatly depends on the technology). The secondary protection device is a grounded gate nMOS transistor reaching its trigger breakdown voltage (called snapback) rapidly before the primary protection circuit turns on. The current through this secondary device causes a voltage drop across the resistor that increases the PAD voltage to a value at which the field oxide transistor triggers, and takes most of the current.

5.3.2 Input Circuitry: Providing "Clean" Input Levels

Input circuitry must provide "clean" or noise-free logic levels to the internal circuitry. An external noisy or ringing input transition may induce multiple switching at the gate output. A solution uses a circuit with a static hysteresis transfer characteristic. The input voltage for which the output responds to a low-to-high or high-to-low transition depends on the output voltage. This circuit is known as an Schmitt trigger. Figure 5.25 shows both the Schmitt trigger transistor-level design and the input–output transfer characteristic. The feedback transistors whose gates are connected to the output provide the hysteresis [2].

The Schmitt trigger circuit in Figure 5.25 has only six transistors, but it has complexity whose explanation will bring together many CMOS concepts [2]. Start by setting $V_{in} = 0$ V and tracking the three nMOS transistors as they change state. When $V_{in} = 0$ V, M_1 and M_2 are off and M_3 is a pass transistor driven fully by the high logic voltage at V_{out}. M_3 will pass V_{DD} to the source of M_2 (drain of M_1) with a weak voltage of $V_{DD} - V_{tn3}$. As V_{in} rises to V_{tn1}, M_1 turns on. M_2 is still off since $V_{GS2} = V_{tn1} - (V_{DD} - V_{tn3})$. As V_{in} rises above $V_{in} = V_{tn1}$, M_1 conducts through M_3 and V_{S2} begins to fall. M_2 will turn on when $V_{in} - V_{S2} = V_{tn2}$. M_2 source and bulk are at different voltages so M_2 will have an elevated threshold voltage, the same as M_3. M_2 conduction now allows a more rapid drop in V_{S2} and V_{out} with

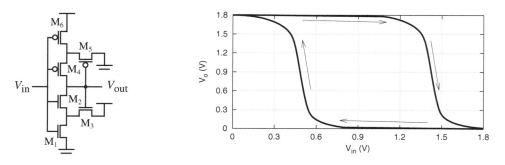

Figure 5.25. (a) Schmitt trigger CMOS design and (b) transfer characteristic.

the onset of transition higher than V_t as in a normal inverter. A similar analysis exists, starting with $V_{in} = V_{DD}$, that watches the transistor actions as V_{in} drops. The p-channel transistors respond to a different level when switching the output voltage to a logic high.

Analytically, the design is examined by equating the saturated state drain current expressions for M_1 and M_3:

$$K_1'(V_{GS1} - V_t)^2 = K_3'(V_{GS3} - V_t)^2 \quad (5.1)$$

We define the switching point of the low- to high-input transition V_{SPH} as the input voltage at which M_2 starts to conduct ($V_{GS2} = V_{t2}$), i.e.,

$$V_{in} - V_{S2} = V_{t2} \quad (2.2)$$

Setting the body effect threshold voltages of M_2 and M_3 equal, from Equation (5.2) the switch point is $V_{SPH} = V_{S3} - V_{tn3}$, giving

$$\frac{K_1'}{K_3'} = \frac{(V_{DD} - V_{SPH})^2}{(V_{SPH} - V_{tn1})^2} = \frac{L_3}{W_3} \frac{W_1}{L_1} \quad (5.3)$$

The conduction control of the pMOS transistors follows a similar analysis, giving

$$\frac{K_6'}{K_5'} = \frac{(V_{SPL})^2}{(V_{DD} - V_{SPL} - V_{tp6})^2} = \frac{W_6}{L_6} \frac{L_5}{W_5} \quad (5.4)$$

The transistor widths and lengths can be designed to achieve a given V_{SPH} and V_{SPL}.

5.3.3 Output Circuitry, Driving Large Loads

IC output circuitry must have strong signal strength to drive other circuits at the PCB level. Large capacitive loads driven at high speed require a large current in a small time. Driving large capacitances is not only an I/O design problem, it also appears within the device when driving long lines or bus lines. This drive is achieved with large transistors that have large input capacitance values because of their size. Specific circuits can drive large loads such as the tapered buffers described in Section 4.2.

Latchup in CMOS Technologies.* Large currents can be triggered by a bipolar mechanism called *latchup*. CMOS technology uses n-type and p-type transistors on the same substrate. Many processes start with a uniformly doped substrate, and construct wells of opposite doping to fabricate both MOS transistor types. This structure has an inherent *pnpn* parasitic bipolar transistor structure shown in Figure 5.26 that is off in normal operation and does not contribute to the circuit behavior. Latchup occurs when a parasitic *pnpn* structure underlying the CMOS structure is turned on, driving large currents and damaging the whole circuit.

The underlying parasitic bipolar transistors are connected with positive feedback, so that once the structure is triggered, the current increases until the device is destroyed. If proper rules are not followed during design or the circuit is operated improperly, then the parasitic bipolar structure may be triggered on, causing severe circuit damage.

---

*This subsection requires the reader to have a knowledge of bipolar (BJT) transistor princicples.

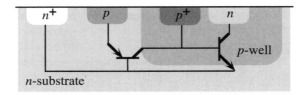

Figure 5.26. Cross section of a CMOS circuit fabricated with a single well and parasitic bipolar devices associated with such a technology.

The CMOS structure with diffused wells in Figure 5.26 shows the parasitic bipolar transistor structure underlying the circuit. The parasitic bipolar devices are connected such that the collector terminal of one device is connected to the base of another in a closed positive-feedback loop.

If an excess of carriers reach the base of some of the parasitic bipolar transistors, the current is amplified at its collector terminal, driving the base of the other bipolar device. This positive feedback connection can increase the current without limit. Figure 5.27 shows the current–voltage characteristic of the parasitic bipolar structure within a CMOS single-well process. Once the structure is triggered (the voltage goes beyond V_{trig}), lowering the voltage does not decrease the current because of the positive feedback. The only way to cut the current through the device is to completely switch off the power supply of the circuit.

Latchup is prevented by proper design that avoids activating a parasitic structure, since this cannot be eliminated. One latchup mechanism uses hot electrons from saturated MOS devices, causing holes to be injected into the substrate. If those holes are not properly collected at substrate and bulk contacts, they may diffuse and cause a voltage drop within the substrate (or well) that is enough to turn on a parasitic bipolar device. High substrate currents are another latchup source. Design strategies to avoid latchup are beyond the scope of this book and can be found in [1]. The modern trend toward SOI technologies and power supplies lowered to around $V_{DD} = 1$ V lessen the threat of latchup

5.3.4 Input–Output Circuitry: Providing Bidirectional Pins

Microprocessors, microcontrollers, programmable logic (FPGA), and memories use bidirectional (i.e., input–output) pins. Depending on the circuit design, certain pins are logic inputs for some operations and logic outputs for others. Bidirectional pins reduce the overall circuit pin count. These pins must have proper protections for the gates that will process the inputs, and also provide enough driving capability when acting as outputs.

Figure 5.28 shows a commonly used design to control bidirectional I/Os. When the control signal OE is low, the logic from inside the circuit (data out) is driven onto the output PAD through the strong output transistors. When the control signal input/output (OE) is high, both output devices are off, and the PAD acts as an input to the circuit.

5.4 SUMMARY

This chapter raises the level of transistor integration, showing how primitive CMOS complementary combinational logic gate designs are built from Boolean algebra equations. More compact circuits that have different power dissipation and speed properties illustrate

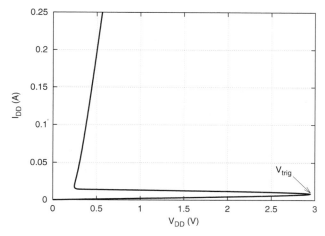

Figure 5.27. Current voltage characteristics of a parasitic bipolar structure underlying a CMOS single-well process.

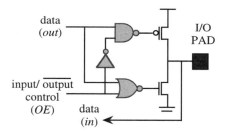

Figure 5.28. A bidirectional I/O circuit.

the popular tri-state gate, pass transistors, and dynamic logic gates. All versions appear in modern CMOS IC design. Sequential or memory-storing circuits partner with combinational logic to build complete ICs. Latches are the first building block, but have transparency properties eliminated by combining latches and transmission gates into flip-flops. Finally, the latchup failure mechanism and important input/output circuits were described.

REFERENCES

1. A. Amerasekera and C. Duvvury, *ESD in Silicon Integrated Circuits,* Wiley, 1995.
2. R. J. Baker, H. W. Li, and D. E. Boyce, *CMOS Circuit Design, Layout, and Simulation,* IEEE Press, 1997.
3. K. Bernstein, K. Carrig, C. Durham, P. Hansen, D. Hogenmiller, E. Nowak, and N. Rohrer, *High Speed CMOS Design Styles,* Kluwer Academic Publishers, 1998.
4. R. H. Krambeck, et al., "High speed compact circuits with CMOS," *IEEE Journal of Solid State Circuits, 17,* 3, June 1982, 614–619.
5. J. Rabaey, A. Chandrakasan, and B. Nikolic, *Digital Integrated Circuits,* Prentice-Hall, 2003.
6. N. Weste and K. Eshraghian, *Principles of CMOS VLSI Design, A Systems Perspective,* 2nd ed., Addison-Wesley, 1993.

148 CHAPTER 5 CMOS BASIC CIRCUITS

EXERCISES

5.1. Given the Boolean function $F = z[\bar{x}yz + x\bar{z}]$, draw the static CMOS transistor schematic.

5.2. Write the Boolean expression F for A, B, and C in the circuit in Figure 5.29.

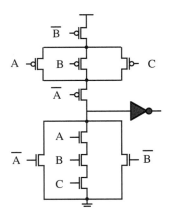

Figure 5.29.

5.3. Draw the static CMOS transistor schematic that performs the Boolean function $F = (g + f) \cdot (m + n)$.

5.4. Draw the CMOS transistor schematic that fulfills the function $F = \overline{[(A \cdot B) + C] \cdot D}$ for both a static and a domino CMOS logic gate.

5.5. Given the schematics of Figure 5.30:
   (a) If it corresponds to the CMOS pull-up network of a static circuit, what is the resultant Boolean expression F?
   (b) If the p- and the n-channel transistors are sized for equal drive current, discuss whether the pull-up will be faster than the pull-down network, or they will they be the same.

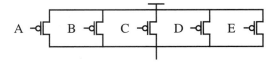

Figure 5.30.

5.6. What Boolean function will the circuit in Figure 5.31 perform?

5.7. Determine the logic function of the circuit in Figure 5.32.

5.8. Given the circuit of Figure 5.33:
   (a) Determine the role of A, B, and C nodes (input or output).
   (b) Determine its Boolean function.

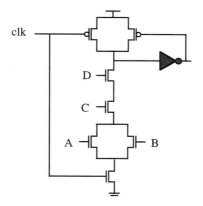

Figure 5.31.

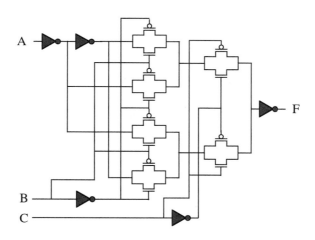

Figure 5.32.

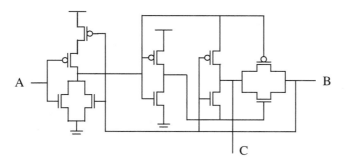

Figure 5.33.

5.9. The circuit of Figure 5.34 has the same function as a basic block used in sequential circuits. Identify the circuit type and the conventional names given to the inputs and outputs. Hint: analyze the equivalent circuit for $y = 0$, and then for $y = 1$.

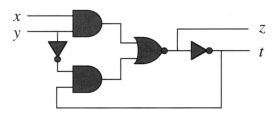

Figure 5.34.

5.10. Figure 5.20(b) shows the transfer properties of a simple static memory circuit. Suppose the input V_i is a short pulse with amplitude 0.6 V_{DD}. If V_0 drives another latch, what is the effect on (a) overall timing, (b) noise sensitivity (margin).

5.11. Identify the function and the input/output conventional node names for the circuit in Figure 5.35.

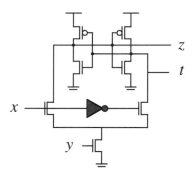

Figure 5.35.

5.12. Combine two circuits of Exercise 5.11 to get a flip-flop.

5.13. (a) What is the difference between EOS and ESD?
(b) If an input protection circuit protects the inner core logic from an ESD assault, but it is damaged, has the protection circuit done its job?

5.14. The DRAM circuits in Figure 5.23 store the bit (voltage) information on a capacitor. Use knowledge from Chapter 2 to determine the affect on refresh frequency if the temperature rises.

5.15. Observe the Schmitt trigger circuit in Figure 5.25(a). Explain how the transfer curves in Figure 5.25(b) behave as the input signal drops from V_{DD} to 0 V. Describe the transistor action.

5.16. If latchup occurs in a CMOS circuit and draws a large current, how do you stop it?

PART II

FAILURE MODES, DEFECTS, AND TESTING OF CMOS ICs

CHAPTER 6

FAILURE MECHANISMS IN CMOS IC MATERIALS

6.1 INTRODUCTION

A single modern IC may have more than a billion transistors, miles of narrow metal interconnect ions, and billions of metal vias or contacts. As circuit entities, metal structures dominate the semiconductor transistors, and their description is as challenging and necessary as that of semiconductors. Transistor oxides have shrunk, and dimensions are now on the order of 5–7 SiO_2 (silicon dioxide) molecules thick. Metal and oxide materials have failure modes that have always been with us, but are now even more significant in the deep-submicron technologies. This chapter addresses these IC failure modes that are caused by failure of materials.

This chapter has three sections:

1. Materials Science of IC Metals
2. Metal Failure Modes
3. Oxide Properties and Failure Modes

We do not seek to overwhelm the reader with mathematical detail, but rather to use visual models and learn the conditions that cause IC materials to fail in time due to interconnect bridges opens or damaged oxides. Circuit failure modes challenge the quality and reliability of large deep-submicron ICs. High-performance ICs now push safety margins closer to expected product lifetimes than before to achieve high clock frequencies. Added concerns are that previously dormant metal and oxide failure modes may now appear at test or during product life.

154   CHAPTER 6 FAILURE MECHANISMS IN CMOS IC MATERIALS

We refer to *intrinsic* material as defect-free, and *extrinsic* material as containing defects. Reliability failure modes are typically identified with failures of intrinsic material, whereas extrinsic material is linked to burn-in yield or production test failures. There are gray areas in this division, since some latent failure modes are related to (extrinsic) defects present during fabrication, but are too difficult to detect at test. Also, violation of fabrication procedures or design rules are known to cause postfabrication failures. Defective thin oxide that leads to rupture of the oxide, or metal stress voiding as a serious metal open failure mechanism are examples that are described later. Parts can fail at the next level of assembly or in the field, but detection of defects causing these serious failures may be almost impossible earlier in the product cycle. Test engineering deals with extrinsic materials, whereas reliability and failure analysis engineers deal with both material types.

Materials science studies the defects of a solid and their relation to the solid's physical properties [2, 3, 4]. The materials foundation of modern ICs were built on this old science. Several material classes exist, but we will only study metals and dielectrics. For many years, Al (aluminum) dominated signal and power interconnections, and SiO_2 was the thin and thick oxide material of the IC. Cu (copper) is now the popular interconnect material and new dielectrics are being developed to replace SiO_2. It is a nontrivial challenge to introduce new materials into the known and controlled CMOS IC manufacturing process.

6.2 MATERIALS SCIENCE OF IC METALS

Interconnect metals are thin films made of small, single crystals called grains. Metal grains are crystals similar to silicon crystals, but their grain surfaces are irregular and not smooth like silicon crystals. Figure 6.1 is a photograph of a polycrystalline Al line with its grain boundaries marked artificially for clarity. The irregular *grain boundaries* are important interface regions that influence the metal resistance against forces that can move atoms and lead to open or bridge metal lines. Metal failures involve extrusions or voids, and so require movement of Al atoms along easy paths, such as grain boundaries. Grain boundaries are about 1–2 atoms wide and are relatively open spaces, allowing easy travel for moving atoms. If a line has large grains and thus fewer grain boundaries, atoms have less opportunity for displacement. The metal in Figure 6.1 has many paths for a mobile metal atom to follow, increasing the likelihood of net Al atom dislocation in the presence of forces in the metal. This chapter looks at these metal forces that derive from electron current, temperature gradients, atomic concentration gradients, and mechanical stress forces.

Figure 6.2 shows two of the 14 possible crystal structures (Bravais lattices). The dots represent the center of the atoms since atomic volume usually extends to neighboring atoms. Figure 6.2(a) shows the corner atoms of a unit crystal with an additional atom

Figure 6.1. Al grains in IC interconnect. (Reproduced by permission of Bill Miller, Sandia National Labs.)

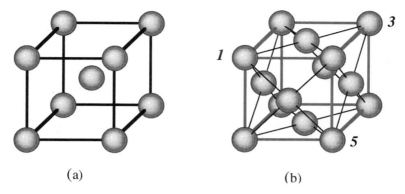

Figure 6.2. (a) Body-centered cubic cell (W). (b) Face-centered cubic cell (Al, Cu).

placed in the middle of the cube. This structure is called a *body-centered cubic* (bcc) cell, and is found in W (tungsten). The dots on the corners are drawn larger than the one in the middle for viewing clarity. Figure 6.2(b) shows a *face-centered cubic* (fcc) cell with an atom placed in the center of each face of the unit crystal. This is the structure of Al and Cu.

The unit crystals in Figure 6.2 can interpret the variable strength of a metal when forces are applied in any direction. Some directions are very resistant to applied force, and other directions are weak. Figure 6.2b numbers corner atoms 1, 3, and 5. If the unit crystal is viewed perpendicular to the plane of these three atoms, and the atoms are enlarged to represent their full diameters, then we see six atoms (Figure 6.3). The corner and face-centered atoms touch, and that surface is called the close-packed plane. The atoms bond strongly in this plane, and lateral forces that try to pull the atoms apart find it difficult to do so. This plane is called the 111 plane, following crystal convention. A force perpendicular to the close-packed plane dislodges atoms more easily because nearest neighbor atoms in this direction are further apart, and bonding strength is less. A fortunate result of laying down Al interconnections on the substrate of an IC is that the 111 texture is energetically favorable. Deposited atoms fill up available space next to the substrate just as oranges do when dumped on the bottom of a grocery fruit bin. This forms a crystal in the 111 plane that is the densest plane, so that more atoms can get closer to the substrate. This places the close-packed plane parallel to the majority of forces that act on the horizontal plane (especially electron flow), providing a natural resistance to dislocation of atoms.

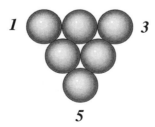

Figure 6.3. Close packed plane of fcc metal.

Metal crystals are imperfect. They have defects that can be intentional, such as alloyed metals, or unintentional from contamination by other atoms or thermal dislocations in a pure metal. Figure 6.4 illustrates three important metal crystal defects. Small atoms, such as B (boron) or H (hydrogen), can squeeze between larger metal atoms and form *interstitials*. Their small size allows easy diffusion within the host material. Some atoms can replace a host metal atom and are called *substitutionals*. Copper forms substitutionals in Al, and is intentionally alloyed with Al to strengthen the metal. A third defect appears naturally due to thermal vibration in otherwise perfect crystals, and is called a *vacancy*.

The vacancies in Figure 6.4 are locations in the metal lattice where an atom is missing. This is a natural condition of the high-frequency, thermal-induced vibrations of atoms. At any instant, there is a probability that an atom has vibrated out of its host position. There are no vacancies at 0 K, but they increase exponentially as temperature rises. Vacancy existence is a condition for movement of metal atoms. When metal atoms move within the lattice, such as in Figure 6.4, they must have a place to go. If an atom jumps into a vacancy, then it creates a vacancy at the location it left. Motion of vacancies is similar to the hole motion concept in semiconductors, except that vacancies don't have an electrical charge.

Metals have line, area, and volume defects besides the atomic defects in Figure 6.4. Volume defects are large metal voids or precipitates. A common line defect called an *edge dislocation* is drawn in Figure 6.5. The regular array of metal atoms is missing a few atoms in one of the crystal planes. Stress forces appear in the region of the circle. The top three atoms in the circle are in compression, and the bottom two atoms are in tension. The metal wants to relieve this energy. The compressive and tensile forces encourage a movement of atoms along the slip plane. The significance of these defects is that the metal is less resistant to holding its structure in the face of external forces (described later). Its strength is less than that of a perfect metal. When stresses are relieved to lower energy states, such as in heat annealing of metal, then conductivity increases. Electrons move easier in a crystal without alloy elements or other defects, although alloy elements are intentionally added to provide mechanical strength.

Grain boundaries are area defects. Figure 6.6 is a photograph of a narrower metal line than the one in Figure 6.1. The arrows locate points where three grain boundaries merge to form a *triple point*. If Al atoms move in a single grain boundary that merges with two other grain boundaries, then at the merger, more atoms can leave that point than enter (or vice versa). This is called a *flux divergence site*. For example, if Al is moving from right to left in Figure 6.6, then the first triple point will show more atoms entering the point than

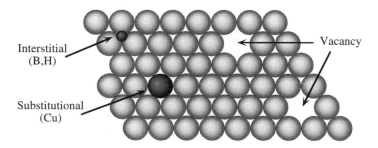

Figure 6.4. Imperfections in a metal crystal. Cu is not drawn to scale; it is about two times larger than Al.

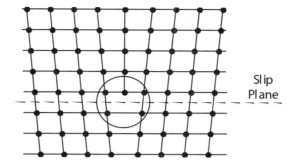

Figure 6.5. Edge dislocation and slip plane.

leaving, causing compressive buildup of atoms. At the second triple point, more atoms will leave than enter, causing tensile forces. The opposite polarity stress sets up a stress gradient.

The vertical grain boundary shown toward the left in Figure 6.6 is typical for metal lines of <0.5 μm, in which grains are large compared to the metal width. The very small lines in deep-submicron ICs have mostly these vertical or *bamboo* structures. You will notice that whereas an Al atom may move laterally in Figure 6.6 along longitudinal grain boundaries, it virtually stops when it encounters an orthogonal grain boundary. Metal lines with bamboo structures are generally stronger than those with triple points

The last topic in this capsule view of materials science concerns the physical gradients that can drive metal atoms. The link between physical gradients and particle (atoms) motion is subtle. Consider four potential gradients that drive metal atoms:

1. dC/dx Concentration gradient of atoms—diffusion
2. dT/dx Temperature gradient in solids—thermotransport
3. dV/dx Voltage gradient in solid—electromigration
4. $d\sigma/dx$ Stress gradient in solid—stress voiding

The concentration gradient causes net atom motion because all atoms are in thermally induced motion. The direction of motion is random, so the region of denser atoms will have more atoms diffusing toward the region of lower density than vice versa. A net dislocation of atoms occurs. Similarly, a temperature gradient has a region of higher-energy (chemical potential) atoms, and the more energetic ones can move to the lower chemical potential energy, similar to the motion induced by a concentration gradient. The voltage and stress gradient forces are discussed in detail since they relate to the electromigration and

Figure 6.6. Al metal and (marked) grain boundaries. (Reproduced by permission of Bill Miller, Sandia National Labs.)

stress voiding metal failures discussed later. Fick's diffusion laws and the concept of chemical potentials underlie all four forces.

Diffusion can be measured for the many atomic elements under diverse conditions such as solid, liquid, or gas phases, or for different structural conditions. Diffusivity denotes the measurement of diffusion, and is defined by Einstein's relation: $D = \mu \cdot kT$ (cm²/s), where μ = particle mobility and $k = 1.38 \times 10^{-23}$ Joules/K (Boltzmann's constant). D is exponentially related to Kelvin temperature (T), an activation energy (E_a), and a material-dependent constant (D_0) by

$$D = D_0 e^{-E_a/kT} \qquad (6.1)$$

Figure 6.7(a) shows the exponential diffusivity (D) relation to temperature, and Figure 6.7(b) plots the commonly presented straight line of $\ln(D)$ versus T^{-1}.

Generic curves are shown for diffusivity (Figure 6.7b) when the metal atoms are in a GB (grain boundary) or diffusing within the crystal lattice. GBs have a diameter that is about twice the size of an atomic diameter, and at low temperatures (right-hand side of x-axis), the grain boundaries have a higher diffusivity than the tightly packed crystal lattice region. As temperature rises, the diffusivity of atoms in the lattice increases and surpasses that of grain boundaries. Metallurgists estimate that the crossover temperature is between 0.4 T_{mp} to 0.6 T_{mp} (T_{mp} = metal melting point). Al melts at $T_{mp} = 660°C = 933$ K, so that the estimated temperatures at which diffusivity crossover occurs are between 100°C to 287°C.

Modern high-performance ICs have average package temperatures above 100°C and IC hot spots of even higher temperatures. The junction temperature of an IC is defined as the temperature of the silicon substrate, and it is a crucial parameter of reliability-prediction procedures and burn-in testing. The measured junction temperature of a 1 GHz 64-bit RISC microprocessor implemented in 0.18 μm CMOS technology was reported as 135°C at $V_{DD} = 1.9$ V [1]. This microprocessor had 15.2 million transistors packed in the 210 mm² chip area. High temperature allows higher diffusivity of metal atoms, which can lead to shorter failure times.

Review. Metal studies began over 100 years ago, and their application to IC thin metal films (interconnections) has a solid knowledge base. Metals are polycrystalline, with

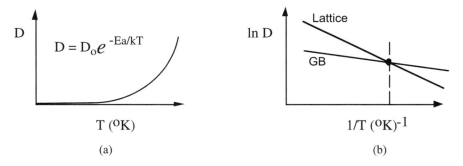

Figure 6.7. Diffusivity of grain boundary and lattice with temperature.

grain boundaries that separate the grains and influence the quality of the metal. Metal defects and high temperature make it easier for atoms to move within a metal, and reduction of these effects requires strong effort by industry. Concentration, voltage, temperature, and mechanical-stress gradients will move metal atoms within an interconnect. The movement of atoms leads to open or bridging circuit defects that are the next topics.

6.3 METAL FAILURE MODES

6.3.1 Electromigration

Electromigration is the net movement of metal under the influence of electron flow and temperature. A metal line will fail if sufficient current density and high temperature are applied. Metals can form a void or may form an extrusion that projects from one of the surfaces of the metal. This failure mechanism is called *electromigration* (EM). The abundant knowledge about aluminum (Al) failures is presented first, followed by a description of copper (Cu) failure mechanisms.

Figure 6.8 is an unusual photograph of two electromigration failure sites in a wide, unpassivated Al line. Here, Al atoms presumably exited the voided region and moved to the extruded bulging region on the left. Electromigration almost stopped the IC industry in the 1960s until methods were found to control electromigration. Electromigration studies began over 35 years ago, and much is known about this failure mechanism.

Electrons are believed to transfer a small but sufficient momentum to thermally active metal atoms forcing those atoms, out of their lattice sites, and moving them under diffusion in the same direction as the electrons. Figure 6.9 illustrates an Al metal line with electrons moving from right to left and colliding with Al atoms. If the thermal energy of Al is at a level such that a small nudge from many electrons dislodges it, then it will move if there is a vacancy to move into. That critical energy is called a saddle point. A small tensile stress is

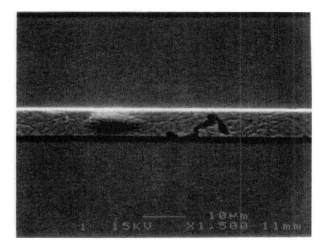

Figure 6.8. Electromigration SEM photo in a wide, unpassivated metal line. (Reproduced by permission of Joe Clement, Sandia National Labs.) Electron current and Al atom motion is from right to left.

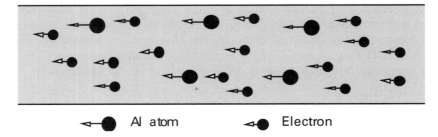

Figure 6.9. Representation of moving electrons and metal atoms during electromigration.

created where an atom is knocked from its lattice site, while downstream, the displaced metal atom creates compressive forces and possible extrusions. The region between the compressive and tensile stresses is under a stress gradient. Figure 6.8 shows these concepts, where the electron current (j) direction displaces Al atoms in the region of the void, and those atoms pile up downstream, forming an extrusion through passivation.

The variables that affect electromigration are clues used to reduce the threat, and we will present a model of electromigration using these variables. A flux J is the number of particles (atoms in this case) crossing a unit area per unit time. Table 6.1 lists the equations and subequations that lead to an expression for the atomic flux J_{em} of an electromigrating metal atom [9]. Equations are given in the middle column and substituted variables are given in the right column.

The final flux expression for electromigrating atoms is

$$J_{em} = \frac{D}{kT}(Z^* \cdot q)\rho \cdot j_e \cdot N \left(\frac{\text{atoms}}{\text{cm}^2 \cdot \text{s}} \right) \qquad (6.2)$$

Equation (6.2) is not typically used to calculate numbers, but to identify variables that influence electromigration. The flux is governed by diffusion and electromigrating atoms, and is said to be under a biased or directed diffusion. The temperature sensitivity of electromigration diffusion in Equation (6.2) is seen in Equation (6.1) showing the exponential temperature dependence.

Table 6.1. Electromigration Flux J_{em} Derivation

Magnitude	Equation	Symbols
Atomic flux J_m	$J_{em} = vN$	v = mean velocity of metal atoms N = concentration of moving metal atoms
Effective charge Q	$Q = Z^*q$	Z^* = effective charge factor q = electron charge
Atom mobility	$\mu = QD/kT$	D = diffusivity kT = Boltzmann's thermal energy
Atom mean velocity	$v = \mu \mathcal{E}$	μ = atom mobility $\mathcal{E}$ = electric field
Electric field	$\mathcal{E} = \rho j_e$	ρ = metal resistivity j_e = electron current density

The RHS of Equation (6.2) has an electron current flux term j_e, whereas the LHS is a flux of atoms. The Z^* term performs a cross coupling in the equation from electron current density to metal atom flux. One qualitative interpretation of Z^* is that it is the ratio of impinging electrons to a single metal atom that is moved. The driving electric field for electrons is the product of resistivity and electron current density across a 1 cm length, and the units are V/cm. Equation (6.2) reinforces that at a given current density and temperature, metal atoms will move along the metal stripe in the same direction as the electrons.

Electromigration Failure Time. A different analysis looks at the electromigration variables that affect the time for a metal line to fail, t_F. James Black made major contributions to understanding electromigration failure [5]. He empirically derived Black's law, which states that the median time to failure (t_F) of a group of Al interconnects is

$$t_F = \frac{A_0}{j_e^2} e^{E_a/kT} \qquad (6.3)$$

A_0 is a technology-dependent constant, T is temperature in Kelvin, j_e is electron current density (A/cm^2), and E_a is the activation energy (eV) for electromigration failure. Metal grain structures vary considerably, so that t_F is a statistical quantity not a predictor of single line failure time. Black's law shows the electromigration failure dependence on j_e^{-2}. When a design is shrunk, current density usually increases, so that designers must work to reduce j_e. An older DC design rule for CMOS technologies kept $j_e < 0.2 - 1$ mA/μm^2. This was a crude estimate with a built-in safety factor. Electromigration design rules for metal dimensions are now taken from circuit simulators that compute average current density and expected temperature in a local interconnection. The waveshape of the current pulses and the dimensions of the metal line needed to assure metal safety are calculated from these variables. Although j_e reduction is important, the exponential temperature term has a more acute effect on t_F.

The Black's law exponent of j_e is given as $n = 2$, as the original equation stated. Subsequent knowledge showed that n can vary from $1 < n < 2$ depending on the metal structure. We will use $n = 2$ for simplicity.

Black's law predicts failure time, and that is useful in test structure studies under accelerated temperature and current density. t_F is typically measured as the median time to failure of many test structures. However, Equation (6.3) is more often used to calculate E_a for the metal when t_F is measured. E_a is the accepted parameter to assess the electromigration quality of a metal technology. If t_F is measured over a temperature range, then E_a can be calculated from the slope of a plot of ln (t_F) versus (1/kT). If E_a is in the low range of $E_a \approx$ 0.4 to 0.6 eV, then the quality of the metal is poor, usually indicating an abundance of grain boundary paths. If $E_a \approx 0.8$ eV to 1.0 eV, then metal quality is good and indicative of bamboo structures. The calculated upper limit on E_a is that of a pure Al crystal, where $E_a \approx 1.48$ eV. A metal interconnect line made of a pure, single Al crystal would be the best structure possible, but this has not yet been practical.

■ **EXAMPLE 6.1**

Use Black's law [Equation (6.3)] to estimate the reduction in useful product life if a metal line is initially run at 55°C at a maximum line current density of 0.6

MA/cm², and then run at 110°C and 2 MA/cm². Use $E_a = 0.7$ eV and Boltzmann's constant of k = 86.17 μeV/K.

Black's law can be written for $T = 55°C = 328$ K and $j_e = 0.6$ MA/cm². Then divide by Black's law for $T = 110 °C = 383$ K and $j_e = 2$ MA/cm². A_0 cancels, and we get the electromigration acceleration factor A_{EM}

$$A_{EM} = \frac{t_{F1}(328\text{ K})}{t_{F2}(383\text{ K})} = \frac{j_{e2}^2}{j_{e1}^2} = e^{E_a/k[(1/328)-(1/383)]} = \frac{(2.0\text{ MA})^2}{(0.6\text{ MA})^2} e^{0.7/k[(1/328)-(1/383)]} = 389.4$$

∎

Typical practice rounds A_{EM} off to a single digit, or $A_{EM} = 400$. The result is that the hotter part has an estimated useful life shortened by a factor of 400. If nominal life is projected for 15 years for the cool part, then the hot part has a predicted life of about 2 weeks. This example is realistic since modern high-performance ICs have package temperatures of 100 °C, and the dies are even hotter. Designers push the j_e limits, since this increases clock frequency performance. This is one example where reliability margin and improved performance can be traded off.

Self-Exercise 6.1

An IC part has an operating temperature 10°C above its specification due to mounting in a package with poor thermal impedance. If metal $E_a = 0.8$ eV and normal use temperature is 85°C, (a) what percentage must the metal current density be reduced to maintain the same expected time to fail (t_F), and (b) if the part is run 10°C hotter, what is the reduction in lifetime.

How Metals Fail in Electromigration. Metals need an imperfection or defect in the structure to begin the failure process. Unfortunately, metals have unavoidable vacancies and irregular grain boundary patterns that can initiate electromigration. The ultimate failure may be an open circuit, or the metal may exert pressure at a site and break the passivation layer, and possibly form a defective bridge. Figure 6.10(a) shows an electromigration failure caused by extrusion of the Al through the passivation material, forming a bridge to another interconnection. Figure 6.10(b) shows a commonly seen open circuit notching characteristic of narrow metal lines (< 0.5 μm).

Failures commonly occur at a flux divergence site, and Figure 6.11 sketches two examples. Figure 6.11(a) shows a region of increased granularity that provides many more paths for Al to move in. The left side of the granular region undergoes tension as atoms leave this location and are easily transported in the granular region. The right border of the granular region undergoes compression as atoms find fewer travel paths and are stuffed into a volume smaller than their natural spacing. The tension and compression regions set up a stress gradient that actually acts to retard the left-to-right motion of Al atoms under electromigration. That is an important failure relief concept that will be developed later.

Figure 6.11(b) shows temperature differences along a line that can lead to flux divergence. The hot region has a higher concentration of vacancies, and Al atoms will diffuse more readily here than in the cold region. The J_{m1} flux tends to cause voiding and tension in and near the hot region, and compression as the moving atoms leave the hot region and approach the cold region. Since the mechanism for motion is diffusion, the cold region

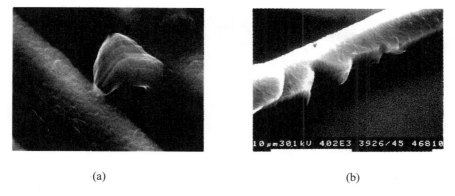

Figure 6.10. (a) Extrusion and bridging defect from EM. (b) Narrow metal line electromigration. (Reproduced by permission of Rod Augur of Philips Semiconductors; reprinted with permission from "Diffusion at the Al/Al oxide Interface during Electromigration in Wide Lines," *J. Appl. Physics, 79,* 6, 15 March 1996, pp. 3,003–3,010. Copyright 1996, American Institute of Physics.)

relatively stops the moving atoms (J_{m2}). The compression can be large enough to rupture the passivation material, allowing extrusion of the metal into the ruptured region [Figure 6.10(a)].

Another serious flux divergence site appeared when W (tungsten) was adapted as a via material in multilayer Al metal ICs. Tungsten has a high melting point, stronger atomic bonds, and high resistance to electromigration. Figure 6.12(a) sketches a metal interconnect with W vias and a TiN (titanium nitride) liner over the top and bottom of the Al line. Electrons entering the structure at the upper right direct Al atoms toward the W via. There is no net motion of atoms in the W, so the Al atoms pile up, causing compression at the top. Since electrons move unimpeded through the W via, they can dislodge Al atoms on the other side of the via, leading to voiding at the bottom (Figure 6.13). The left-side via experiences the opposite effect—compression on the bottom and tension (voiding) at the top.

Figure 6.12(b) shows a metal structure with Al vias. Here, Al atoms will pass through all structures. Unless there are local flux divergence sites along the path, there will not be voiding or compression. W replaced Al as the via material several years ago when reduced scaling made it difficult to retain Al. Cu (copper) interconnections with Cu vias unfortu-

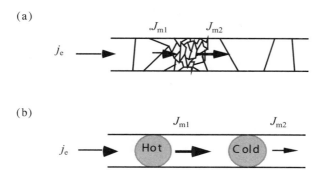

Figure 6.11. Flux divergence sites due to (a) granularity differences, (b) temperature gradient.

164   CHAPTER 6 FAILURE MECHANISMS IN CMOS IC MATERIALS

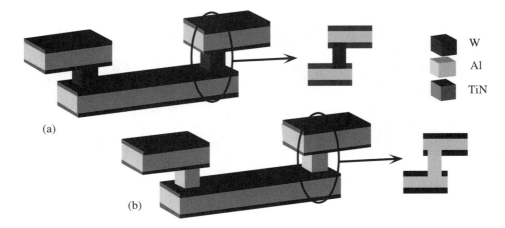

Figure 6.12. Metal–via structures (a) Al–W–Al. (b) Al–Al.

nately have a thin barrier metal interrupting the Cu flow at the bottom of the via. This flux divergence site is a common location for Cu electromigration voiding.

The TiN metal adjacent to the Al is a safeguard against breaks that might occur in the Al (Figure 6.13). If an open circuit forms in the Al, then current will pass through the TiN shunt path. This open-circuit protection does not prevent extrusions that occur horizontally and form bridges. However, barrier metals on the Al are a good retardant to electromigration voids, and one reason why electromigration is not more prevalent. Barrier metals such as TiN safely shunt current around an Al void location, but resistivity of barrier metals can be 30–40 times higher than Al. Voltage drops are typically small, but surprisingly large amounts of heat may be generated. Self-Exercise 6.2 illustrates this.

Figure 6.13. Void at the bottom of the blocking tungsten plug in metal-1 aluminum–copper alloy due to accelerated electromigration stressing [7].

Self-Exercise 6.2

The Al void in Figure 6.14 is 2.5 μm long and 1 μm wide, and the resistance of the TiN conduction path is 3 Ω. The current shunts through the 2.5 μm of TiN and then back to the Al. Calculate the heat flux (Watts/cm²) at the TiN surface, assuming that all heat goes out the bottom surface when: (a) current is 100 μA and (b) current is 10 mA. (c) What are voltage drops across the shunted void? (The problem assumes only one side of the Al is shunted by TiN.)

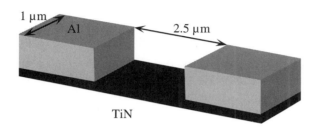

Figure 6.14. Al metal line with a void.

Current Polarity and Pulse Frequency. Three current waveform types exist in metal: pure DC, unipolar pulse currents with an average DC value, and bidirectional (bipolar) currents. DC currents exert the most electromigration stress, and are the typical condition for process characterization and long-term reliability studies. With the exception of off-state leakage, pure DC currents do not appear in CMOS circuits that are fully static and fully complementary. Fully static means that the circuit can operate at zero clock frequency, and fully complementary means that there are equal numbers of *p*MOS and *n*MOS transistors per gate. DC currents may appear if pull-up or pull-down resistors are used, or in any design that allows a continuous path from V_{DD} to V_{SS}. A significant reliability implication is that CMOS DC currents are also caused by most bridge defects and certain open-circuit defects. If these defect-induced currents exceed electromigration design rules in a given interconnect, electromigration may unexpectedly occur.

Unipolar pulses have an average DC current that occurs in drain or source terminals. Positive current always enters the source of a *p*MOS transistor (dotted line in Figure 6.15) and exits the drain, whereas current always enters the drain of an *n*MOS transistor and exits the source. The transistor gate voltage turns source–drain current on and off. The average (DC) current is used as the stress parameter for electromigration.

Bipolar current (solid line in Figure 6.15) occurs in the interconnections from a gate output terminal to the next logic gate input. Current enters the load gate interconnection, charging the capacitance during pull-up. The current reverses direction during pull-down. Interconnects carrying bipolar current have a low susceptibility to electromigration because damage caused on the forward current phase tends to heal on the reverse current phase. Metal atoms move in one direction during pull-up, and then reverse their direction during pull-down.

Blech Effect and Electromigration. The force between atoms at a given temperature is similar to the force between balls attached by a mechanical spring. If you push against a crystal of Al atoms, they compress and exert a back force. That back force acts on the com-

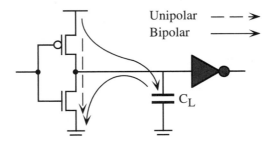

Figure 6.15. Unipolar and bipolar currents.

pressed atoms, tending to eject them from the stressed space. If tension is put on a crystal, then the opposite effect occurs as the region will tend to pull atoms back into the tensile space. Ilan Blech found an interesting and beneficial relation between electromigration and stress gradients [6]. He discovered that for a fixed metal line length, there was a current density below which electromigration would not occur. Conversely, for a given current density, there was a line length below which electromigration failure also would not occur. This so-called Blech effect is important in containing electromigration in deep-submicron ICs. We will derive the Blech effect flux equations as we did earlier for electromigration in Table 6.1.

Stress analysis uses mechanical concepts. We will derive the Blech effect starting with the definition of stress (σ):

$$\sigma = \frac{F}{A} \tag{6.4}$$

where F is the force across a material with area (A). A material or atom with high stress (σ), but no stress gradient ($d\sigma/dx = 0$) has no driving force to move it. However, when a stress gradient is present, then the force on an atom with atomic volume Ω is

$$F_\sigma = \Omega \frac{d\sigma}{dx} \tag{6.5}$$

This expression is subtle, but quite important for understanding stress voiding in metals. The equation predicts that applying equal high stress throughout a material has no displacement influence on particles, molecules, or atoms, but a difference in stress across a material will tend to move an atom. How is this so?

Figure 6.16 shows a unit cube representing a small solid. Initially, the block has no net force on its sides, but when a force F is applied to the right-hand face, the x-dimension compresses by Δx. If the left-hand side is constrained, we can derive the relation between the force, cube dimensions, pressure (p), and induced stress σ.

The energy statement (w) including pressure (p) and volume (v) is

$$dw = F \cdot \Delta x = \Delta p \cdot \Delta v \tag{6.6}$$

or

$$F = \Delta v \frac{\Delta p}{\Delta x} \tag{6.7}$$

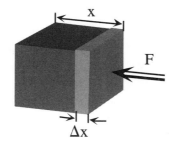

Figure 6.16. Element volume of particle or atom under stress forces.

Since stress $\sigma = p$, and the smallest volume is an atom of volume Ω, then

$$\Delta V = \Omega \text{ (smallest unit volume, approximately an atomic volume)} \quad (6.8)$$

and in the limit

$$F = \Omega \frac{d\sigma}{dx} \quad (6.9)$$

If a stress gradient exists in the metal, then Equation (6.9) provides the force on atoms to move them with the stress gradient. Equation (6.9) allows us to derive the flux equation using an electromigration-induced stress gradient, and then an equation for the Blech observation. The initial flux equations are

$$J_\sigma = vC = (\mu F)C = \frac{D}{kT} FN \quad (6.10)$$

Substituting Equation (6.9) into Equation (6.10) gives the flux term for metal under a stress gradient J_σ:

$$J_\sigma = \frac{D}{kT} \Omega \frac{d\sigma}{dx} N \quad (6.11)$$

The diffusion mechanism is evident in the stress effect. Electromigration simultaneously creates a tensile force and a compressive force as an atom is displaced. Blech recognized that the electromigration force and an oppositely directed stress gradient force may achieve a flux balance, thus stopping the net flow of atoms. Those two electromigration (J_{em}) and stress driven (J_σ) fluxes are repeated:

$$J_{em} = \frac{D}{kT}(Z^* \cdot q)\rho \cdot j_e \cdot N$$

$$J_\sigma = \frac{D}{kT} \Omega \frac{d\sigma}{dx} N \quad (6.12)$$

At equilibrium, the absolute values of the fluxes are equal, $J_\sigma = J_{em}$.

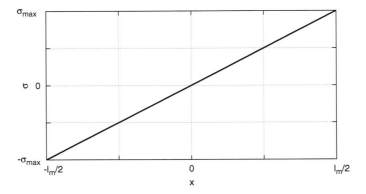

Figure 6.17. Stress versus distance curve for a thin film of metal.

Figure 6.17 shows a stress versus distance curve for a passivated metal. The stress gradient is assumed to be due to metal atom migration from the LHS of the curve to the RHS. σ_{max} is the maximum stress that the passivation can take before it cracks.

We can rearrange Equations (6.12) as

$$j_e dx = \frac{\Omega d\sigma}{Z^* q\rho} \qquad (6.13)$$

This balance equation requires one last concept. The balance ends when the passivation ruptures at a distance l_m under the high pressure of a maximum stress σ_{max}. If dx is integrated from $x = -l_m/2$ to $+l_m/2$ and stress from $-\sigma_{max}$ to $+\sigma_{max}$ we get

$$\int_{-l_m/2}^{l_m/2} j_e dx = \int_{-\sigma_{max}}^{\sigma_{max}} \frac{\Omega}{Z^* q\rho} d\sigma \qquad (6.14)$$

or

$$l_{max} j_e = \frac{2\Omega \sigma_{max}}{Z^* q\rho} \qquad (6.15)$$

The labor needed for this derivation is worth the result. The RHS of Equation (6.15) is a constant for a given technology. If we raise j_e, then l_{max} must drop and vice versa. Equation (6.15) explains Blech's law. Experimental measurements find the $l_{max} \cdot j_e$ product is about 1,000–3,000 A/cm for unpassivated metal. An example and exercise will illustrate this.

■ EXAMPLE 6.2

An Al interconnect has a length of 100 μm. If the $(l_{max} \cdot j_e)$ product is 3,000 A/cm, what current density limit should be assigned to prevent electromigration?

$$l_{max} \cdot j_e = 3{,}000 \text{ A/cm} = (100 \text{ μm})(j_e)$$

So if $j_e < 3$ mA/μm², electromigration will not occur in this unpassivated line. ■

Self-Exercise 6.3

If the ($l_{max} \cdot j_e$) product is 3,000 A/cm, what maximum length should the IC interconnect lengths be if designers keep effective current densities in all lines at less than 1.2 mA/μm^2.

Electromigration and High Frequency. Lines carrying pulsed currents show more resistance to electromigration as the frequency increases. Figure 6.18 shows the increase in t_f for a 50% duty cycle as current pulse frequency increases from DC [16]. During the off-portion of the cycle, a back stress is exerted on the metal from atoms moved during the on-cycle. As pulse frequency increases, average line temperature rises, increasing back-diffusion efficiency and healing during the off-state. This is an unusual case of metal reliability increasing with temperature.

6.3.2 Metal Stress Voiding

A bad "discovery" was made in the early 1980s when certain metal lines pulled apart, forming open circuits even if the IC was not powered. This failure mechanism, called *stress voiding,* (or *stress-induced voiding*) was linked to the TCE (thermal coefficient of expansion) differences of metal and the passivation materials surrounding it. When deposited metal is taken to 400°C or higher for a passivation step, the metal expands and tightly bonds to the passivation material. When cooled to room temperature, enormous tensile stresses arise from the differences in the TCE of the metal and passivation material. The passivation material basically does not move, so the metal bonded to the passivation material undergoes extreme tensile stress. These stresses are parallel to the metal line, and can pull lines apart if stress gradients appear. The time required to do this varies with the quality of the metal. It can happen during the fabrication process itself, or can take weeks to years for voiding to appear. Figure 6.19 shows photographs of two stress void failures.

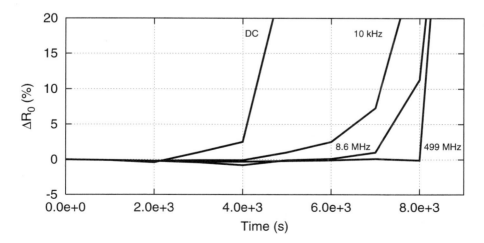

Figure 6.18. Time to failure versus current pulse frequency [16].

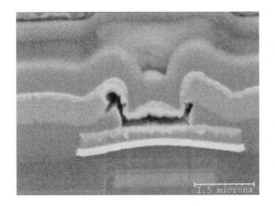 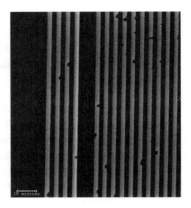

Figure 6.19. Stress void photos. (Reproduced by permission of Bill Filter, Sandia National Labs.)

Stress void analysis uses concepts from materials science, physics, and mechanical engineering. It is a mechanical failure mechanism that involves no electron current, but often it is the most prevalent metal failure mode in modern ICs. We will examine unpassivated and passivated metal line responses to the large mechanical forces that they undergo in normal IC environments. Passivation layers provide electrical isolation and protection to the metals, but the unavoidable problem lies in the sharp differences of thermal expansion coefficients. Al has a TCE = α = 23.6 × 10^{-6} parts per °C, and silicon dioxide has an α = 0.5 × 10^{-6} parts per °C. This means that a 1 meter Al line will expand by 24 μm for each degree of elevation in temperature, and SiO_2 by 0.5 μm per degree.

When SiO_2 is deposited and tightly bonded to Al at 400°C, there are no thermally generated stresses between the materials. However, when the two bonded materials cool to room temperature, enormous lateral stresses are generated, since the SiO_2 moves little while the Al strains to contract. A related problem is when Al reacts with a metal such as Ti. If both materials are passivated before the reaction, then the formation of $TiAl_3$ occurs with an approximate volume reduction of 5%. This is another high-stress-generating mechanism. We will calculate the stresses on Al for an unpassivated and a passivated metal line, developing simple equations that predict the stresses and strains in modern IC metals. We will then use numerical examples to show these enormous values, and conclude with methods to reduce the probability of stress void occurrence.

We begin with a simple example of metal stress forces, using a metal rod suspended in air and pulled at its ends. Figure 6.20(a) shows a metal rod with an applied force F, an area A, and a stress $\sigma = F/A$. When F is applied, the rod has a uniform stress along its axial length, and stretches a small amount, ΔL. The amount of stretching is called strain ε, and its measurement is normalized with respect to its original unstressed length L, or $\varepsilon = \Delta L/L$ [Figure 6.20(b)]. When metal is pulled in air, lengthening corresponds to a decrease in the diameter or circumference of the metal. The length increase is compensated for by a diameter reduction in the lateral walls, so that the volume remains constant. The surface atoms have no restraining force to prevent them from moving inward. Accurate strain measurements take this area change into account.

Figure 6.20(b) shows a generic stress–strain curve. Material in the linear region can be stretched to a point that upon release of the force returns the material to its original length.

6.3 METAL FAILURE MODES

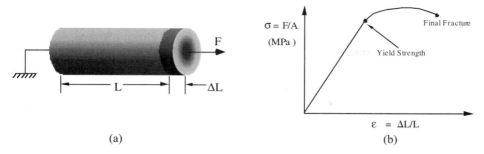

Figure 6.20. (a) Measurement setup for stress and strain. (b) Stress (σ) versus strain (ε) curve.

Atoms in this region are acting in accordance with the atomic spring model. The slope of this straight line is called Young's modulus, Y_m. Y_m characterizes a material's resistance to an applied force, and also allows calculations of strain given the stress (or vice versa). Stress is related to strain by a simple but important relation:

$$\sigma = Y_m \varepsilon \qquad (6.16)$$

The stress point at which the material enters the nonlinear or plastic deformation region is called the yield strength of the material. Metal stretched beyond the yield strength will not return to its original length. It has undergone plastic deformation, after which atoms no longer act as springs, but slide past each other. Finally, the material ruptures. The pressure unit is Pascals (Pa) (Newtons per meter^2), where 1 MPa $\approx$ 146 lb/in^2. Al alloys have a yield strength of about 95 MPa (about 14,000 psi) and a Young's modulus of about 71.5 GPa.

Temperature and length are typically linear, and related by the thermal coefficient of expansion (TCE) where

$$\alpha = \frac{\Delta L/L}{\Delta T} = \frac{\varepsilon}{\Delta T} \qquad (6.17)$$

Stress, strain, and temperature are combined from these equations to give

$$\sigma = Y_m \varepsilon = Y_m \alpha \Delta T \qquad (6.18)$$

Equation (6.18) is an important link between stress, temperature, and the material constants Y_m and α. An example will show how to estimate the stress forces on an unpassivated Al line.

■ EXAMPLE 6.3

Given an unpassivated Al line surrounded by air [except at the ends, where forces are applied (Figure 6.21)]. Let the length at 430°C be $L + \Delta L$, and the length at 30°C be L. When the metal shrinks as temperature drops from $T = 430$°C to 30°C, what stress is required to hold the ends at the $T = 430$°C dimension ($L + \Delta L$). Assume $\alpha_{Al} = 23.6 \times 10^{-6}/$°C and $Y_m = 71.5$ GPa.

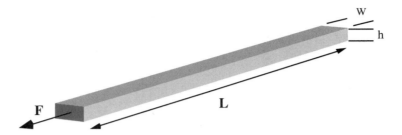

Figure 6.21. Unpassivated Al line.

The fractional change in the Al line is equal to the strain ε, where

$$\varepsilon = \frac{\Delta L}{L} = \alpha \Delta T = 23.6 \; 10^{-6} \times (430 - 30) = 0.00944$$

Since $Y_m = 71.5$ GPa and

$$Y_m = \frac{\sigma}{\varepsilon}$$

then

$$\sigma = \varepsilon \, Y_m = 0.00944 \times 71.5 \text{ GPa} = 675 \text{ MPa}$$

■

This 675 MPa stress greatly exceeds the yield and fracture strength of Al, therefore the TCE forces would tear the Al line apart. Most Al metal interconnections do not pull apart, so what protects the thin metal lines? The answer lies in the unique properties that passivation brings to these material systems. Surprisingly, even higher stress forces are generated in passivated metals than for the case of Al in air.

> **Self-Exercise 6.4**
>
> Assume a 100 µm long unpassivated Al metal line (Figure 6.22) with width of 1 µm, height of 0.4 µm, and yield strength of 95 MPa.
>   (a) How many pounds of force F are required to cause the metal to enter the plastic deformation region? How many kg?
>   (b) How far can you pull the 100 µm line before it goes into plastic deformation?

We next look at the more relevant situation in which the metal is bound and constrained by a dielectric. We are indebted to Bill Filter of Sandia National Labs for his stress void lectures at the University of New Mexico, giving his insights and examples of stress void calculations. Unpassivated metal atoms on the surface of the line do not have a bond to another atom on the surface plane, since there is only air at the surface. As the metal line is pulled at the ends, all atoms feel a tension, but the surface atoms, having no

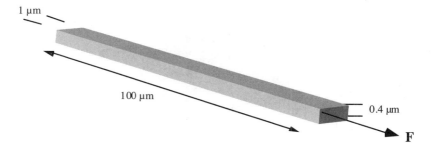

Figure 6.22. Al line.

restraint at the surface plane, tend to move inward, reducing the section of the metal line. The inward displacement provides some stress relief, but the stress remains high.

When a metal line is passivated, its surface atoms bond strongly to the fairly rigid passivation material and, basically, these metal atoms cannot move, even under very large stresses. The stresses are larger when a metal is passivated, and unless there are defects present in the metal, it maintains its shape. A key issue is that the surface atoms that could move inward under tensile stress now remain in position, and a stress exists at the surface of the metal that is orthogonal to the stress applied at the ends of the line. The stress equation is modified from that of Figure 6.20 to reflect this increase, using a parameter called Poisson's ratio. Equation (6.19) shows this modification for passivated metal stress [18]:

$$\sigma = \frac{Y_m \varepsilon}{1 - 2v} \qquad (6.19)$$

where v is called Poisson's ratio; $v = 0.35$ for aluminum. When two materials are bonded, the thermally induced strain must consider the differences in their TCE, so that using Equations (6.16)–(6.18),

$$\varepsilon = \frac{\Delta L_2 - \Delta L_1}{L_2} = \frac{L\alpha_2 \Delta T - L\alpha_1 \Delta T}{L} = \Delta \alpha \, \Delta T \qquad (6.20)$$

where $L_1 = L_2 = L$ for constrained metal. When we substitute Equation (6.19) into Equation (6.20) we get

$$\sigma = \frac{\Delta \alpha \Delta T Y_m}{1 - 2v} \qquad (6.21)$$

A simplified example shows the even larger stress calculation for a passivated metal.

■ EXAMPLE 6.5

Given a passivated Al line [Figure 6.23)], if $\alpha_{Al} = 23.5 \times 10^{-6}/°C$, $\alpha_{SiO_2} = 0.5 \times 10^{-6}/°C$, $v = 0.35$, and $Y_m = 71.5$ GPa, calculate the stress on the Al at its ends after the material drops from 430°C to 30°C. In other words, what is the required

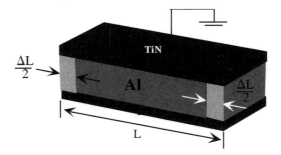

Figure 6.23. Passivated Al.

stress on the ends needed to maintain the longitudinal dimension when the metal and passivation material cool to 30 °C?

The fractional change in the Al line is equal to the strain ε, where

$$\varepsilon = \Delta\alpha\, \Delta T = (23.6 \times 10^{-6} - 0.5 \times 10^{-6})(430 - 30) = 0.00924$$

Since

$$\sigma = \frac{\varepsilon Y_m}{1 - 2v}$$

then

$$\sigma = \frac{(0.00924)(71.5 \text{ Gpa})}{1 - 2 \times 0.35} = 2{,}202 \text{ MPa}$$

∎

The stresses in this calculations are enormous, far exceeding the Al yield and fracture strengths. Again, why doesn't Al instantly pull apart? The answer lies in the ability of the passivation molecules to bond to the metal atoms. They can withstand stresses of GPa's, whereas Al–Al bonds permanently deform at less than a hundred MPa. Al is a soft metal.

How does this impact an integrated circuit? Stresses of hundreds of MPa have been measured in passivated Al lines. The everyday ICs that we use, such as wristwatch ICs, and those in personal computers, pocket calculators, etc., are subject to these large stresses throughout their product life. What is the concern? It is the threat of stress voiding, and that is final subject of this section.

How do stress voids appear in this high-stress environment? Voids in the metal do not just appear spontaneously. Voids form only in a stress gradient. Equation (6.22) is the flux equation for materials under a stress gradient:

$$J_\sigma = \frac{D}{kT}\Omega \frac{d\sigma}{dx} N \qquad (6.22)$$

Net atomic displacement at room or operational temperature can only happen if $d\sigma/dx$ is not zero. If the *whole* material has an equal stress of 2.2 MPa, then net atomic displacement does not happen. Metals and their passivation material can lie in these large stresses, and nothing much happens unless a gradient occurs. That is what causes stress voids.

A void requires an imperfection in the metal, such as a small nucleation at the metal–passivation material boundary. The surface of the metal atoms of a void nucleus have zero stress, so that a large stress gradient forms in the metal. The metal atoms now want to move and relieve the stress gradient. The last condition for destructive stress void growth is a diffusion path and sufficient temperature for these metal atoms to move. A neighboring grain boundary unfortunately satisfies this condition. In summary, there are three conditions for a stress void to occur:

1. A large stress must be present in the metal.
2. A defect must be present to convert the stress to a stress gradient.
3. A diffusion path and sufficient temperature must be present that allow the void to grow.

The flux J_σ in Equation (6.22) is proportional to the stress gradient, but atomic travel is still controlled by the diffusivity of the metal. These two factors combine to give an interesting temperature dependence to stress void sensitivity. Figure 6.24 shows a family of acceleration factor curves relative to room temperature as a function of processing temperature. The top curve has a deposition temperature of 435°C and the bottom curve of 300°C. Each curve has a peak with zero acceleration factors at high temperatures and near zero at low temperatures. At high temperatures, the TCE-induced stress differences are small, so little net metal atomic motion occurs, even though diffusivity is large. At low ambient temperatures, the TCE-induced stress is very large, but diffusivity is small, and, again, little net motion of the metal occurs. A peak occurs at which diffusivity and stress gradients

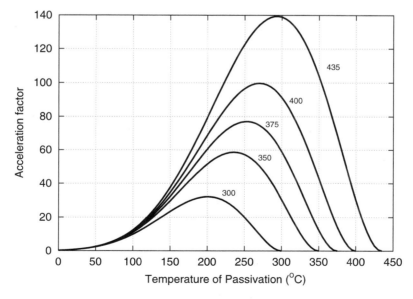

Figure 6.24. Stress void sensitivity versus test or ambient temperature. (Reproduced by permission of Bill Filter, Sandia National Labs.)

combine for maximum effect. The peak sensitivity shifts at lower process temperature, and the magnitude of the peak sensitivity is lower. A warning from Figure 6.24 is that modern high-performance ICs have die temperatures above 100°C.

Stress Voiding and Electromigration Comparisons. Electromigration and stress voiding have distinctions that are important when seeking the root cause of a failure (Table 6.2). Electromigration requires circuit operation that provides the necessary current density and elevated temperature, and it worsens with temperature. Stress voiding needs only a moderate temperature; it has a temperature peak in sensitivity, and it requires no power. When small stress voids nucleate during the cool-down or quenching process, then lines become more sensitive to electromigration during circuit operation. The small stress voids finitely reduce the metal area, thus increasing j_e.

6.3.3 Copper Interconnect Reliability

Cu appeared in 1998 as a substitute for Al in some high-performance ICs. Cu resistivity is about 1.9 μΩ · cm compared to Al–Cu alloy resistivity of about 3 μΩ · cm. This is about a 37% reduction in interconnect resistivity and RC time constant. Cu has interesting properties besides resistivity. Its melting point is 1358 K, considerably higher than Al at 933 K. This means that Cu has stronger bonds than Al, and should be more resistant to atomic motion by electromigration or stress voiding. Lee et al. reported a Cu Young's modulus of 110 GPa compared to 70 GPa for Al [10]. They also reported a Blech threshold of 3700 A/cm for passivated Cu lines.

It was originally thought that Cu would never electromigrate or have stress void failures, but that is not the case. Cu electromigration activation energies can show $E_a \approx 0.8$ eV, about the same as bamboo Al. Why the similarity if Cu–Cu bonds are stronger? Metal

Table 6.2. Comparison of Electromigration and Stress Voiding Failure Mechanisms

Property	Electromigration Failure	Stress Voiding Failure
Powered On	Necessary	Not necessary
Defect Class	Bridges and opens	Only opens
Ambient Temperature	Gets exponentially worse as temperature rises	Has a temperature peak sensitivity to failure
Passivation Strength	Blech Length gets longer as passivation hardness increases, making metal more durable	TCE-induced stress gets worse as hardness increases; thus, more vulnerable to failure
Initial Event	Displaced atoms cause tension and compression regions; stress gradient formed from displaced atoms	Stress gradient appears due to void nucleation before metal atoms are displaced
Atomic Displacement Mechanism	Diffusion	Diffusion
Final Failure Mode	Displaced atoms cause voiding (opens) and compression (bridging)	Displaced atoms cause voiding

migration in Al bamboo structures primarily takes place at the Al_2O_3–passivation interface. Cu electromigration also takes place at the Cu–passivation interface, and Cu granularity is higher than Al. Al bonds much tighter to the passivation than does Cu, so that Cu migration at its interface is larger. Improvements in Cu processing have raised its measured activation energies.

The Cu advantage in RC time constant can be taken from the R or C. For example, if a Cu and Al metal line have the same geometry, then the Cu line resistance is about 37% lower than the Al with the C unchanged. However, if the height of the Cu metal line is reduced, then its resistance increases, but its sidewall capacitance drops. The load C_L and crosstalk capacitance are smaller for Cu, and the IC power dissipation drops, as we saw in Chapter 4:

$$P = C_L V_{DD}^2 f_{clk} \tag{6.23}$$

Cu has a unique reliability risk not found in Al and that is the high diffusivity of Cu in SiO_2 and Si. Cu lines must be bound with a thin ($\approx$ 150 Å) barrier metal liner such as TaN. Cu can ruin transistor *pn* junctions if it is not contained. TaN liners bind three surfaces of the Cu line: the bottom and the two sides. A partial containment solution uses W (tungsten) at the first metal layer that connects drain, source, and gate contacts. This further separates Cu from the transistor *pn* junctions. Although W resistivity is higher, the signal lines are kept short so that IC performance is not compromised.

A Cu process uses the dual-damascene process, which is quite different from the Al metal sputtering process. The metal regions are first etched as trenches in the dielectric. Then a thin barrier metal layer is deposited on the bottom and two sidewalls of the trench. Next, a thin Cu seed layer is deposited in the trenches using PVD or CVP techniques followed by a Cu electroplating that fills the trenches and upper dielectric surface. The surface is then polished flat using the chemical mechanical planarizing (CMP) technique. This removes the excess Cu on the top surface, leaving Cu interconnects in the trenches.

The quality of the Cu interconnects depend critically on the properties of the seed layer. A rough Cu seed layer promotes small grain sizes, and this degrades the ability of Cu to resist EM fluxes. The dual-damascene process is more complex, with its required barrier protection metal liner as an essential part of Cu interconnect integrity. Cu is also a strong contaminant of the ICs in a fabrication facility if it spreads to the equipment in the lab. Originally, Cu was thought to be a perfect metal, without electromigration or stress void reliability risk. However, studies show otherwise [8, 10, 13, 15].

Song et al. noted that Cu is a noble metal that is less reactive than Al [15]. This may seem to be an advantage, but Cu forms weaker bonds with the surrounding dielectric than Al. Therefore, we see higher EM fluxes occurring at the Cu-to-dielectric interface, and even see lateral (intrametal layer) breakdown leakage paths.

The high via aspect ratios make liner dimension and liner continuity integrity challenging for the billions of vias that may populate an IC. The total metal length of a modern IC is on the order of several kilometers. The initial via etch through the dielectric must be taken just to the level of the bottom metal and no more. A shallow etch will leave a thin layer of dielectric in the serial path of the via. A 90 nm technology can use minimum vias on the order of 90 nm in diameter, subject to via–metal interconnect line design rules. The via fabrication challenges translate to lower IC yield, more test escapes, and increased reliability concerns.

Many via electromigration failures occur at the bottom of the vias, where the liner intersects and interrupts the Cu via path [8]. The subsequent Cu flux divergence identifies an electromigration weak spot. This site is similar to the electromigration voiding found under a tungsten via in Al systems. Also, an initial defect-induced voiding in the Cu could electromigrate to a complete open in the via. This forces the via current to pass through the parallel path of the liner. Although this offers protection, failure analysis showed that about 20% total via voiding could lead to excessive heat in the high-resistive liner, and even lead to a thermal opening of the liner itself.

The evolution of Cu interconnects to low-k dielectrics will impact Cu reliability. Lee et al. studied Cu with the low-k dielectric SiLK™ and found that t_{50} values were 3 to 5 times lower for Cu–SiLK™ than for Cu–oxide materials [10].

Doong et al. reported design rules for stress-induced voiding (SIV) in a 130 nm Cu–damascene technology [19]. The interconnect variables were width of the metal lead, via location in a wide metal lead, and width of the other connecting metal. They found that SIV was more severe on a via fed by a wide metal line than on one fed by a narrow line. Design rules are a key ingredient to preventing stress voiding in an IC.

6.4 OXIDE FAILURE MODES

Transistor gate oxides made of SiO_2 (silicon dioxide) are the beating hearts of a MOS transistor. Gate control of channel charge depends on the dimensions and quality of this oxide. Although SiO_2 appears in different parts of an IC, this section specifically uses the word oxide to refer to the thin dielectric material that separates the transistor gate from the channel substrate. Financial penalties for poor quality oxides are longer time to market and customer dissatisfaction. Oxide thickness in the 1970s was about 750 Å, and now oxide dimensions are below 20 Å. Gate oxide electric fields at the turn of this century were higher than burn-in field strengths in the early 1990s. Test, field failure, and burn-in are just three examples of why we must understand the chemical and electronic nature of oxides. We will look at the chemical structures of the thin oxide and then two oxide failure mechanisms: wearout and hot carrier injection. We will close with a description of a relatively recent pMOS transistor oxide reliability concern called negative bias temperature instability (NBTI).

Figure 6.25 is a remarkable TEM (transmission electron microscope) photograph of a MOS capacitor structure, showing the atoms of the single-crystal Si material, the non-crystalline or amorphous SiO_2 thin oxide molecules, and the polysilicon gate material above [24]. Imperfections cannot be avoided when the amorphous SiO_2 surface abuts the Si crystal. The interface is the site of numerous dangling bonds in which Si atoms or SiO_2 molecules have unshared bonds, leading to charges that are readily filled if an electron, hole, or hydrogen atom (H^+) is near. Process steps use or generate hydrogen and water that can bond with the unfulfilled states at the interface. The transistor threshold V_t is altered by these charge exchanges, with an important impact on speed.

Figure 6.26 shows the molecular orientation of SiO_2 molecules. A Si atom (open circle) appears to bond to four O atoms (shaded circles), but since each O atom also bonds to another Si atom, the chemical ratio is one Si to two O atoms, or SiO_2. McPherson and Mogul described the oxide structure in which each SiO_2 tetrahedron molecule forms rigid 109° angle bonds between Si and O [41]. Significantly the bonding between tetrahedrons is not rigid, but bond angles form from about 120° to 180°, with a mean of about 150°.

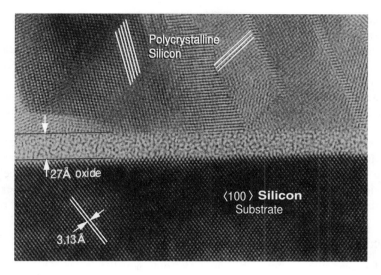

Figure 6.25. MOS capacitor cross section. (Reproduced by permission of Doug Buchannan, IBM Corporation.)

The bond angle weakens as the angle deviates from the mean. The variable bond strength is one source of the statistical behavior of oxide wearout and breakdown. Another weak bond occurs when an O atom is absent, allowing two Si atoms to bond to each other (Figure 6.26). These weaker (strained) bonds are more susceptible to rupture, leaving sites for holes, electrons, or atoms such as hydrogen to attach to.

The dangling bonds, the variable bond strength of SiO_2–SiO_2 molecular angles, and the absence of O atoms in the normal pairing leads to defects in the oxide called *traps*. A trap is an oxide defect, and the electronic charge on that trap is called a *state*. Traps can

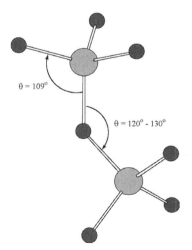

Figure 6.26. Si and O bond geometry in SiO_2 [41].

exist after the processing steps, or can be created when bonds are broken by energetic particles such as electrons, holes, or radiation.

Traps lying at the Si–SiO$_2$ border are called *interface traps*. Interface traps can rapidly exchange charge with channel carriers, since they are in close proximity to the channel. A trap 25 Å into the oxide will exchange channel charge in about one second. The oxide depth of the trap from the interface determines the exchange rate with the channel. Each trap that is 2.5 Å deeper in the oxide increases charge tunneling time by about one decade [30]. Border traps are those that lie deeper than interface traps, but less than 50 Å deep. Fixed oxide traps lie deeper than 50 Å and, basically, do not exchange charge with the channel. Most oxide dimensions are now less than 50 Å, so that these deep traps are less relevant to modern failure mechanisms. Charge exchange between the channel carriers and the oxide traps has a negative influence on transistor performance. The next section builds on these physical descriptions, describing oxide wearout and subsequent rupture.

6.4.1 Oxide Wearout

Good oxides wear out and rupture if a charge is continuously injected. This has nothing to do with defects from the fabrication, and the actual failure mechanism has eluded good people doing expensive experiments for many years. Each time a logic circuit has a voltage put across its gate oxide, a small amount of charge is injected into the oxide. The question is how long will it take for a normally operating transistor to wear out and rupture. That time must exceed expected product lifetime, since miscalculation could have severe consequences if premature oxide wearout caused ICs to fail during product life. Oxide wearout time decreases as oxide stress increases, and a concern is that the voltages and electric fields of thin oxides will cause premature oxide wearout. Oxide field strength is the force that accelerates electrons across the oxide. The example below illustrates the rising oxide field occurring for deep-submicron transistors in their use condition. The values are typical for the succession of technologies since the late 1980s.

■ **EXAMPLE 6.6.**

Calculate the oxide field strengths in V/cm for the following technologies: (1) 5 V and T_{ox} = 300 Å, (2) 5 V and T_{ox} = 200 Å, (3) 3.3 V and T_{ox} = 100 Å, (4) 2.8 V and T_{ox} = 60 Å, (5) 2.5 V and T_{ox} = 40 Å, and (6) 1.2 V and T_{ox} = 20 Å.
1 Å = 10^8 cm, so:

1. $\mathcal{E}_{ox}$ = 5 V/300 × 10^{-8} cm = 1.7 MV/cm
2. $\mathcal{E}_{ox}$ = 5 V/200 × 10^{-8} cm = 2.5 MV/cm
3. $\mathcal{E}_{ox}$ = 3.3 V/100 × 10^{-8} cm = 3.3 MV/cm
4. $\mathcal{E}_{ox}$ = 2.8 V/60 × 10^{-8} cm = 4.7 MV/cm
5. $\mathcal{E}_{ox}$ = 2.2 V/40 × 10^{-8} cm = 5.5 MV/cm
6. $\mathcal{E}_{ox}$ = 1.2 V/20 × 10^{-8} cm = 6 MV/cm

■

Significant tunneling of electrons through the gate oxide can occur when the oxide thickness becomes less than about 40 Å. As T_{ox} goes to 20 Å and 15 Å, the tunneling current is worse. These increased gate currents are reliability and power concerns in modern ICs.

Most of the research on oxide reliability has used MOS capacitor structures. Some early work on transistor gate oxide shorts showed that the gate capacitance could store sufficient energy ($\frac{1}{2}\,CV^2$) so that when a breakdown rupture occurred, this energy was released into the small, weakened oxide site, causing severe local damage. The silicon on either side of the oxide became temporally molten and joined; i.e., the polysilicon gate material physically bonded to the silicon substrate. An n-doped polysilicon gate joined with the p-well (nMOSFET) or n-well (pMOSFET) to form parasitic diodes or resistors. As transistors were scaled to modern technologies, power supply voltages dropped from 5–10 V to 1.0–1.2 V. Gate dimensions scaled from channel lengths of 1–5 μm to 90–130 nm. The gate area scaled by $(0.7)^2$ for each technology node so that gate capacitance scaled on the order of 2^7 as we went from 1.0 μm to 130 nm technologies. This dropped the gate capacitance by a factor of over 100 and V_{DD}^2 by about 20–25. The stored gate capacitance then became sufficiently small so that the violent thermal ruptures were replaced with the more gradual and subtle breakdowns that are described next.

What causes oxide wearout? The answer lies in which technology we work with. The older-technology oxides greater than 40 Å thick have a breakdown model quite different than the oxides we now build (below 30 Å). Oxides less than 30 Å thick are known as the ultrathins. Ultrathin oxide breakdown shows a distinct soft breakdown. Soft breakdown results in an irreversible damage to the oxide. Its most significant effect is an increase in noise of the gate voltage or current. Figure 6.27(a) shows this breakdown for oxide thicknesses from 2.4 nm to 5.5 nm [52]. The oxides were stressed with a constant current and the 5.5 nm oxide shows a precipitous drop in gate voltage at 75 s when a stressing gate current is applied. The 2.4 nm gate oxide did not change gate voltage with the oxide damage event, but shows an increase in noise.

The noise plotted in Figure 6.27(b) shows a four orders of magnitude increase after soft breakdown. Noise increase is the only certain evidence of the irreversible damage to ultrathin oxides. The noise associated with soft breakdown is thought to be trap-assisted conduction through a small conducting path in the oxide. The electrons hop noisily from trap to trap. In contrast, hard breakdowns in the thick oxides of older technologies showed severe gate voltage or current changes (Figure 6.27(a)). A hard breakdown was defined as a thermal event that merged the material above and below the oxide. The physical touch-

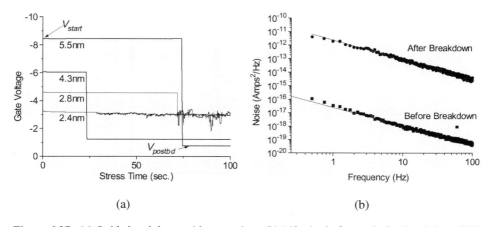

Figure 6.27. (a) Oxide breakdown with stress time. (b) $1/f$ noise before and after breakdown [52].

ing of two differently doped materials can create diodes, or resistors if the doping is of opposite polarity.

The normal functioning of a transistor with an ultrathin oxide is not as effected as those with thicker oxides following rupture [49]. The ultrathin wearout and breakdown model shows that rupture is primarily related to the gate voltage V_G and the amount of charge driven through the oxide (fluence). Evidence for the voltage model is shown in Figure 6.28, which plots the log of time to breakdown (T_{BD}) versus V_G. The interpretation is that electrons tunnel through the gate oxide, accelerating in the oxide field. The oxide electric field is constant across the oxide, but the internal oxide voltage drops as the electron reaches the anode of the structure. The relation of oxide rupture to gate voltage implies that the electron travels through the oxide without interaction, achieving a maximum kinetic energy before striking the anode, where it causes bond breakage. The likely weak bonds are H–Si and H–O. One subsequent damage mechanism is believed to be release of a hydrogen ion that reenters the oxide, causing trap damage. This is the anode hydrogen release model (AHR). The other damage mechanism is thought to be creation of a hole that migrates back into the oxide. This is the anode hole injection model (AHI). H^+ and a hole feel the attractive pull of the oxide electric field, causing trap damage as they enter and interact with the oxide molecules. There is evidence that both AHI and AHR contribute to wearout and breakdown [37].

Oxides do not breakdown after a single hole or electron are reinjected into the oxide. Oxides have a wearout and a breakdown phase. The wearout is believed to be the continuous addition of damage sites (traps) distributed throughout the oxide. When a statistical distribution of these traps is critically aligned in a vertical path supporting an increase in conduction, then a thermally damaging current goes through the oxide. This model is known as the *percolation model* of wearout and breakdown [28]. Figure 6.29 sketches such a statistical distribution of traps. The path in the middle of the figure indicates a trap distribution that is sufficiently close to form a breakdown percolation path.

Ultrathin breakdown has been characterized into three stages:

1. Slow defect generation within the oxide (wearout) until a defect path links the gate terminal to the substrate (percolation model)

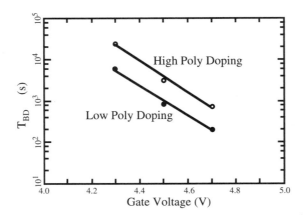

Figure 6.28. TBD as function of V_G [37].

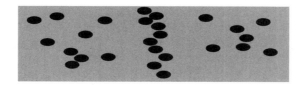

Figure 6.29. Percolation model of wearout and breakdown [28].

2. A soft breakdown (SBD) at low voltages that permanently increases gate current (< 100 nA at 1.2 V in 150 Å oxide) and gate noise
3. The appearance of a "hard breakdown" (HBD), showing continuous exponential increase in I_G.

There is evidence that SBD and HBD may be independent events [40, 44].

An ultrathin-oxide, voltage-dependent time-to-breakdown model (T_{bd}) has been proposed [43]. This breakdown model [Equation (6.24)] includes the gate oxide thickness (T_{ox}) and the gate voltage (V_G):

$$T_{bd} = T_0 \cdot e^{\gamma \left(\alpha \cdot T_{ox} + \frac{E_a}{kT_j} - V_G \right)} \tag{6.24}$$

where γ is the acceleration factor, E_a is the activation energy, α is the oxide thickness acceleration factor, T_0 is a constant for a given technology, and T_j is the average junction temperature. Time-to-breakdown physical parameter values were extracted from experiments as follows: $(\gamma \cdot \alpha) = 2.0$ 1/Å, $\gamma = 12.5$ 1/V, and $(\gamma \cdot E_a) = 575$ meV [43]. The voltage model and its supporting data suggest that ultrathin oxide rupture will be a greater concern with the increased electron tunneling (fluence) of thin oxides. V_{GB} decreases with each shrinking technology, but it is still high enough to support electron tunneling and subsequent reentry of high-energy particles into the oxides. However, recent data suggest that the soft ruptures of ultrathin oxides may not pose as serious a reliability threat to actual transistors.

There is general agreement on the ultrathin voltage-driven wearout model and the percolation theory of breakdown, but there is need for ultrathin technology data relating wearout and breakdown to logic circuit failure, not just to oxide capacitors. The ultrathin oxide experiments indicate that reliabilities may not be as risky as for breakdown in older technologies, but we must take care with these conclusions from wearout studies since they are predominantly done on capacitor oxides and to a lesser extent on transistors having drain, channel, and source regions. The studies reported on the effect of transistor oxide rupture on circuit functionality are now reviewed.

The evolution of ultrathin oxide studies from MOS capacitors to MOSFET transistors shows distinct characteristics. A rupture of the older technology gate oxide shorts may or may not cause logic failure in the IC [32, 33, 49]. However, the effect of ultrathin oxide soft breakdown on transistor V_t and g_m was reported as negligible [52]. Figure 6.30 shows the small time-varying changes in V_t and g_m during the pre-soft-breakdown stress and afterward—V_t dropped by 1.3% and g_m increased by 3.1%.

Crupi et al. [27] stressed 24 Å thin oxide transistors, and found breakdown in the overwhelming majority of nMOSFET devices. I_{Doff} was significantly increased, and the $|V_G/I_G|$ ratio showed hard breakdown values in the range from about 1 kΩ to 100 kΩ. High I_{Doff}

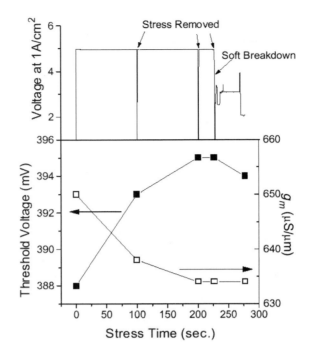

Figure 6.30. Time-varying change in V_t and g_m of an ultrathin oxide during current stress [52].

drain currents from about 1 μA to 1 mA occurred only for breakdown in the in the gate-to-drain region. Soft breakdown with much lower I_{Doff} occurred dominantly in the gate-to-source and gate-to-channel regions of the transistor. Hard breakdowns were not found for any of the three regions in the *p*-channel MOSFETs. The implication is that only the gate-to-drain breakdowns of *n*MOSFETs are serious reliability threats. Although the experiment clearly shows a sensitivity of the *n*-MOSFET, it should be noted that the oxides were protected from harder breakdown by a 1 kΩ series resistor in the gate electrical path. A normal logic IC may show more variation in breakdown hardness from soft to hard categories. Also, the implications for a logic circuit with damaged transistors, such as a NAND gate, were not shown.

Rodriguez et al. measured the effect of ultrathin gate oxide breakdowns on inverter properties [48]. Inverter transfer curves showed weakened logic voltages and, finally, functional failure for inverters that underwent a stronger stress. The weak logic voltage compromises noise margins, and could also cause I_{DDQ} elevation if the weak voltage output is sufficient to turn on downstream load gates. HBD can significantly load a previous logic gate stage to the point of logic failure or severe weakening of noise margin.

Dumin et al. showed an interesting result that the stress on an *n*MOS transistor oxide is greater if $V_G = 0$ V and $V_D = V_{DD}$ [29]. An inverter in the high-output-logic state would show this stress. This contrasts with traditional thinking, which assumed that the gate voltage was set at V_{DD} and source and drain terminals were at ground potential.

***E* versus *E*$^{-1}$ Models for Oxide Breakdown.** The research community debated for several years two oxide breakdown models that pertain to oxides of $T_{ox} > 40$ Å [51]. These are the E-model and the E^{-1} (or 1/E) model. The E- and E^{-1} models are increasingly less relevant with ultrathin technology use, but these models consumed large research resources, and they taught us a great deal about oxide properties. The E-model is thermodynamic, and one interpretation is that the initial event is field emission of an electron in the oxide. When field strength is high enough, an electron can be pulled from an atom in the material. This field emission causes a trap, and when a sufficient number of traps are vertically lined up, the oxide field strength exceeds that needed to rupture it. The E^{-1} model assumes an initial preferential tunneling of charge into a spot that has local thinning with respect to neighboring regions. A trap occurs at the thinner spot, resulting in a higher oxide field strength and leading to more tunneling and damage. The damage increases the oxide field in the region of traps until rupture occurs. The difficulty in distinguishing between the two models is that test data must be taken at abnormally high field strengths to accelerate the failures in a reasonable time. Wearout and breakdown data would take months or years to collect at normal use field strengths. The data on high oxide field stress (> 8 MV/cm) overlay almost exactly for the E- and E^{-1} models, and that was the problem.

Figure 6.31 compares time to breakdown, t_{BD}, found when plotting the data with the E-model or E^{-1} model. Although t_{BD} predictions are virtually identical in the high field region, a wide discrepancy is seen for projections of data to user conditions between 2–5 MV/cm. Experiments done at lower field strengths and elevated temperatures show that high-temperature breakdowns occurred with the same mechanism as those at lower temperatures. Suehle and colleagues then used this observation to fit t_{BD} data to the E-model over a broad range, including the user region [50].

6.4.2 Hot Carrier Injection (HCI)

This second major oxide failure mechanism occurs when the transistor electric field at the drain-to-channel depletion region is too high. This leads to the hot carrier injection (HCI)

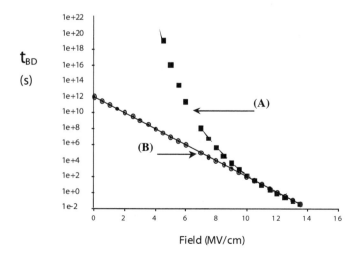

Figure 6.31. Time to breakdown t_{BD} for E^{-1} model plot (A) and E model plot (B) using extrapolation of data from high-oxide fields (8–10 MV/cm) [51].

effects that can alter circuit timing and high-frequency performance. HCI is a systematic failure resulting in a decline in the maximum operating frequency (F_{max}) of the IC. It seldom leads to catastrophic failure. The typical parameters affected are: I_{Dsat}, transistor transconductance (g_m), threshold voltage (V_t), weak inversion subthreshold slope (S), and increased gate-induced drain leakage (GIDL).

HCI can happen if the power supply voltage is higher than intended for the design, the effective channel lengths are too short, there is a poor oxide interface or poorly designed drain–substrate junctions, or overvoltage accidentally occurs on the power rail. Figure 6.32 sketches an nMOS transistor cross section showing the drain depletion field. The horizontal electric field in the channel $\mathscr{E}_{ch}$ gives kinetic energy to the free electrons moving from the inverted portion of the channel to the drain. When the kinetic energy is high enough, electrons strike Si atoms around the drain–substrate interface causing impact ionization. Electron–hole pairs are produced in the drain region and scattered. Some carriers go into the substrate, causing an increase in substrate current I_{SUB}, and a small fraction have enough energy to cross the oxide barrier and cause damage. It is estimated that an electron needs at least 3.1 eV to cross the barrier and a hole needs 4.6 eV. Even more energy is needed to break bonds leading to trap formation. Typically, damage is creation of acceptor-type interface traps near the drain by electrons with energies of 3.7 eV or higher. A possible mechanism is that a hot electron breaks a hydrogen–silicon bond at the Si–SiO$_2$ interface. If the silicon and hydrogen recombine, then no interface trap is created. If the hydrogen diffuses away, then an interface trap is created [42].

The energy follows a Boltzmann distribution in which particle thermal energy is $E_t = kT/q$ where k is Boltzmann's constant, T is degrees Kelvin, and q is the electron charge. An electron of 3.1 eV then has an equivalent mean temperature of $T = E_t/k = 3.1$ eV/(86.17 μeV/K) = 36,000 K. This is the basis for the expression "hot electrons." Ambient temperature has an interesting relation to HCI since carrier mobility increases as temperature decreases. Carriers with higher mobility more efficiently create hot holes and electrons, so that HCI increases as temperature is lowered. This property is sometimes used when using HCI reliability test structures to rapidly show damage.

Once a hot carrier enters the oxide, the vertical oxide field $\mathscr{E}_{ox}$ determines how deeply the charge will go. If the drain voltage is positive with respect to the gate voltage, then holes entering the oxide near the drain are accelerated deeper into the oxide, and electrons in the same region will be retarded from leaving the oxide interface. $\mathscr{E}_{ch}$ restricts the damage to oxide over the drain–substrate depletion region, with only a small amount of dam-

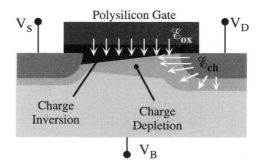

Figure 6.32. Saturated-state nMOS transistor and its internal electric fields $\mathscr{E}_{ch}$ and $\mathscr{E}_{ox}$.

age just outside the depletion region. In practice, the I_{Dsat} parameter is typically used to measure HCI degradation. I_{Dsat} is the transistor parameter that most closely approximates the impact on circuit speed, since it impacts the charge and discharge of load capacitors. Also, the MOSFET current model equations in Chapter 3 showed that I_{Dsat} is a function of V_t. The increased trap density and subsequent charging of the traps alters transistor threshold voltage V_t. Typically, nMOS transistors show increased V_{tn} causing the transistor to slow, and decreased I_{off} and g_m. pMOS transistors typically show the opposite effect: $|V_{tp}|$ decreases, I_{off} and g_m increase, and the transistors switch faster.

Figure 6.33 shows time degradation for a nMOS transistor under a hot-carrier stress. The important circuit speed parameter is the I_{Dsat} parameter that shows only slight degradation in time. This is the parameter that largely controls the oscillation frequency of a circuit such as a ring oscillator or a microprocessor. V_{tn} also changes slightly in time. Other parameters change more readily such as the g_m, I_{Dsat} reverse, and I_{Dsat} forward. The forward and reverse designations refer to normal bias of the drain and source (forward), and reversing the normal bias of the drain and source (reverse). The point is that whereas some transistor parameters change markedly, I_{Dsat} is the overall speed determining parameter, and it changes slowly.

Figure 6.34 shows $I_D = I_{off}$ (off-state leakage current) versus gate voltage for a pMOS transistor subjected to a drain-to-source overvoltage. This nominal 2.8 V transistor had $V_{DS} = -4.5$ V during the time of the measurements. After 1 minute of stress, I_{off} increased over two orders of magnitude. The damage is quick and easily measured, but, surprisingly, such damage does not affect circuit performance to the same degree. Chatterjee et al. reported that a stressed ring oscillator failure, defined as a 5% reduction in oscillation frequency, occurred for a time 100 times longer than the 10% damage criteria for measuring

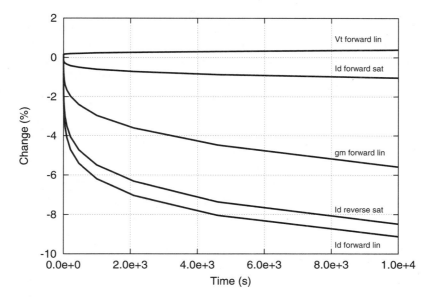

Figure 6.33. Plot of HCI degradation for transistor parameters. (Reproduced by permission of Duane Bowman, Sandia National Labs.)

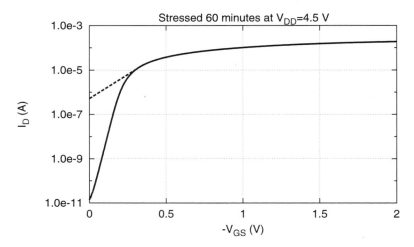

Figure 6.34. Stress time and pMOS transistor damage due to hot-carrier injection in a 0.35 μm technology.

I_{LIN}, the drive current of the transistor in an ohmic bias state [25]. Reasons for this paradox are developed next.

Why does the obvious damage to a transistor by HCI not evoke the same measure of IC performance reduction? One reason is that the dominant speed parameter I_{Dsat} is minimally affected by HCI stress. The oxide damage occurs dominantly in the oxide over the drain-depletion–substrate-depletion region when the transistor is in the saturated bias state, and that only occurs during the logic transition. If HCI damage is present in a nMOS oxide, then V_{tn} is increased only in the drain-depletion region. Charge inversion and V_{tn} are irrelevant in the depletion region during the logic transition. In the transition from V_{DD} to GND, the transistors are in the depletion state for about 75% of the excursion. The damage is in an unusual "don't care" location. If a transistor drain and source are electrically exchanged, then much more damage is observed, since the threshold voltage is critically altered near the source region affecting normal carrier inversion. Another effect offsetting IC performance is that a pMOS transistor undergoes a drop in V_{tp} and operates faster than normal.

The typical HCI effect is reduction in F_{max} (maximum measured operating frequency). A production concern is that ICs with statistically short effective channel lengths will have better F_{max} performance, but higher drain depletion fields. The question is whether F_{max} will degrade by HCI during customer use. Unless HCI is severe, then expected reductions in operating frequency are on the order of 1–3%. Guardbanding at test by raising the F_{max} limit by 5% is one protection. However, all manufacturers must make these determinations from their own parts characterization.

Hot-Carrier Injection and Bias State. Figure 6.35(a) shows the substrate current I_{SUB} versus V_{GS} when V_{DS} is high, holding the transistor mostly in the saturated state. The schematic for this measurement is shown in Figure 6.35(b). Hot-carrier generation in the drain-depletion region causes impact ionization with holes and electrons entering the substrate as well as the oxide. I_{SUB} is larger and more easily measured than I_G, and is often used as a proportionality indicator of hot-carrier generation.

The plot for an n-channel transistor in Figure 6.35(a) peaks near $V_{GS} \approx 0.5\, V_{DS}$. When V_G is below threshold, few carriers exist in the channel, and I_{SUB} is near zero. When V_{GS} approaches and becomes larger than V_{DS}, the transistor enters the nonsaturation state, and the depletion field at the drain disappears. Again no hot carriers are generated. When V_{GS} goes just above V_t then the device is in saturation and free carriers exist in the channel, some of which cause hot-carrier generation. Holes and electrons that enter the gate experience an oxide electric field whose positive field is at the drain. The gate voltage is lower than the drain voltage so that holes are preferentially attracted to the gate.

The left-hand portion of the curve in Figure 6.35(a) is the region in which holes enter the oxide and become gate current. When V_{GS} goes well beyond the peak, the gate voltage rises, and electrons are drawn to the gate. The peak in the I_{SUB} curve is a condition in which both holes and electrons are entering the oxide, but neither with maximum field attraction such as at the ends of the curve. The holes and electrons in the middle portion of the curve tend to cause more interface damage here, in contrast to traps deeper in the oxide.

These bias curves are useful for engineers who design reliability monitor structures for measuring HCI. Wafer-level reliability (WLR) monitor structures are designed for rapid measurement of damage. The HCI is maximized by biasing transistors at these worst-case conditions (i.e., $V_{GS} \approx V_{DD}/2$) and even at lower temperatures. The curves also show that HCI occurs during the logic state transitions, and not during the quiescent states. HCI requires the saturated bias state of the transistor, and that only occurs during the logic gate transition. The terminal polarities of an inverter are more dynamic than the curves in Figure 6.35 since V_{DS} is not constant, but drops as V_{GS} increases. $\mathscr{E}_{ch}$ and $\mathscr{E}_{ox}$ vary in a complex manner over a wide range during a logic transition.

Prediction of Integrated Circuit HCI Reliability. One method of estimating how long an IC will perform in a normal hot-carrier injection environment surprisingly uses

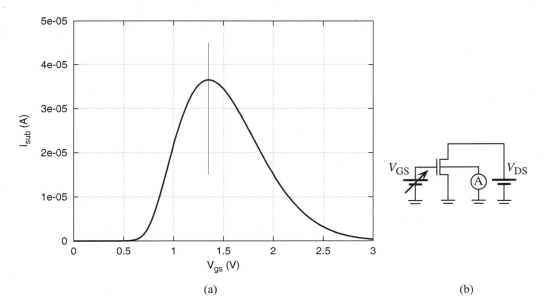

Figure 6.35. (a) n-channel transistor hot-carrier generation curves as function of V_{GS} and I_{SUB}. (b) Schematic for measurement.

data collected from individual transistors. These data are then combined with estimates of the IC duty cycle and other operating parameters. We thank Steve Mittl of IBM, who lectured on these subjects at the University of New Mexico, for the following discussion.

A theoretical model forms the basis of HCI reliability prediction for most companies [34]. The end result is a calculation of transistor lifetime τ as

$$\tau = C' \frac{W}{I_D} \left(\frac{I_{SUB}}{I_D} \right)^{-\phi_{it}/\phi_i} \qquad (6.25)$$

where C' is a constant, W is transistor width, I_D is drain current, I_{SUB} is substrate current, ϕ_{it} is the interface state activation energy (≈ 3.7 eV), and ϕ_i is the impact ionization activation energy (≈ 1.2 eV). The ratio of ϕ_{it}/ϕ_i is a constant of about 3. τ is the defined lifetime degradation due to HCI (i.e., a 10% drop in V_t or g_m, etc.). If we rearrange the terms, we get

$$\frac{\tau I_D}{W} \propto \left(\frac{I_{SUB}}{I_D} \right)^{-\phi_{it}/\phi_I} \qquad (6.26)$$

We can measure τ, I_D, and I_{SUB} for each transistor in the stress experiment, and we know W and $\phi_{it}/\phi_I \approx 3$. If we plot these variables on log-log scale, we get a straight line such as shown in Figure 6.36. The data on the lower right are for transistors that failed in a few seconds under a HCI stress condition.

The slope in Figure 6.36 is constant. The line will be offset for transistors of different quality. The next step is to estimate chip HCI lifetime. The method predicts chip HCI lifetime by estimating I_D and I_{SUB} at use conditions, and then derives τ from those use conditions in Figure 6.36. We then use the conversion equation from DC stress into chip power-on hours (POH)

$$\text{DC stress time} = \text{duty factor} \times \text{switch factor} \times \text{chip POH} \qquad (6.27)$$

where duty factor is (device DC equivalent stress per cycle)/(cycle time). Switch factor is a fraction that may consider that stress only occurs during low-to-high input transitions, and

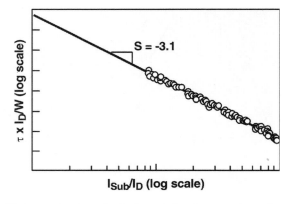

Figure 6.36. Lifetime projection of HCI degradation for several transistors (from [42]).

then you solve for chip POH. Remember that HCI damage occurs only in normal operation when the transistor is in saturation, and that occurs only during the logic transitions. The method is tedious, but does provide HCI estimated lifetimes. Power-on hours are typically 10 years, but will vary with product expectations. Typical HCI reliability goals are about 0.1–1.0 years of DC stress. Practical stresses of test structures are 10–100 hours.

6.4.3 Defect-Induced Oxide Breakdown

Foreign particulates or poor quality oxides typically lead to early breakdown of the oxide for lower applied voltages. These are called gate oxide shorts. The origins of particulate-driven gate shorts are different than those for HCI or wearout, but the end result can be similar to breakdown by wearout. The time to failure t_F is shorter for defect-related gate oxide shorts appearing as early as the production test or from field returns weeks to years later. Manufacturers try to eliminate defect-induced gate shorts by stressing the ICs at high voltages for short time periods (burn-in). Chapter 7 will analyze the electronic effects that occur when gate oxide shorts are present in an IC.

6.4.4 Process-Induced Oxide Damage

Plasma and ion implantation process techniques are manufacturing steps used for etching or doping some areas of the circuit during semiconductor device fabrication. During these steps, devices may be exposed to charges collected by aluminum or polysilicon conducting lines (called "antennas") connected to the gate terminals of MOSFETs. As a result, gate oxides can be damaged, affecting V_t, the subthreshold slope, and the transconductance of transistors. The stress currents to which these charges give rise may eventually cause oxide breakdown. For oxides that do not experience breakdown, oxide trapped charges and interface states can be created. These damaging processes can be avoided by providing alternative conducting paths to the charge collected during these fabrication steps and thus protecting the gate oxides. Often, reverse-biased diodes are tied to the "antennas" to provide a charge path to ground.

6.4.5 Negative Bias Temperature Instability (NBTI)

A recent oxide reliability issue has appeared that impacts short-channel pMOS transistors with their p-doped polysilicon gates. It is called negative bias temperature instability (NBTI), and it is a wearout mechanism with positive charge buildup at the channel interface of pMOS transistors. It causes threshold voltage absolute magnitude increase and reduction in I_{Dsat}. The damage has been referred to as caused by "cold holes" [38], and is identified as getting worse with scaling of oxide thickness. The "instability" refers to the time variation in V_{tp} and I_{Dsat} [21]. The affected pMOS transistor has higher V_{tp} than a transistor in normal inversion.

Several mechanisms have been proposed for NTBI but, presently, the accepted one is a hole-assisted electrochemical reaction [38]. Interface states are formed when a hole breaks a silicon bond, forming trivalent silicon in the interface region. Hydrogen is believed to be the spun-off element, and as H diffuses away, a positive charge is left on the trivalent Si atom. H_2O and holes are believed to be reactants. The exact nature of the reaction is still not well understood. The necessary components for NBTI are holes, hydrogen (either as H_2, H^+, or H_2O moisture), high temperature (> 100°C), and oxide voltage [38].

NBTI occurs in the pMOS transistor when $V_{in} = 0$ V, but NBTI shows some recovery phenomena when $V_{in} = V_{DD}$, especially at high temperatures such as at burn-in. Chen et al. reported that for inverter experiments, NBTI was less during dynamic stressing than that of DC stressing, and that NBTI damage was overestimated for the DC studies [26]. NBTI is a significant problem in pMOSFET performance for advanced technologies [38]. Circuit design and lifetime projections must consider the competing degradation from hot-carrier injection.

6.5 CONCLUSION

Metal and oxide failure mechanisms were described that showed the relation between material properties and potential IC failure. Electromigration and stress voiding are constant challenges for deep-submicron transistor ICs. Oxide wearout, hot-carrier injection, oxide ruptures due to defects, and NBTI are other materials concerns. Engineers in the CMOS IC industry need to understand these reliability failure mechanisms.

ACKNOWLEDGMENTS

We thank Bill Filter, Joe Clement, Bill Miller, Ted Dellin, Dave Monroe, Duane Bowman and Steve Yazzie of Sandia National Labs, and Shannon Hawkins of Boeing Aircraft Corp. for critical reviews and suggestions. Bill Filter's stress void lectures and Joe Clement's electromigration lectures in the University of New Mexico Microelectronics Reliability class were particularly beneficial. We also thank Steve Nelson and Rod Augur of Philips Semiconductors, Steve Mittle, Jim Lloyd, Tim Sullivan, and Bob Rosenberg of IBM, Cleston Messick of Fairchild Semiconductors, Tim Turner of Keithley Instruments, and Don Pierce and Eric Snyder of Sandia Technologies, who lectured in the University of New Mexico Microelectronics Reliability Series Lectures and influenced us.

BIBLIOGRAPHY

Materials Science

1. J. Ahn, H.-S. Kim, T.-J. Kim, H.-H. Shin, Y.-Ho Kim, D.-U. Lim, J. Kim, U. Chung, S.-C. Lee, and K.-P. Suh, "1GHz microprocessor integration with high performance transistor and low RC delay," in *Proceedings of International Electron Devices Meeting (IEDM)*, pp. 28.5.1–28.5.4, December 1999.
2. R. A. Flinn and P. K. Trojan, *Engineering Materials and Their Applications,* Houghton Mifflin, 1986.
3. Annual MRS Symposium Proceedings (1991–present), Vols. (225, 265, 309, 338), titled *Materials Reliability Issues in Microelectronics,* from the Materials Research Society.
4. R. E. Reed-Hill and R. Abbaschian, *Physical Metallurgy Principles,* PWS-Kent, 1997.

Electromigration

5. J. Black, "Mass transport of aluminum by momentum exchange with conducting electrons," in *International Reliability Physics Symposium,* pp. 148–159, April 1967.

6. I. Blech and E. Meieran, "Electromigration in thin aluminum films," *Journal of Applied Physics, 40,* 2, 485–491, 1968.
7. J. Clement, "Electromigration modeling for integrated circuit interconnect reliability analysis," *IEEE Transactions on Device and Materials Reliability, 1,* 1, 33–42, March 2001.
8. J. Gill, T. Sullivan, S. Yankee, H. Barth, and A. von Glasow, "Investigation of via-dominated multi-modal electromigration failure distributions in dual damascene Cu interconnects with a discussion of the statistical implications," in *International Reliability Physics Symposium (IRPS),* pp. 298–304, April 2002.
9. A. Goel and Y. Au-Yeng, "Electromigration in the VLSI interconnect metallizations," in *32nd IEEE Midwest Symposium on Circuits and Systems,* pp. 821–824, 1989.
10. K-D. Lee, X. Lu, E. Ogawa, H. Matsuhashi, V. Blaschke, R. Augur, and P. Ho, "Electromigration study of Cu/low k dual damascene interconnects," in *International Reliability Physics Symposium (IRPS),* pp. 322–326, April 2002.
11. B. Li, T. Sullivan, and T. Lee, "Line depletion electromigration characteristics of Cu interconnects," in *International Reliability Physics Symposium (IRPS),* pp. 140–145, April 2003.
12. J. Lloyd, "Topical review: electromigration in thin film conductors," *Semiconductor Science Technology, 12,* 1,177–1,185, 1997.
13. E. Ogawa et al., "Stress-induced voiding under vias connected to wide Cu metal leads," in *International Reliability Physics Symposium (IRPS),* pp. 312–321, April 2002.
14. D. Pierce and P. Brusius, "Electromigration: A review," *Microelectronics Reliability, 37,* 7, 1,053–1,072, 1997.
15. W. Song et al., "Pseudo-breakdown events induced by biased-thermal-stressing of intra-level Cu interconnects-reliability and performance impact," in *International Reliability Physics Symposium (IRPS),* pp. 305–311, April 2002.
16. D. Pierce, E. Snyder, S. Swanson, and L. Irwin, "Wafer-level pulsed DC electromigration response at very high frequencies," in *International Reliability Physics Symposium (IRPS),* pp. 198–206, April 1994.

Stress Voiding

17. *American Institute of Physics (AIP) Conference Proceedings #263,* "Stress-Induced Phenomena in Metallization," (1991 to present, every two years).
18. M. F. Ashby and D. R. H. Jones, *Engineering Materials I: An Introduction to Their Properties and Applications,* 2nd ed., Butterworth-Heinemann, 1996.
19. K. Doong et al., "Stress-induced voiding and its geometry dependence characterization," in *International Reliability Physics Symposium (IRPS),* pp. 156–160, April 2003.
20. S. Rauch and T. Sullivan, "Modeling stress-induced void growth in Al–4 wt% Cu lines," *Proceedings of SPIE, 1,* 805, 197–208, 1993.

Oxide Reliability

21. W. Abadeer and W. Ellis, "Behavior of NBTI under ac dynamic circuit conditions," in *International Reliability Physics Symposium (IRPS),* pp. 17–22, April 2003.
22. M. Alan and R. Smith, "A phenomenological theory of correlated multiple soft-breakdown events in ultra-thin gate dielectrics," in *International Reliability Physics Symposium (IRPS),* pp. 406–411, April 2003.
23. G. Barbottin and A. Vapaille, *Instabilities in Silicon Devices, Vol. 2: Silicon Passivation and Related Instabilities,* North-Holland, 1989.
24. D. Buchanan, "Scaling the gate dielectric: Materials, integration and reliability," *IBM Journal of Research and Development, 43,* 3, 245–264, 1999.

25. P. Chatterjee, W. Hunter, A. Amerasekera, S. Aur, C. Duvvury, P. Nicollian, L. Ting, and P. Yang, "Trends for deep submicron VLSI and their implications for reliability," in *International Reliability Physics Symposium (IRPS)*, April 1995.
26. G. Chen et al., "Dynamic NBTI of PMOS transistors and its impact on device lifetime," in *International Reliability Physics Symposium (IRPS)*, pp. 196–207, April 2003.
27. F. Crupi, B. Kaczer, R. Degraeve, A. De Keersgieter, and G. Groeseneken, "Location and hardness of the oxide breakdown in short channel n- and p-MOSFETs," in *International Reliability Physics Symposium (IRPS)*, pp. 55–59, April 2002.
28. R. Degraeve, B. Kaczer, A. De Keersgieter, and G. Groeseneken, "Relation between breakdown mode and breakdown location in short channel nMOSFETs," in *International Reliability Physics Symposium (IRPS)*, pp. 360–366, May 2001.
29. N. Dumin, K. Liu, and S.-H. Yang, "Gate oxide reliability of drain-side stresses compared to gate stresses," in *International Reliability Physics Symposium (IRPS)*, pp. 60–72, April 2002.
30. D. Fleetwood, "Fast and slow border traps in MOS devices," *IEEE Transactions on Nuclear Science, 43,* 3, 779–786, June 1996.
31. C. R. Helms and B. E. Deal, *The Physics and Chemistry of SiO_2 and the Si–SiO_2 Interface,* Plenum, 1988.
32. C. Hawkins and J. Soden, "Electrical characteristics and testing considerations for gate oxide shorts in CMOS ICs," in *International Test Conference (ITC)*, pp. 544–555, November 1985.
33. C. Hawkins and J. Soden, "Reliability and electrical properties of gate oxide shorts in CMOS ICs," in *International Test Conference (ITC)*, pp. 443–451, September 1986.
34. C. Hu et al., "Hot-electron induced MOSFET degradation–Model, monitor and improvement," *IEEE Transactions on Electron Devices, ED-32,* 375–385, February 1985.
35. V. Huard, F. Monsieur, and S. Bruyere, "Evidence for hydrogen-related defects during NBTI stress in p-MOSFETs," in *International Reliability Physics Symposium (IRPS)*, pp. 178–182, 2003.
36. Proceedings and Tutorials, *International Reliability Physics Symposium (IRPS),* Annual Spring Conference Sponsored by IEEE.
37. J. McKenna, E. Wu, and S-H Lo, "Tunneling current characteristics and oxide breakdown in p+ poly gate pFET capacitors," in *International Reliability Physics Symposium (IRPS)*, pp. 16–20, April 2000.
38. G. La Rosa, "NBTI challenges in pMOSFETs of advanced CMOS technologies," *Tutorial in International Reliability Physics Symposium (IRPS),* Section 241, April 2003.
39. Y. Leblebici, *Hot-Carrier Reliability of MOS VLSI Circuits,* Kluwer Academic, 1993.
40. B. Linder, J. Statis, D. Frank, S. Lombardo, and A. Vayshenker, "Growth and scaling of oxide conduction after breakdown," in *International Reliability Physics Symposium (IRPS),* pp. 402–405, April 1998.
41. J. McPherson and H. Mogul, "Disturbed bonding states in SiO_2 thin-films and their impact on time-dependent dielectric breakdown," in *International Reliability Physics Symposium (IRPS),* pp. 47–56, April 1998.
42. S. Mittl, IBM Corp. Hot-carrier injection lecture given at the University of New Mexico in March 2000.
43. F. Monsieur, E. Vincent, D. Roy, S. Bruyere, G. Pananakakis, and G. Ghibaudo, "Time to breakdown and voltage to breakdown modeling for ultra-thin oxides ($T_{ox} < 32$ Å)," *Proceedings of IEEE International Reliability Workshop (IRW),* pp. 20–25, October 2001.
44. F. Monsieur et al., "Evidence for defect-generation-driven wearout of breakdown conduction path in ultrathin oxides," in *International Reliability Physics Symposium (IRPS),* pp. 424–431, April 2003.
45. E. Nicollian and J. Brews, *MOS Physics and Technology,* Wiley, 1982.

46. P. Nicolean, W. Hunter, and J-C Hu, "Experimental evidence for voltage-driven breakdown models in ultrathin gate oxides," in *International Reliability Physics Symposium (IRPS)*, pp. 7–15, April 2000.
47. V. Reddy, "Introduction to semiconductor reliability," *Tutorial at International Reliability Physics Symposium (IRPS)*, Section 111, April 2002.
48. R. Rodriguez, J. Statis and B. Linder, "Modeling and experiemntal verification of the effect of gate oxide breakdown on CMOS geometry," in *International Reliability Physics Symposium (IRPS)*, pp. 11–16, April 2003.
49. J. Segura, C. De Benito, A. Rubio, and C. Hawkins, "A detailed analysis of GOS defects in MOS transistors: testing implications at circuit level," in *International Test Conference (ITC)*, pp. 544–550, October 1995.
50. J. Suehle and P. Chaparala, "Low electric field breakdown of thin SiO_2 films under static and dynamic stress," *IEEE Transactions on Electron Devices, 44*, 5, May 1995.
51. J. Suehle, "Ultrathin gate oxide reliability: Physical models, statistics, and characterization," *IEEE Transactions on Electron Devices, 49*, 6, 958–971, June 2002.
52. B. Weir et al., "Ultra-thin dielectrics: They break down, but do they fail?," in *International Electron Device Physics Symposium (IEDM)*, pp. 41–44, December 1997.

EXERCISES

6.1. Why would a Cu alloy in an Al base interconnect (a substitutional interconnect) increase the resistivity over pure Al?

6.2. When ICs increase their operating temperature from about 85°C to 120°C, how does this impact metal reliability?

6.3. When electromigration time to fail is plotted against the width of metal samples, a typical curve looks like the one sketched in Figure 6.37. From your knowledge of electromigration and grain boundary models, why does t_F increase for very narrow and very wide interconnects?

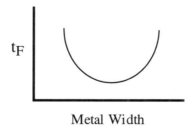

Figure 6.37. Time to failure versus interconnect width.

6.4. Aluminum metal lines using tungsten vias have increased sensitivity to electromigration at the Al/W interface due to the inert nature of W to electromigration. Do copper interconnect systems using Cu vias form a perfect interface, thus avoiding the W/Al problem?

196 CHAPTER 6 FAILURE MECHANISMS IN CMOS IC MATERIALS

6.5. The line that interconnects a 2NAND gate drive circuit to a load 2NOR gate input has certain electromigration design rules. If an *n*-channel transistor in the 2NOR gate connected to this line acquires a gate-to-source oxide rupture, discuss the electromigration risk.

6.6. Two walls in an airtight room move slowly inward and stop. The room pressure is now 10 atmospheres. There is no net motion of air molecules under this high stress. Why?

6.7. Why does electromigration failure rate decrease at higher clock frequencies?

6.8. An unpassivated interconnect line has a Blech constant of 3500 A/cm and a Blech length of 95 μm. If the same structural interconnect is passivated, the Blech constant goes to 6500 A/cm. How is the Blech length effected?

6.9. Given for an Al interconnect that $\alpha_{Al} = 23.5 \times 10^{-6}/°C$, $\alpha_{SiO2} = 0.5 \times 10^{-6}/°C$, $Y_m = 71.5$ GPa, and $\nu = 0.35$. Compare the stress at 30°C when the passivation deposition temperature is lowered from 430°C to 400°C to 300°C.

6.10. Cu with its higher melting temperature of 1085°C was expected to offer total protection from electromigration. What prevented Cu from being the perfect electromigration metal?

6.11. Copper interconnects require barrier metal lines to confine Cu. Compare the via-resistive effect of a Ta liner that occupies 10% of the damascene space to that of a pure Cu via. The dimensions of a cylindrical interconnect are given in Figure 6.38 ($\rho_{Cu} = 1.7$ μΩ · cm and $\rho_{Ta} = 200$ μΩ · cm).

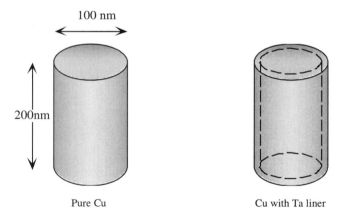

Figure 6.38.

6.12. An oxide dielectric has a shorter time to failure if the oxide has greater area and thinner dimensions. Use the percolation model to explain why.

6.13. Explain what effect large load capacitance has on hot-carrier injection.

6.14. Assume that an aluminum line of 0.5 μm thickness and 0.5 μm width is overlaid with TiN of 0.1 μm thickness. When the Al opens, it leaves a void of 0.5 μm length. The current density is $J = 20$ MA/cm². If the resistivities are $\rho_{Al} = 2.66$ μΩ · cm and $\rho_{TiN} = 2.66$ μΩ · cm, what is the power dissipation in the TiN section of the break? Give your answer in power per unit area (Watts/cm²), where area is the bottom face of the TiN.

6.15. Electromigration T_{50} data: two metals evaluated from two different processes. Boltzman's constant = 1.38×10^{-23} eV/°K.
   (a) Find the thermal activation energy for both metals and state which metal is better.
   (b) Give reasons why the quality might be different for (A) and (B) in Figure 6.39.

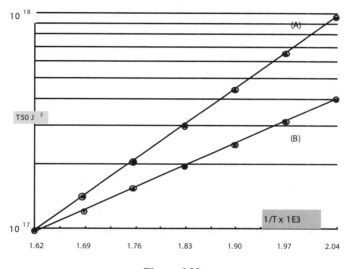

Figure 6.39.

6.16. An overall stress acceleration factor of 5×10^5 is desired for a particular defect in a qualification test. The thermal activation energy is 1 eV, normal temperature is 55°C, normal voltage is 5 V, stress voltage is 7 V, and the oxide voltage acceleration constant B = 400 Å/T_{ox}, where T_{ox} = 100 Å. Calculate the stress temperature T_s in °C to provide the acceleration factor of 5×10^5.

6.17. A company had been using an oxide defect thermal activation of $E_a = 0.3$ eV and a calculated failure rate of 500 FITs (1 FIT = 10^{-9} fails/hour). From new extrapolated data, they found that $E_a = 0.6$ eV. What would the failure rate calculation be if $E_a = 0.6$ eV? The experimental temperatures were $T_1 = 55$ °C and $T_2 = 125$ °C.

6.18. An aluminum stress experiment is conducted in which $T_1 = 27$ °C, $T_2 = 227$ °C, and $J_1 = J_2 = 10^7$ A/cm². The average times to fail are: $t_{F1} = 5000$ hour and $t_{F2} = 15$ hour.
   (a) Calculate the Al activation energy E_a.
   (b) If J_2 increases to 3×10^7 A/cm², calculate the expected failure time t_{F2}.

CHAPTER 7

BRIDGING DEFECTS

7.1 INTRODUCTION

The previous chapter showed failure mechanisms resulting in shorts between IC conducting paths. A bridge or shorting defect is an unintentional connection between two or more circuit nodes. Bridges in ICs induce abnormal electrical behaviors that depend on certain circuit parameters and the resulting circuit topology. The major bridge defect variables are

- Ohmic or nonlinear
- Intragate-connections across transistor internal nodes
- Connections across the I/O nodes of separate logic gates
- Power rail to ground rail
- Combinational or sequential resulting circuit topology
- Interconnect material types—metal, polysilicon, diffusion region
- Critical resistance—transistor drive strength and *W/L* ratios

Ohmic bridge defects can be metal slivers bridging two interconnections (Figure 7.1(a), large amounts of material shorting more than one interconnect (Figure 7.1(b)), or certain forms of transistor gate oxide shorts. Gate oxide short defects are ruptures of the transistor thin oxide that electrically connect the gate to the silicon structures underneath. Gate shorts are well controlled in some fabrication processes and a plague in others. Bridging defects in memory cells or flip-flops may or may not show responses different than those in combinational circuits. Power rail shorts between V_{DD} and GND are common and though they do not involve signal paths of the IC, they need to be recognized and con-

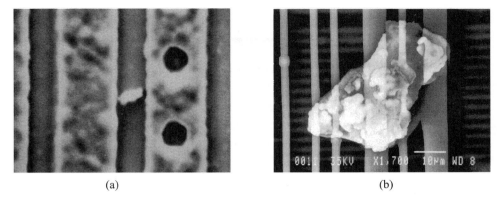

Figure 7.1. (a) Metal sliver. (b) Metal blob. (Reproduced by permission of Jerry Soden, Sandia National Labs.)

trolled. Low-power and battery-powered products cannot sustain predicted life if defect-induced power supply leakage occurs.

This chapter characterizes bridge defect behavior at the circuit level, showing how to calculate signal node voltages and power supply currents that are altered. We introduce the parameters related to bridging defects, and then analyze their effect at the circuit level for combinational and sequential circuits.

7.2 BRIDGES IN ICs: CRITICAL RESISTANCE AND MODELING

7.2.1 Critical Resistance

Critical resistance may be the most important concept in bridge defect electronics. It relates the defect resistance to the electrical properties of the surrounding circuitry and its induced logic behavior [18]. The critical resistance is the defect resistance value above which the circuit will not functionally fail. Contamination particles like the ones shown in Figure 7.1 cause shorts between two or more lines, resulting in bridging resistance between the shorted nodes.

Figure 7.2(a) shows one inverter (I_1) driving a second inverter (I_2) with an ohmic defect bridge that shorts the output of I_1 (V_2) to the V_{DD} rail. When the input to I_1 is at 0 V, the

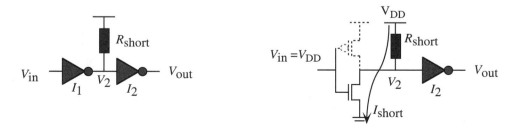

Figure 7.2. Power rail to signal node bridge defect.

nMOS transistor is off, whereas the pMOS is on and V_2 is pulled to a logic one state. Since there is no voltage drop across the bridging defect, no current passes through the short. In this situation, it is said that the defect is not activated, since it has no effect on the circuit in this logic state. When the input gate is at logic one, the pMOS transistor shuts off and the nMOS transistor turns on. The pMOS transistor is not conducting, and current is drawn through the resistor (I_{short} in Figure 7.2(b)) with a subsequent voltage drop from the V_{DD} rail. The defect is said to be activated for the logic one input signal.

When the defect is activated in Figure 7.2, there is a competition between the nMOS transistor of the first inverter that pulls node V_2 to ground and the short resistance that pulls this node to V_{DD}. The final node voltage will depend on the relationship between the value of the short resistance and the current drive of the inverter nMOS transistor. Figure 7.3(a) plots a set of voltage measurements at this node versus the input voltage V_{in} for different resistance values obtained for an inverter chain taken from [18]. Figure 7.3(b) shows the power supply current versus the input voltage of the circuit for each resistance value. Figure 7.3(a) shows that the higher the resistance value, the closer the transfer voltage curve to the fault-free one (note that a defect of infinite resistance would be equivalent to a fault-free circuit). When the voltage at the input is at V_{DD}, the nMOS transistor transconductance takes its maximum value, and the inverter output is reduced for each decreasing defect resistance value.

When the defect resistance is 1 kΩ, the lowest possible voltage at the I_1 inverter output is around 4.3 V. Clearly, the circuit will exhibit a logic error for this defect resistance value, and also for any other short resistance in the 0–5 kΩ range. For larger defect resistance values, the voltage in node V_2 is correct but logically weak. The successive gates can recover the full logic response so that the static logic behavior remains correct, but the logic gate whose input is near the defect has little protection against electrical noise spikes. Since the critical resistance is defined as the defect resistance value beyond which the logic operation is correct, this example shows a value of about 5 kΩ. All bridges except power rail shorts have a critical resistance whose value depends on the strength of the contending transistor(s) across that bridge.

The impact of a given bridging defect on the logic behavior of a circuit does not depend only on the resistance value and the transconductance of the transistors that compete

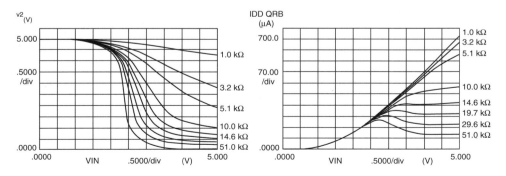

Figure 7.3. (a) Voltages and (b) consumption current for different resistance values of an internal node to V_{DD} short for an inverter chain [18].

with the shorting resistance. The transfer characteristic of the gates driven by the weak node will determine the impact of the defect, since they will interpret the logic value that corresponds to each intermediate voltage. The driver gate may have a symmetrical V_{out} versus V_{in} transfer curve, or it could be skewed to the left or right depending upon the individual p- and n-channel transistor drive strength (W/L ratio).

Functional failure is defined by a parameter called the logic threshold voltage V_{TL} and it is measured on the inverter voltage transfer curve at the point where $V_{in} = V_{out}$. V_{TL} is easily found as the intersection of a 45° line from the origin and the V_{out} versus V_{in} transfer curve. V_{TL} is typically about $V_{DD}/2$ which is a convenient parameter to define failure for the critical resistance calculations that follow. Some examples illustrate critical resistance calculations.

■ EXAMPLE 7.1

For the defective circuit in Figure 7.2(a), $K_n = 75$ μA/V², $W/L = 4$, $V_{DD} = 1.5$ V, $V_{tn} = 0.4$ V, and the logic threshold voltage is 0.75 V. Find the current and voltage V_2 when $R_{short} = 100$ Ω and (b) $R_{short} = 4$ kΩ.

(a) Figure 7.2(b) shows the equivalent circuit for analysis. The pMOS transistor is removed and the input activates the defect. The saturated state equation for the transistor is

$$I_{Dn} = K_n \frac{W}{L}(V_{DD} - V_{tn})^2 = 75 \frac{\mu A}{V^2} 4(1.5 - 0.4)^2 = 363 \ \mu A$$

KVL gives

$$V_{DS} = V_2 = V_{DD} - I_{Dn}R_{short} = 1.5 - (363 \ \mu A)(100) = 1.46 \ V$$

Check the bias state:

$$V_{GS} < V_{DS} + V_{tn} \quad 1.5 \ V < 1.46 + 0.4$$

The transistor is saturated, so the answer is correct and $V_2 = 1.46$ V. This is an error voltage since $V_{in} = 1.5$ V should produce V_2 of about 0 V, or at least below the logic threshold voltage $V_{TL} = 0.75$ V.

(b) R_{short} increases to 4 kΩ and the saturated state equation again gives $I_{Dn} = 363$ μA, as in (a) and

$$V_{DS} = V_2 = V_{DD} - I_{Dn}R_{short} = 1.5 - (363 \ \mu A)(4 \ k\Omega) = 48 \ mV$$

Check the bias state:

$$V_{GS} < V_{DS} + V_{tn} \quad 5 \ V < 1 + 0.4$$

The transistor is in the nonsaturated state so we must try again by combining the nonsaturated state equation:

$$I_{Dn} = 75 \frac{\mu A}{V^2} 4[2(V_{in} - V_{tn})V_2 - V_2^2] = 300 \frac{\mu A}{V^2}[2(1.5 - 0.4)V_2 - V_2^2]$$

with KVL:

$$I_{Dn} = \frac{V_{DD} - V_2}{R_{short}} = \frac{1.5 - V_2}{4\ k\Omega}$$

Solving both equations simultaneously:

$$V_2 = 0.492\ V,\ or\ 2.54\ V$$

the solution is

$$V_2 = 0.492\ V$$

V_2 is below the logic threshold value of 0.75 V; therefore, a correct logic value would be read. 1 kΩ is above the critical resistance, but this defective circuit has lost significant noise immunity at the V_2 node, making it vulnerable to logic upset by noise spikes. Also, $I_{DDQ} = (1.5 - 0.492)\ V/4\ k\Omega = 252\ \mu A$ is considerably above the normal quiescent power supply current value for the gate. ■

Critical resistance was evident in these two examples. The 100 Ω defect resistance was small, and V_2 failed its intended logic value. The larger 4 kΩ resistance did not cause V_2 to fail. The critical resistance R_{crit} lies between 100 Ω and 4 kΩ. The next example calculates its exact value.

■ EXAMPLE 7.2

Calculate the exact value of the critical resistance $R_{short} = R_{crit}$ for the circuit of Figure 7.2(a), using $K_n = 75\ \mu A/V^2$, $W/L = 4$, $V_{tn} = 0.4\ V$, $V_{DD} = 1.5\ V$, and the logic threshold voltage $V_{TL} = 0.75\ V$.

This problem is not as bad as it looks. The logic threshold of $V_{TL} = 0.75\ V$ defines the V_2 voltage point where failure just occurs. If $V_0 = 0.75\ V$, then the nMOS transistor is in the nonsaturated state since

$$V_{in} > V_0 + V_{tn} \qquad 1.5\ V > 0.75\ V + 0.4\ V$$

The nonsaturated equation gives

$$I_{Dn} = 75\frac{\mu A}{V^2}\ 4[2(1.5 - 0.4)0.75 - 0.75^2] = 326\ \mu A$$

$$R_{crit} = \frac{V_{DD} - V_0}{I_{Dn}} = \frac{(1.5 - 0.75)\ V}{326\ \mu A} = 2.3\ k\Omega$$

The circuit will functionally fail if the bridge defect resistance is below 2.3 kΩ. ■

■ EXAMPLE 7.3

Calculate the critical resistance for the bridge defect in Figure 7.4, given $K_p = 300\ \mu A/V^2$, $K_n = 500\ \mu A/V^2$, $W/L = 1$ $V_{tn} = 0.35\ V$, $V_{tp} = -0.35\ V$, $V_{DD} = 1.2\ V$, and the logic threshold for high/low distinction is $V_{TL} = 0.6\ V$.

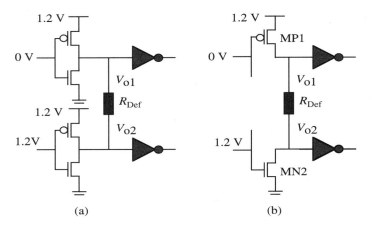

Figure 7.4.

Figure 7.4(b) shows the equivalent circuit when the off-transistors are removed. We must first choose which node attached to R_{Def} will fail first. If V_{o1} drops below 0.6 V, then it fails, or if V_{o2} rises above 0.6 V, then it fails. The clue is that MP1 has weaker drive strength than MN2 and, therefore, cannot win the contest with MN2. The drain at MP1 will fail first. At failure, we can assign $V_{DP1} = V_{TL} = 0.6$ V. We then know all node voltages except the drain of MN2. Since $V_{GP1} = 0$ V, $V_{DD} = 1.2$ V, and $V_{DP1} = 0.6$ V, you can verify that MP1 is in the nonsaturated state. The drain current through both transistors and the defect is

$$I_{DP1} = 300 \frac{\mu A}{V^2}[2(1.2 - 0.35)0.6 - 0.6^2] = 198 \ \mu A$$

We can find the voltage V_{o2} and R_{crit} is found by Ohm's law. We know that $V_{o2} < 0.6$ V, since we deduced that it had not failed. MN2 is then in the nonsaturated state since

$$V_{in} > V_{o2} + V_{tn} \quad 1.2 \text{ V} > 0.6 \text{ V} + 0.35 \text{ V}$$

The nonsaturated equation allows calculation of V_{o2}:

$$I_{DN1} = 500 \frac{\mu A}{V^2}[2(1.2 - 0.35)V_{o2} - V_{o2}^2] = 198 \ \mu A$$

where the valid solution is $V_{o2} = 279$ mV.
R_{crit} is

$$R_{crit} = \frac{V_{o1} - V_{o2}}{I_{DN1}} = \frac{(0.6 - 0.279) \text{ V}}{198 \ \mu A}$$

$$R_{crit} = 1.62 \text{ k}\Omega$$

Self-Exercise 7.1.

For the circuit and problem in Figure 7.4, all parameters are the same except
(a) Calculate R_{crit} if $K_p = 150$ μA/V² and $K_n = 250$ μA/V².
(b) What is the ratio of K_p/K_n when R_{crit} goes to zero; what does that mean?

These problems deepen insight into critical resistance properties. The single transistor bridged to a power rail has a critical resistance that is dependent upon the current drive strength of the transistor. When the defect resistance connects a pull-up to a pull-down transistor, the R_{crit} is a function of the current drive mismatch and the current drive strengths. The signal node with the weaker current drive will fail first. The weaker a transistor's current drive, the larger the resistance needed to protect that node from failure.

R_{crit} increases as the mismatch in pull-up and pull-down current strength gets larger, but R_{crit} goes to zero when the pull-up and pull-down transistor current strengths are equal. The product significance is that real IC bridging defects have a range of resistance or impedance values, and many ICs will still function for bridge defects with quite low resistances. It is not known a priori whether a particular bridge will fail or pass the IC when voltage-based tests are applied.

Different logic combinations affect R_{crit} since pull-up and pull-down strengths vary with the logic input signals. Figure 7.5 shows a 3NAND contending with a 2NAND. The critical resistance values were simulated from a standard cell library for various input combinations, and results are in Table 7.1. The pMOS transistor widths were double the nMOS transistor widths. The minimum R_{crit} of 150 Ω occurred when a single pMOS transistor contended with two nMOS transistors in series. The pull-down was only slightly stronger than the pull-up strength for this logic state. The maximum R_{crit} of 1750 Ω occurred for the worst-case contention of three n-channel series pull-down transistors (weak) against two parallel p-channel pull-ups (strong). This example shows the strong influence of input logic states on electrical response in the presence of a bridging defect. These bridging defects cause weak node voltages or logic failure, but the quiescent power supply current I_{DDQ} is always elevated.

Self-Exercise 7.2

Two inverters have a bridge defect connection at their output terminals. The nominal $K'_n = 100$ μA/V² and inverter current drives (K'_n, K'_p) are matched within 10% of the worst case. What is the range of R_{crit} if $V_{tn} = 0.5$ V, $V_{tp} = -0.5$ V, $V_{TL} = 0.75$ V, and $V_{DD} = 1.5$ V?

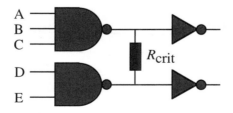

Figure 7.5. Bridging defect between two NAND gate outputs.

Table 7.1. Critical Resistance as a Function of Logic State

A	B	C	D	E	$R_{crit}(\Omega)$
0	0	0	1	1	950
0	0	1	1	1	725
0	1	1	1	1	150
1	1	1	0	1	1250
1	1	1	0	0	1750

■ EXAMPLE 7.4

Use the short channel nMOS transistor curves from Figure 3.34(b) in Figure 7.2(a) to graphically find the critical resistance when that transistor has a bridge defect tied between its drain and V_{DD} rail.

We repeat the transistor curves in Figure 7.6 for convenience. The supply voltage for this technology is 2.5 V, and we will assume that the value for the gate logic threshold voltage is $V_{TL} = V_{DD}/2 = 1.25$ V. To compute the critical resistance value, we must find the resistor load curve that intersects the drain current curve with $V_{GS} = V_{DD}$ at $V_{DS} = 1.25$ V.

We plot a vertical line at $V_{DS} = 1.25$ V and look at the intersection with the topmost drain current curve (corresponding to $V_{GS} = V_{DD}$). The intersection point ($V_{DS} = 1.25$ V, $I_{DS} = 12.8$ mA), gives the drain current that will pass through the nMOS device and the resistor short for the critical resistance value. We can compute this critical resistance from the slope of the line joining the $V_{DS} = V_{DD}$, $I_{DS} = 0$ point to the intersection point found ($I_{DS} = 12.8$ mA), or from the Ohm's law using the obtained current value, i.e:

$$R_{crit} = \frac{0.5 V_{DD}}{I_D(V_{GS} = V_{DD}, V_{DS} = V_{DD}/2)} = \frac{1.25 \text{ V}}{12.8 \text{ mA}} = 97.6 \, \Omega$$

■

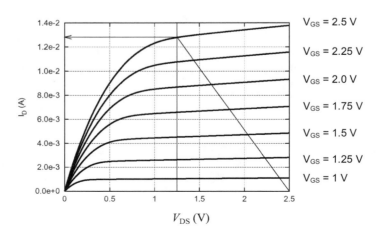

Figure 7.6. Transistor curves and resistor load to compute the critical resistance.

7.2.2 Fault Models for Bridging Defects on Logic Gate Nodes (BF)

We saw that when a bridge defect is activated, it causes intermediate voltages at the shorted nodes. An accurate description of the induced behavior may use analog-based circuit simulators, such as SPICE, to determine the operation region of the affected transistors, the intermediate voltages at the short nodes, and the induced power supply current increase. A test and diagnosis goal is to predict the faulty behavior of the circuit to find appropriate circuit stimuli (called test vectors) that can expose the bridging fault (BF).

Most circuit test pattern generators derive test vectors from a logic description (net list) of the IC. Unfortunately, logic circuit simulators cannot describe the behavior induced by a BF since the altered node voltages may not fit defined logic values. Analog simulators take too long to calculate the state of large circuits, and in many cases a detailed circuit analysis by analog simulators is not required. Several methods try to overcome this difficulty using logic fault models. A logic fault model is an approximation at the logic level of the behavior induced by a defect within an IC.

Generally, fault models try to overcome the gap between the logic description and the analog behavior induced by defects in the circuit. Since logic fault models are widely used in industry to calculate test vectors, we describe some bridging fault models intended for bridge defects at the logic gate I/O level. These models include:

- Stuck-at
- Pseudo stuck-at
- Logic-wired AND/OR
- Voting
- Biased voting

A brief view of each logic model will be given with more details found in [1, 8, 10, 21].

Stuck-at Fault Model (SAF). The stuck-at model, further discussed in Chapter 10, is the simplest and most used logic fault model in the industry. It came from the bipolar transistor IC era, and was accurate for that technology. Many authors showed that stuck-at faults are inadequate for CMOS technology, but its substitution by more accurate models has been slow, given the SAF's easy computational efficiency and its established practice.

Stuck-at fault models applied to bridging defects at the logic level assume that one signal node is permanently tied to the power rail or to ground, and is therefore referred to as stuck-at–1 or stuck-at–0. As shown in previous examples, the SAF model can only quantify detection of low-resistance (sub-R_{crit}) BFs between a signal node and the power/ground rails. Literally, the SAF models a zero Ohm bridge defect to one of the power rails. The inefficiency of the SAF model in detecting BFs motivated the search for more accurate logic fault models.

Pseudo Stuck-at Fault Model. This fault model, initially proposed in [7], exploits the leakage current mechanism induced by bridging faults to simplify the computation effort for ATPG and increase fault coverage. Schematically, this fault model targets stuck-at faults at primitive logic gate inputs, and considers a defect to be detected if its effect is propagated to the output of such a gate. The effect of the fault does not need to be propagated to the IC primary outputs, since it will be detected at the circuit power supply pin by measuring the quiescent current. The advantage of this model is at the computational level, since only small changes to conventional stuck-at tools are required to adopt this fault model.

Logic-Wired AND/OR Model. Logic-wired models for BFs were taken from bipolar technologies (ECL and TTL), in which defective shorted logic gate outputs are logically equivalent to a logic OR or logic AND gate. This behavior appears in technologies where one of the logic levels is always stronger than the other. Therefore, when two nodes are shorted, the stronger node overrides the weaker. Although this logic fault model is more versatile than the stuck-at model, it is not well suited for CMOS technologies, since CMOS ICs have no logic value always stronger than the other. We know that the voltage at a shorted node depends on the relative sizing of the gates involved, the bridge resistance, and the input logic states. An experimental study of wired-logic BFs showed poor correlation between this fault model and real defects [3]. A detailed analysis of logic wired fault models for BFs is in [1].

Voting Model. The voting model was a step forward in modeling BFs [2]. It observes that when shorted nodes are driven to opposite voltages, there is a competition between conducting pMOS and nMOS transistors. The model assumes that the set of drivers having the largest driving strength (i.e., those driving more current) will decide the final logic value.

The voting model does not account for the nonzero bridging resistance value or the logic threshold voltage of the fan-out gates. Nonzero BFs allow the output of the shorted gates to be at different voltages (because of the voltage drop at the bridge), and therefore be interpreted as different logic values for the subsequent gates. Maxwell showed that two logic gates whose inputs are shorted could interpret the bridge-defect-related analog voltage differently [8].

Biased Voting. This model overcomes the limitations for circuits with variable logic gate thresholds [8]. The biased voting model finds the conductances of the transistors involved in the bridge, taking into account the particular voltage of the bridged node. The voting model assumes a fixed initial voltage to calculate the conductance of the involved transistors. Therefore, the biased voting model accounts for the nonlinear transistor characteristics in calculating the driving strength of a device. Additionally, it takes into account the different thresholds of the logic gates connected at the shortened outputs.

Mixed Description. Since the electrical behavior induced by a bridge defect is analog, an accurate model of the induced behavior uses a mixed description. The whole circuit is described logically, except for the fault site that is described with an analog simulator. This method, described in [13], joins the accuracy of analog simulators for the defect site with the efficiency of simulating large circuits with logic-based tools.

7.3 GATE OXIDE SHORTS (GOS)

This section describes a form of intratransistor bridge defect caused by hard transistor oxide breakdown from particles or oxide imperfections. Chapter 6 described oxide wearout and rupture, and hot-carrier degradation of oxide material that inherently had no defects. The particle-induced oxide failures described here are found at production test, or during the infant mortality phase of the product life cycle.

Gate oxide shorts (GOS), or gate shorts, have troubled MOS technology since its beginning in the mid-1960s. The thin oxide of the MOSFET is the control region in which a transistor modulates charge population in the channel. A gate oxide short is a rupture in the thin silicon dioxide (SiO_2) between the polysilicon gate and any of the silicon structures beneath the oxide. The undamaged regions of the thin oxide generally still show normal charge inversion. In some cases, the transistor may still support functionality, although I_{Dsat} may be degraded.

Figure 7.7 shows two forms of gate oxide shorts. Figure 7.7(a) shows thermal filament growth on the gate edge, caused by high overvoltage on the gate. The electric field is higher on the edge of the gate, causing breakdown, and filament growth between gate and source. Figure 7.7(b) shows a gate short to the *p*-well caused by a small particle. The electrical response is different for the two types of gate shorts.

7.3.1 Gate Oxide Short Models

Figure 7.8 shows an inverter cross section in an *n*-well technology. Gate drain/source oxide shorts have simple electrical models, depending on the relative doping of the shortened terminals. There are six places where a gate short acquires a distinct parasitic connection when the gate material merges with the substrate material. Since gate oxide shorts connect the gate polysilicon with the drain, source, or bulk of the device, the electrical properties of the contact depend on the doping type of the terminals being shorted. If the gate and diffusion are of the same doping type, then the electrical model is a resistor between both terminals. If the shorted region has opposite doping, the electrical model is a *pn* junction diode. A detailed description of all the possibilities is given next.

n*MOS Transistor Gate–Drain/Source Oxide Shorts.* An Ohmic connection forms in an *n*MOS transistor when the rupture is from the *n*-doped polysilicon gate to the *n*-doped drain or source (Figure 7.8) [4, 19]. The result is similar to that of an external resistor connected between the gate to drain/source terminals. Figure 7.9 shows an I-V

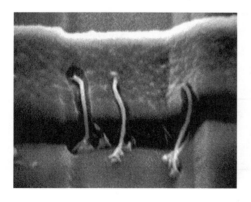
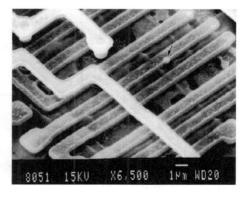

(a) (b)

Figure 7.7. (a) Thermal filaments across gate to source. (b) Particle-induced gate short from gate to *p*-well [4, 19].

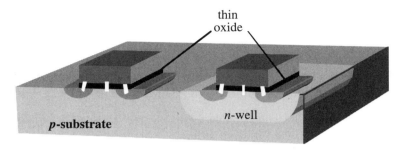

Figure 7.8. Possible GOS defects in a CMOS inverter.

curve (a) taken with probes placed across the gate and source terminals. This signature is an Ohmic gate short between the *n*-doped gate to *n*-doped source region. These forms of gate shorts can result from weak oxides or, preferentially, because electric fields are higher at the edges of the gate structure than in the middle. Typical resistances for *n*MOS transistor gate drain/source shorts formed in the lab ranged from about 1 kΩ to 20 kΩ, putting them above most critical resistances. The gate short resistance in Figure 7.9, curve (a), is about 20 kΩ.

nMOS Transistor Gate–Substrate Oxide Shorts. An *n*MOS transistor gate short between the *n*-doped polysilicon gate to substrate doping results in a diode with its cathode, or negative end, at the gate. Under normal positive biasing on the gate, this parasitic diode is reversed-biased and might never conduct, but its depletion region behaves as an additional "parasitic" drain diffusion region, taking electrons from the channel. The dam-

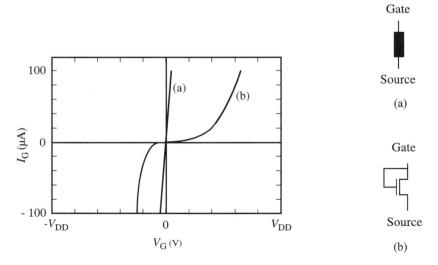

Figure 7.9. I-V curves for different parasitic elements forms of *n*MOS transistor gate oxide shorts. (a) Gate to source. (b) Gate to *p*-well short [sketched from 4].

age-induced depletion region is surrounded by an inversion layer of electrons in the non-damaged portion of the transistor. A parasitic nMOS transistor forms, with its gate and drain connected (Figure 7.9(b)).

Chapter 3 showed that a transistor with its gate connected to its drain is always in the saturated state when $V_{GS} > V_t$. Therefore the I_G versus V_G curve is always that of a saturated nMOS transistor. Curve (b) in Figure 7.9 shows this quadratic curve for a gate short between the n-doped polysilicon gate and the p-well. Quadrant I is the normal bias operation of the transistor. Quadrant III shows the parasitic diode response in forward bias as the gate voltage is reversed. The defect can be modeled with two parameters describing the defect position within the transistor and its effective resistance [19].

> *Self-Exercise 7.3*
>
> An inverter has an nMOS transistor with a gate short whose I-V properties are shown in Figure 7.9, curve (a). Given K_n = 200 μA/V^2, K_p = 150 μA/V^2, W/L = 1, V_{TL} = 1.25 V, V_{tn} = 0.5 V, V_{tp} = –0.5 V, and V_{DD} = 2.5 V, show by calculation whether the circuit will functionally fail. Hint: you must include the relevant driving transistor.

p*MOS Transistor Gate–Drain/Source Oxide Shorts.* Short-channel technology changed the pMOS transistor polysilicon gate doping from n-doped arsenic to p-doped boron to reduce short-channel effects. We will review responses of n-doped polysilicon gates taken from measurements, and relate them to expected behavior in p-doped polysilicon gates.

pMOS transistor gate shorts form diodes when the rupture is across an n-doped polysilicon gate to a p-drain or p-source. Such a diode gate short at the source clamps the pMOS transistor gate voltage at one diode drop below V_{DD} when a preceding pull-down transistor attempts to pull the gate node to a logic zero. A gate short to the drain is a little more complicated since the diode acts as a nonlinear feedback element from the drain (output node) to the gate (input node). A p-doped polysilicon gate short to the drain/source diffusion regions causes an Ohmic short.

p*MOS Transistor Gate–Substrate Oxide Shorts.* When the defect appears between the n-doped polysilicon gate and substrate of a pMOS transistor, the GOS is a low-resistive Ohmic contact to the device substrate since they have the same doping type [19]. However, when power is applied, the total pMOS transistor structure combines with the defect to form a parasitic pnp (bipolar) transistor. The parasitic GOS resistance allows base terminal current injection to the pnp transistor. The gate current acts as the base current of a bipolar transistor, and the resulting device characteristics mix MOSFET and bipolar transistor current characteristics. When the gate voltage drops toward logic zero, the pMOS transistor now provides an impedance path to the previous logic gate nMOS transistor. This action raises the pMOS transistor gate voltage, weakening its correct signal. Figure 7.10 draws the parasitic bipolar structure.

An important circuit effect occurs within a single-well CMOS structure. The parasitic bipolar device biased by the defect is connected to other parasitic bipolar devices inherent to the structure that may cause latchup (See Chapter 5). Figure 7.11(a) shows a GOS defect in a pMOS transistor causing light emission. Figure 7.11(b) shows the subsequent

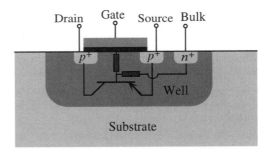

Figure 7.10. Electrical equivalent of a gate–substrate GOS for a *p*MOS transistor.

measured latchup behavior at the circuit level. When sufficient current passes through the latchup structure, a negative I-V slope region rapidly locks the structure into a high-current state.

General Electrical Model Equivalents for Hard Gate Oxide. All combinations of transistor type, defect location, and polysilicon doping type lead to a generalized gate oxide rupture model with 12 subcircuits. The electrical principles are: gate–drain/source shorts form diodes or resistors, depending on whether the relative doping type is the same (short) or opposite (diode). For gate–substrate shorts, defects connecting regions of different doping type create a parasitic MOSFET, whereas shorts between same doping type regions activate the parasitic bipolar transistor. Figure 7.12 summarizes the parasitic electrical elements. The *n*MOS transistor equivalent defect circuits are shown in the first row, and the *p*MOS transistors in the second row.

The *p*-doped polysilicon gate to *n*-well short creates a *pn* junction diode with its anode at the *p*MOS transistor gate. This situation is similar to the *n*MOS transistor gate to *p*-well short, where a parasitic MOSFET is formed. In the *p*MOS transistor gate short, the con-

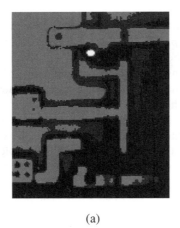
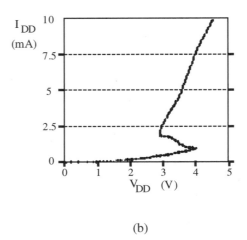

(a)   (b)

Figure 7.11. (a) A *p*MOS transistor GOS emitting light. (b) Latchup current response at the circuit level.

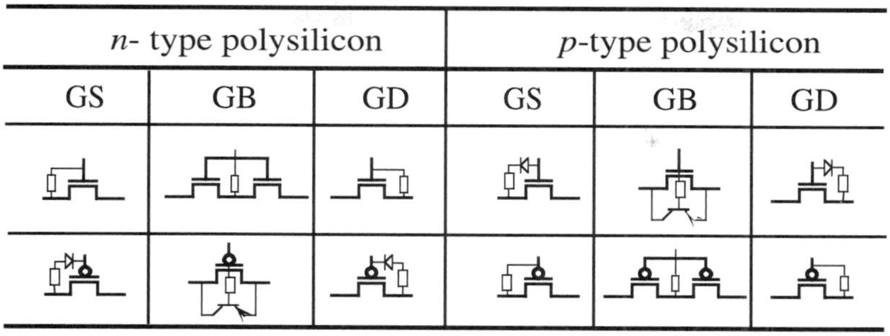

n- type polysilicon			p-type polysilicon		
GS	GB	GD	GS	GB	GD

Figure 7.12. Generalized electrical model for gate oxide short defects.

nection forms a parasitic pMOS transistor that is always in the saturated state when gate drive is greater than threshold voltage. In all gate short cases, the power supply quiescent current, I_D, is elevated, and logic voltages are either weakened or erroneous. ICs that have GOSs, but still pass functional testing are reliability risks [5].

Soft Gate Oxide Shorts in Ultrathin Oxides.
Recent studies show that wearout and breakdown in the ultrathin oxides below 30 Å shows different properties than for the thicker oxides in large-channel-length transistors [23, 24, 25, 26]. The older oxides were well characterized, emphasizing the elevation of I_{DDQ} and the reliability risk when gate shorts passed through the test process. Ultrathin oxides show a soft breakdown in addition to the hard breakdowns of thicker transistor oxides. Soft breakdown is an irreversible damage to the oxide, the most significant effect of which is the increase in noise of the gate signal. I_{DDQ} is not elevated for the soft gate–substrate ruptures of ultrathin oxides. The noise can show up to four orders of magnitude increase after soft breakdown, and this is the only certain evidence of the irreversible damage to ultrathin oxides.

Degraeve et al. found that uniformly stressed 24 Å gates oxides have a uniform area breakdown in transistors [23]. Breakdown over the channel had a high resistance from 10^5–10^9 Ω, whereas breakdowns over the drain and source had resistances of 10^2–10^4. Since the drain and source gate overlap area was much smaller than the area over the channel, most of the breakdowns occurred over the channel. By measuring breakdowns in 200, 180, and 150 nm transistors, they found that the percentage of gate to drain and source breakdowns increased as the transistors became smaller. The drain and source gate overlap area becomes a larger fraction of the transistor gate area. The gate-to-drain or source breakdowns were likely to cause hard failure, whereas the gate-to-channel breakdown showed no failure effects. A 41-stage ring oscillator with seven transistor gate ruptures continued to function with a 15% decrease in oscillator frequency [25]. Part of the frequency reduction was due to hot-carrier damage during a prestress. Significantly, the ring oscillator did not fail despite having several gate oxide breakdowns present.

Henson et al. also found that gate oxide breakdown was more severe at the drain and source than over the channel [24]. They fabricated three channel-length transistors at 4.65 μm, 1.3 μm, and 0.45 μm, where the percentages of gate-to-drain–source breakdowns were 30%, 25%, and 75%. The data show that failure may occur for gate-to-drain or source breakdowns, and the shrinking size of the transistors will begin to shift a greater percentage of breakdowns from the channel region to the critical drain–source region.

Detection of softly ruptured ultrathin oxide at production does not appear possible at this time, nor is the reliability status clear. The normal functioning of the transistor with an ultrathin oxide is not as effected as the killer ruptures of the thicker oxides. The recent ultrathin oxide experiments indicate that test escape and subsequent reliabilities may not be as risky as for breakdown in older technologies. These different properties of the transistor oxide demand more studies at the circuit level to assess the implications of test escapes.

Percolation paths are high-impedance paths in the oxides connecting the gate to the substrate regions. This negates the hard breakdown diode and low resistance gate-diffusion paths. It also replaces parasitic MOSFETs with parasitic bipolar transistors with the path percolation path forming a high-resistance base lead. There is need for more research of these types of shorts at the logic gate level.

Three Other Transistor Node Shorts. There are also drain-to-bulk, source-to-bulk, and gate-to-well (substrate) shorts. These can form *pn* junctions with soft or hard breakdowns between diffusion regions and the well (substrate), or from drain to source. Drain–source shorts can arise from channel punchthrough, or mask particle-related shortened L_{eff}, causing increased leakage.

7.4 BRIDGES IN COMBINATIONAL CIRCUITS

At the circuit level, we distinguish between intragate and intergate BFs. Intragate BFs are shorts between internal nodes of a gate, whereas intergate shorts are between two or more logic gates. This distinction is important for automatic test pattern generation (ATPG) tools since the circuit description (at the gate level or at the transistor level) determines which set of BFs is targeted. The behavior induced by BFs in combinational and sequential circuits is different. BFs in combinational circuits are of two types: nonfeedback bridging faults and feedback bridging faults.

7.4.1 Nonfeedback Bridging Faults

A nonfeedback bridging fault is a short that does not create a logic path between a gate output and a node that can determine the value of any of its inputs. This distinction is made because nonfeedback bridging faults in combinational circuits cannot induce sequential behavior. This is the simplest bridging fault, and corresponds to the BFs discussed to this point. When the defect is activated, the shorted nodes assume intermediate voltages whose values depend on the bridge resistance and the current drive strength of the transistors involved. The logic behavior depends on the critical resistance of the shorted nodes and the logic threshold of fan-out gates.

When the defect is activated, the resulting logic value may or may not be correct, but the power supply current will be elevated. This significantly impacts the coverage achieved with different voltage-based test techniques, as explained in Chapter 10.

7.4.2 Feedback Bridging Fault

A feedback bridging fault is a short in which some logic input combinations to the circuit create a logic path from a gate output to one of its inputs. Specifically, a BF between two

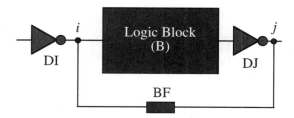

Figure 7.13. Feedback bridging fault.

nodes i and j is a feedback bridging fault if the logic value of j depends on the logic value of i. Node i is called *predecessor,* while node j is called *successor.* Figure 7.13 shows such a bridge between inverters DI and DJ. The output of DJ is connected to the output of inverter DI, and this may determine the logic state of DJ.

Feedback bridging faults are complex, since they can induce sequential behavior in a combinational circuit. The behavior induced by a feedback-bridging fault depends on the logic connecting the shorted gates (Block B in Figure 7.13). For a given input to the circuit, three situations may appear:

Case A. The input to the circuit is such that the logic value of node j is independent of the logic value of node i.

Case B. The logic value of node j is equal to the logic value of node i.

Case C. The logic value of node j is opposite to the logic value of node i.

The first case is equivalent to a nonfeedback bridging fault since the logic values of the shorted nodes are not related. The second case creates a noninverted feedback bridging fault, whereas the third is referred to as an inverted feedback bridging fault. The two latter cases are discussed after an example.

■ EXAMPLE 7.5

Determine the logic values of A, B, and C in Figure 7.14 that categorize the circuit as Case A, Case B, or Case C. Assume that the NAND gate is always stronger than the inverter.

We redraw the circuit to better identify the role of each gate in the circuit (Fig-

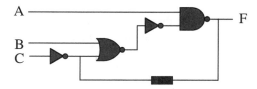

Figure 7.14. A combinational circuit with a BF.

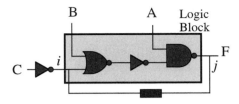

Figure 7.15.

ure 7.15). Node i is the output of the first inverter, whereas node j is the output of the NAND gate (predecessor and successor, respectively).

Case A. When A = 0, the NAND gate output is $j = 1$, independent of any other input to the circuit. When B = 1, the output of the NOR gate is 0 for all value of i, thus making the value of j again independent of i. These two conditions cover 6 of the 8 possible combinations.

Case B. In the fault free case, j can be put in terms of i: $j = \overline{A} + (\overline{B} \cdot i)$. There is no condition under which $j = i$.

Case C. From the expression derived in the previous case, when A = 1 and B = 1, then $j = \overline{i}$. This covers the two remaining cases. ∎

Noninverted Feedback Bridging Fault. These faults were traditionally thought to be redundant, since the shorted nodes are at the same logic level. This is only true when the strength of the predecessor gate (DI in Figure 7.13) is much stronger than the driving capability of the successor gate. If the successor gate (DJ) has a stronger driving capability than the predecessor gate (DI), then the value of node i is determined by DJ. If both drivers have the same logic output, then there is no conflict, and the defect is not activated. If DI makes a transition against the stronger DJ gate, then node i will not change and a competition between gates will appear. This causes an intermediate voltage and a quiescent supply current elevation. The situation can be described as a latched state.

A subtle effect appears when gates DI and DJ have similar driving capabilities, with DI slightly stronger than DJ. This case causes a significant delay because of a transient behavior appearing while the drivers are competing to set the final node value.

Inverted Feedback Bridging Faults. Inverted feedback BFs appear when the logic values of the shorted gates are opposite. Two behaviors may occur, depending on which gate is stronger. When the gate driving the predecessor (gate DI in Figure 7.13) is stronger than the successor (gate DJ), then the defect causes a logic error and current elevation, since driver DI overrides the output of DJ.

If the successor gate is stronger than the predecessor one, then the defect causes oscillation in the circuit. The period of oscillation is related to the delay of the logic connecting the predecessor output to the successor (logic block B in Figure 7.13).

Figure 7.16 shows experimental data from a test circuit. When the input was at logic zero, then the predecessor dominated and no voltage oscillations were observed. When the circuit input was set to a high logic value, the successor driver was dominant, causing oscillations.

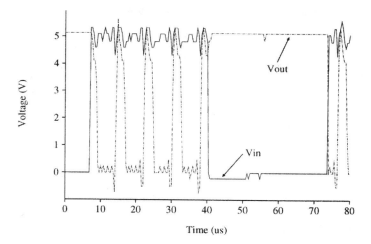

Figure 7.16. Oscillating behavior caused by an inverted feedback BF.

7.5 BRIDGES IN SEQUENTIAL CIRCUITS

The electrical behavior induced by BFs in sequential circuits may differ from combinational ones, since the conditions to detect a BF in combinational circuits may not hold for sequential ones. This depends on the circuit topology and defect site location. Lee introduced the concept of *control loops* to designate groups of transistors that control each other to create memory elements [6]. A control loop is said to be in a *floating state* when it is in a memory state, so that its value cannot be changed by the driving logic. If the memory element is driven by its preceding logic, then it is said to be in a *forced state*. In some situations, a control loop in a floating state may prevent quiescent current elevation since the internal nodes cannot be forced to the values required to elevate this magnitude. Since these defects swap memory values, they may be detected with logic-voltage-based testing, depending on the circuit implementation (shown in Chapter 10). The next section discusses the effect of BFs in flip-flops as the basic memory element in sequential circuits, and also BFs in semiconductor memories.

7.5.1 Bridges in Flip-Flops

The behavior induced by BFs in flip-flops depends on the circuit implementation. Flip-flops can be designed with a standard NAND gate configuration, but when using CMOS technology, tri-state inverter gates or pass gate transistor-based designs are often used. The induced BF behavior differs with the design.

Rodriquez et al. showed that flip-flops implemented with tri-state inverters and pass gate transistors may exhibit behavior in which the defect is not current-testable because of the floating state control loop [15]. These BFs change the logic state of the memory element and are therefore logically testable. Sachdev proposed a design modification using an additional inverter to make these cells current-testable [16].

Metra et. al. showed that flip-flops may have BFs that do not change the logic behavior of the gate, and do not elevate the quiescent current of the circuit [9]. The defect impacts the timing parameters of the cell, changing the set-up or hold time of the cell.

7.5.2 Semiconductor Memories

Bridging defects in SRAMs were analyzed by several researchers [12, 16, 22]. Bridging defects at SRAM cells may or may not cause quiescent current elevation. The main reason for not elevating the quiescent current is similar to the effect in flip-flops, and is due to floating control loops. BFs between cells are not generally current-testable because the cells that are not accessed do not maintain their logic value, and do not create a logic conflict required to elevate quiescent current. These defects may be detected using logic testing.

Sachdev showed that internal timing restrictions inherent to asynchronous operations in the memory may prevent current elevation for a desired period of time [17]. This significantly impacts current-based test techniques since the time window required to measure the quiescent current is not guaranteed. The proposed solution requires design modifications to improve current-testability coverage.

A detailed analysis of the behavior induced by gate oxide shorts in SRAMS was given in [20]. Although gate oxide shorts (GOS) are inherent to a single transistor, they may involve more than one single SRAM cell. GOS defects are responsible for many failures in RAM cells such as data retention (a memory cell looses its value a short time after it was written), coupling faults (more than one cell is accessed simultaneously), sense amplifier recovery (the memory sense amplifier does not respond at the normal frequency of operation), and stuck-at behavior.

Sachdev analyzed BFs in DRAMs [17]. DRAM cells made of a storage capacitor (typically 50 fF) need refreshing because of leakage currents. BFs in these cells cause logic errors, but the quiescent current is not elevated since there is no voltage stress or drive. Moreover, since the charge stored in a DRAM cell is very small, even a high-impedance defect can seriously affect the storage capability since there is no feedback structure to compensate for the defect-induced leakage.

7.6 BRIDGING FAULTS AND TECHNOLOGY SCALING

Technology scaling does not have a direct impact on the bridging defect's inherent characteristics. The physical mechanisms for these defects that appear during IC manufacturing or operation generally remain invariant. The same holds for the electrical properties of the short.

Since the effect of a short on IC behavior is related not only to the defect itself, but also to the surrounding circuitry characteristics, the impact of these defects is expected to change with technology scaling. The main parameters affected are the critical resistance and the quiescent current detectability of the defect. As technology scales down, the driving strength of transistors increases. Therefore, following the analysis made at the initial part of the chapter, the critical resistance values are expected to decrease, thus making logic testing even less effective in detecting BFs.

On the other hand, traditional quiescent current testing based on a single pass/fail threshold (this technique will be introduced in detail in Chapter 10) is becoming less effective in submicron technologies due to elevated current leakage values and variability.

From this perspective, BFs are expected to be a challenge in submicron technologies since more sophisticated techniques based on monitoring the delay induced by this defect or variations in current leakage will be required.

7.7 CONCLUSION

This section described the many variables affecting the response of a circuit with a bridge defect. The critical resistance concept is perhaps the most important, as it explains why flagrant visual bridging defects may not cause a functional test of the circuit to fail. Bridges between logic gate transistor nodes and bridges between signal nodes of two or more gates have slightly different responses. Gate oxide shorts are a major source of the intratransistor node bridges. They have linear and nonlinear properties depending on relative doping levels and location of the gate short. The bridge models developed to assist the test vector generation problem were described. Finally, the behavior of bridging defects in memory circuits was explained. Bridging defects cause I_{DDQ} elevation and either a correct but weakened node voltage or functional failure occurs. An exception to I_{DDQ} elevation is found in certain forms of memory bridges.

REFERENCES

1. M. Abramovici, M. Breuer, and A. Friedman, *Digital Systems Testing and Testable Design*, IEEE Press, 1993.
2. J. Acken and S. Millman, "Accurate modelling and simulation of bridging faults," in *Custom Integrated Circuits Conference (CICC)*, pp. 17.4.1–17.4.4, 1991.
3. R. Aitken, "Finding defects with fault models," in *IEEE International Test Conference (ITC)*, pp. 498–505, October 1995.
4. C. Hawkins and J. Soden, "Electrical characteristics and testing considerations for gate oxide shorts in CMOS ICs," in *IEEE International Test Conference (ITC)*, pp. 544–555, November 1985.
5. C. Hawkins and J. Soden, "Reliability and electrical properties of gate oxide shorts in CMOS ICs," in *IEEE International Test Conference (ITC)*, pp. 443–451, September 1986.
6. K. J. Lee and M. Breuer, "Design and test rules for CMOS circuits to facilitate I_{DDQ} testing of bridging faults," *IEEE Transactions on Computer-Aided Design*, 11, 5, 659–670, May 1992.
7. Y. K. Malaiya and S. Y. H. Su. "A new fault model and testing technique for CMOS devices," in *IEEE International Test Conference (ITC)*, pp. 25–34, October 1982.
8. P. Maxwell and R. Aitken, "Biased voting: A method for simulating CMOS bridging faults in the presence of variable gate logic thresholds," in *IEEE International Test Conference (ITC)*, pp. 63–72, October 1993.
9. C. Metra, M. Favalli, P. Olivo, and B. Ricco, "Testing of resistive bridging faults in CMOS flip-flop," in *European Test Conference*, pp. 392–396, 1993.
10. S. Millman, E. McCluskey, and J. Acken, "Detecting bridging faults with stuck-at test sets," in *IEEE International Test Conference (ITC)*, pp. 773–783, October 1990.
11. T. Miller, J. Soden, and C. Hawkins, "Diagnosis, analysis, and comparison of 80386EX I_{DDQ} and functional failures," *IEEE I_{DDQ} Workshop*, Washington DC, October 1995.
12. Naik, F. Agricola, and W. Maly, "Failure analysis of high density CMOS SRAMs using realistic defect modeling and I_{DDQ} testing," in *IEEE Design & Test of Computers*, pp. 13–23, June 1993.

13. J. Rearick and J. Pate, "Fast and accurate CMOS bridging fault simulation," in *IEEE International Test Conference (ITC),* pp. 54–60, October 1993.
14. R. Rodriguez-Montanes, E. Bruls, and J. Figueras, "Bridge defects resistance measurements," in *IEEE International Test Conference (ITC),* pp. 892–899, September 1992.
15. R. Rodriguez and J. Figueras, "Analysis of bridging defects in sequential CMOS circuits and their current testability," in *European Design and Test Conference,* pp. 356–360, 1994.
16. M. Sachdev, "I_{DDQ} and voltage testable CMOS flip-flop configurations," in *IEEE International Test Conference (ITC),* pp. 534–543, October 1995.
17. M. Sachdev, "Reducing CMOS RAM test complexity with I_{DDQ} and voltage testing," *Journal of Electronic Testing: Theory and Applications, 6,* 2, pp. 191–202, 1995.
18. J. Segura, V. Champac, R. Rodríguez, J. Figueras, and A. Rubio, "Quiescent current analysis and experimentation of defective CMOS circuits," *Journal of Electronic Testing: Theory and Applications, 3,* 337–348, 1992
19. J. Segura, C. De Benito, A. Rubio, and C. Hawkins, "A detailed analysis of GOS defects in MOS transistors: Testing implications at circuit level," in *IEEE International Test Conference (ITC),* pp. 544–550, October 1995.
20. J. Segura, C. De Benito, A. Rubio, and C. Hawkins, "A detailed analysis and electrical modeling of gate oxide shorts in MOS transistors," *Journal of Electronic Testing: Theory and Application, 8,* 229–239, 1996.
21. T. Storey and W. Maly, "CMOS bridging fault detection," in *IEEE International Test Conference (ITC),* pp. 842–850, October 1990.
22. H. Yokoyama, H. Tamamoto, and Y. Narita, "A current testing for CMOS static RAMs," in *IEEE International Workshop on Memory Technology, Design And Testing,* pp. 137–142, August 1993.
23. R. Degraeve, B. Kaczer, A. De Keersgieter, and G. Groeseneken, "Relation between breakdown mode and breakdown location in short channel NMOSFETs," in *International Reliability Physics Symposium (IRPS),* pp. 360–366, May 2001.
24. W. Henson, N. Yang, and J. Wortman, "Observation of oxide breakdown and its effects on the characteristics of ultra-thin *n*MOSFET's," *IEEE Electron Device Letters, 20,* 12, 605–607, December 1999.
25. B. Kaczer, R. Degraeve, G. Groseneken, M. Rasras, S. Kubieck, E. Vandamme, and G. Badenes, "Impact of MOSFET oxide breakdown on digital circuit operation and reliability," in *IEEE International Electron Device Meeting,* 553–556, December 2000.
26. J. Segura, A. Keshavarzi, J. Soden, and C. Hawkins, "Parametric failures in CMOS ICs—A defect-based analysis," in *IEEE International Test Conference (ITC),* October 2002.
27. J. Suehle, "Ultrathin gate oxide reliability: Physical models, statistics, and characterization," in *IEEE Transactions on Electron Devices, 49,* 6, 958–971, June 2002.
28. B. Weir et al., "Ultra-thin gate dielectrics: they break down, but do they fail?," in *International Electron Device Meeting (IEDM),* pp. 73–76, 1997.

EXERCISES

7.1. A 2 kΩ defect resistance connects the inverter output to ground (Figure 7.17). K'_n = 200 μA/V², K'_p = 100 μA/V², V_{tn} = 0.6 V, V_{tp} = –0.6 V, and the logic threshold voltage is V_{TL} = 1.5 V. Will the circuit logically fail?

7.2. (a) As V_{DD} is reduced, explain why the critical resistance increases and how this relates to more or less defect detection sensitivity by voltage-based testing.
 (b) As temperature is reduced, carrier mobility increases and normal circuits run

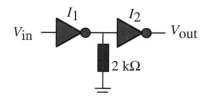

Figure 7.17.

faster. Assume that the defect does not change resistance, What affect would lowering temperature have on the values of critical resistance. Is the defect easier to detect at low temperature for any given value of V_{DD}?

7.3. Figure 7.1(a) shows a metal sliver between two metal interconnections.
   (a) Calculate the sliver resistance if the height, width, and length are 0.5 μm, 0.4 μm, and 1 μm. Assume that an Al sliver has a resistivity of 3.4 μΩ · cm.
   (b) A physical measurement of the resistance between the two metals interconnects gives a resistance of about 500 Ω. Why is there a discrepancy between the calculated and measured resistance?
   (c) What is the impact on detection of this defect by a conventional voltage test that tests for functionality?

7.4. Repeat the calculation for the critical resistance of the circuit shown in Figure 7.4(a) when the pull-up and pull-down drive strengths are closer. Let K_p = 300 μA/V² and K_n = 310 μA/V² when W/L = 1, V_{tn} = 0.35 V, V_{tp} = –0.35 V, V_{DD} = 1.2 V, and V_{TL} = 0.6 V. Comment on the significance of this.

7.5. (a) An integrated circuit is suspected of having a hard gate oxide short in an n-channel transistor. With measurements at the IC pin level, can you distinguish whether this short is to the channel or diffusion regions (drain, source)?
   (b) Repeat (a) for an ultrathin-oxide transistor.

7.6. (a) When a nMOSFET hard gate short occurs between a n-doped gate to a p-doped substrate or a pMOSFET p-doped gate to a n-doped substrate, then a parasitic MOSFET appears. When power is applied to an IC with such a short, explain why these parasitic transistors are in either the saturated or nonsaturated state bias.
   (b) What is the effect when a soft-gate oxide rupture occurs on an ultrathin oxide?

7.7. (a) Assume that all transistors in the gates of Figure 7.14 have $K_n(W_n/L_n)$ = $K_p(W_p/K_p)$, and V_{tn} = –V_{tp}, except for the NOR gate that has V_{TL} = $V_{DD}/3$ for the input driven by the inverter. If the defect resistance is negligible, determine which input combinations will lead to oscillations. *Hint: the circuit needs to be in Case C.*
   (b) If the transistor sizes of the inverters are doubled with respect to (a), and for all devices V_t is 20% of V_{DD}, compute the critical resistance that will lead the inverter override the NAND output.

7.8. Compute the critical resistance trend for the circuit in Exercise 7.4 when scaling the technology if V_{DD} is reduced by 16%, V_t's are reduced by 28.55%, and the transistor drive current (K_n and K_p) increases by 15%.

7.9. An IC shows no functional failure, but I_{DDQ} is abnormally elevated for every test vector measurement. What type of defect might this indicate? What are the reliability implications?

7.10. A functionally failing IC shows abnormal I_{DDQ} for certain test vectors. What might you conclude about this IC?

CHAPTER 8

OPEN DEFECTS

8.1 INTRODUCTION

Open circuit defects are unintended breaks or electrical discontinuities in IC interconnect lines occurring in metal, polysilicon, or diffusion regions.* Figure 8.1 shows two CMOS open circuit defects. Figure 8.1(a) is a via not well bonded to the metal liner in the via hole. Figure 8.1(b) shows locations in the metal where two vias are missing. This caused an open circuit to two inverter logic gates. Open defects have greater behavioral complexity than bridge defects.

The major open circuit defect variables are

- Size of the open defect. Is the crack wide or narrow?
- Defect location:
  —Open gate to a single transistor
  —Open drain or source
  —Open to a logic gate input affecting a CMOS complementary transistor pair
- Open on a metal line driving several gates
- Capacitive coupling of open node to surrounding circuit nodes

Deep-submicron CMOS technologies use metal line widths of 130 μm or less and via height-to-width ratios of more than 5:1. These dimensions, when coupled with IC via

---

*In the context of this chapter, an open is a complete disconnect. Resistive or weak opens are described in Chapter 9 as extrinsic parametric failures.

CMOS Electronics: How It Works, How It Fails. By Jaume Segura and Charles F. Hawkins
ISBN 0-471-47669-2 © 2004 Institute of Electrical and Electronics Engineers

Figure 8.1. (a) A resistive open with poor bonding between the via metal and via liner (reproduced by permission of Bruce Draper of Sandia National Laboratories), (b) Missing vias (arrows) [6].

counts from hundreds of millions to over a billion and total metal lengths of several kilometers, make via- and contact-related open defects more probable than before. Open defects are unavoidable, and their detection is sometimes nearly impossible. We will start by modeling the behavior of a floating node within an IC, and then analyze its impact on circuit behavior.

8.2 MODELING FLOATING NODES IN ICs

The main effect of an open IC signal line is that one circuit node is no longer driven by any gate, but may be left in a floating or high-impedance (high-Z) state. The node does not have a conducting path to V_{DD} or ground through a low impedance connection. The voltage on the floating node depends on the properties and topology of the surrounding circuitry. Two primary variables determine the final voltage value of a floating node: (1) the size of the crack and (2) the amount of charge present at the floating node.

The size of the crack determines if electrons can tunnel across the open, thereby controlling the amount of charge injected from the original driver (the gate that should drive the node in the fault-free circuit) toward the floating node. The charge at the floating node also depends on the capacitive coupling to the surrounding nodes, and the charge at the gate and drain terminals of the transistors to which the node may be connected. It is important to emphasize that the complete problem is often a complex combination of all these factors. We will describe each effect and then summarize with a model that includes all the effects.

8.2.1 Supply–Ground Capacitor Coupling in Open Circuits

Capacitor voltage dividers appear in some MOSFET open circuits. Figure 8.2(a) shows an open on a metal line over a field oxide (floating node V_2). Part of the floating metal line runs over the IC substrate (tied to ground), and part of it runs over the well area (connected to V_{DD}). The metal–oxide–semiconductor structure is a capacitor, so that the floating

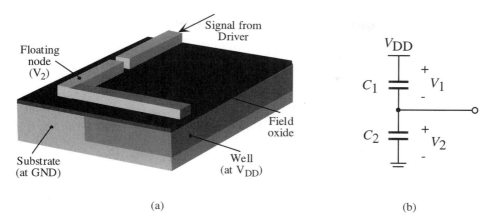

Figure 8.2. (a) Open crack in a metal line and (b) its electrical equivalent: a capacitor voltage divider.

node is capacitatively coupled to ground and supply. The values of the coupling capacitors to V_{DD} and to ground depend on the length of the metal track running over the well and the substrate.

The equivalent circuit of the floating metal has two capacitors in series (Figure 8.2(b)). The voltages V_1 and V_2 are functions of V_{DD} and the values of C_1 and C_2. From Chapter 1, we review

$$C_1 = \frac{Q}{V_1} \quad \text{and} \quad C_2 = \frac{Q}{V_2} \tag{8.1}$$

then

$$V_{DD} = V_1 + V_2 = \frac{Q}{C_1} + \frac{Q}{C_2} = Q\left(\frac{1}{C_1} + \frac{1}{C_2}\right) \tag{8.2}$$

Substituting

$$V_{DD} = V_2 C_2 \left(\frac{1}{C_1} + \frac{1}{C_2}\right) \tag{8.3}$$

or

$$V_2 = V_{DD}\left(\frac{C_1}{C_1 + C_2}\right) \tag{8.4}$$

In general, the voltage at the floating node can be expressed as

$$V_2 = \alpha V_{DD} \tag{8.5}$$

where α is a constant ranging between 0 and 1.

Self-Exercise 8.1

In Figure 8.2, (a) If $V_{DD} = 2.8$ V, $C_1 = 10$ fF, and $C_2 = 18$ fF, calculate V_2. (b) If $V_2 = 0.768$ V, $C_2 = 25$ fF, and $C_1 = 11$ fF, calculate V_{DD}.

The circuit in Figure 8.2(b) is used in open circuit defect analysis. For example, when the transistor gate is open, it is in a high-impedance state but the parasitic capacitive environment can often be reduced to a form of Figure 8.2(b). The defective circuit responds to transistor gate voltages set up by the capacitor voltage divider.

8.2.2 Effect of Surrounding Lines

Typically, an IC metal line is surrounded by other conducting lines. When two conducting lines at different metal levels cross over, there is a parasitic capacitance that couples both nodes. The value of this capacitance depends on the metal lines' intersecting area, which in this case is not large. A similar situation occurs for two conducting lines running parallel on the same metal level for a given distance, but in this case the coupling capacitance is bigger.

Figure 8.3(a) shows two metal lines running adjacent on the same metal level (V_m and V_F nodes), while a third metal line (V_3 node) on a higher level crosses both metals. One of the lower metal lines has an open defect, and it is floating. The coupling capacitance from the adjacent line in the same metal level will be higher than the coupling capacitance to the upper metal, so that the influence of the adjacent line is much stronger.

The equivalent electrical circuit for the floating node in Figure 8.3(a) is shown in Figure 8.3(b). Neglecting the coupling influence of the metal-2 line (which is much smaller than the influence of the metal-1 line), the voltage of the floating node (V_F) will depend on the logic value of the metal-1 line (V_m). An analysis similar to that developed previously leads to

$$V_F = V_{DD}\left(\frac{C_1 + C_m}{C_1 + C_2 + C_m}\right) \quad \text{if } V_m = H$$
$$V_F = V_{DD}\left(\frac{C_1}{C_1 + C_2 + C_m}\right) \quad \text{if } V_m = L \tag{8.6}$$

In general, when n nodes with voltage V_i are each coupled to the floating metal line, the final voltage is expressed as

$$V_F = \alpha V_{DD} + \sum_{i=1}^{n} \alpha_i V_i \tag{8.7}$$

where α and α_i are constants ranging between 0 and 1.

In a real IC, the signal nodes that influence the floating node experience logic transitions, thus changing the floating voltage with time. We can compute the floating voltage after a transition of the influencing metal, using charge conservation. If the influencing metal (V_m node) makes a high-to-low transition, then the initial voltage at the floating line before the transition (V_{Fin}) is computed from

$$(C_m + C_1)(V_{DD} - V_{Fin}) = C_2 V_{Fin} \tag{8.8}$$

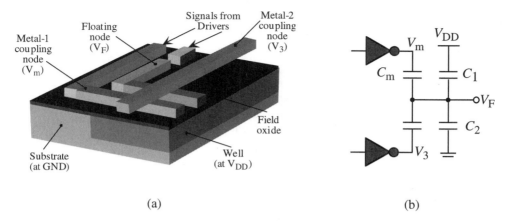

Figure 8.3. (a) Two metal-1 signal lines crossing one metal-2 line. The coupling capacitance of each metal-1 line to the other will be higher than the coupling capacitance to the metal-2 line. (b) Equivalent electrical model with the signal drivers shown as inverters.

which gives

$$V_{Fin} = \left(\frac{C_m + C_1}{C_m + C_1 + C_2}\right) V_{DD} \tag{8.9}$$

Once the transition is finished, the final voltage at the floating line (V_{Fend}) is computed from

$$C_1(V_{DD} - V_{Fend}) = (C_m + C_2) V_{Fend} \tag{8.10}$$

which gives

$$V_{Fend} = \left(\frac{C_1}{C_m + C_1 + C_2}\right) V_{DD} \tag{8.11}$$

Subtracting Equation (8.9) from Equation (8.11), the floating voltage change ΔV_F is

$$\Delta V_F = \left(\frac{C_m}{C_m + C_1 + C_2}\right) V_{DD} \tag{8.12}$$

Equation (8.12) shows that the influence of the surrounding metals depends proportionally on the coupling capacitance value.

■ **EXAMPLE 8.1**

Consider the dynamic NAND gate shown in Figure 8.4, where part of the metal-1 interconnect layout is shown (the interconnect between metal-1 and the transistor gate runs in polysilicon and is not illustrated). Determine the output value in the evaluation phase (clock = 1) if transistor N2 input makes a transition from 0

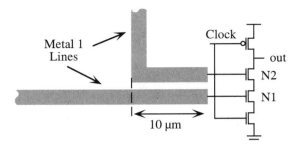

Figure 8.4. Dynamic NAND gate.

to V_{DD} when the metal-1 input to N1 has an open (a) 5 μm away from the polysilicon, (b) 280 μm away from the polysilicon. Discuss the results in both cases.

The metal lines shown are at the minimum distance. Consider that the input gate transistor capacitor and the polysilicon-substrate capacitor are negligible. The circuit is fabricated with a technology that has $V_{tn} = 0.35$ V, $V_{DD} = 1$ V, a metal-1 to bulk capacitance of 0.05 fF/μm, and a metal-1 to metal-1 capacitance of 0.35 fF/μm for minimum distance metal-1 lines.

(a) The metal-1 substrate capacitance of the metal portion connected to the N1 gate is $C_{ms} = 5$ μm × 0.05 fF/μm = 0.25 fF, whereas the capacitance to the metal line driving the N2 transistor input is $C_{mm} = 5$ μm × 0.35 fF/μm = 1.75 fF. Assuming that the gate of N1 is initially grounded, the final voltage of the N1 gate input node will be given by the voltage of a capacitor divider between $V_{Dmetal\text{-}1\ to\ metal\text{-}1}$ (N2 input) and ground.

$$V_{GN1} = V_{DD}\left(\frac{C_{mm}}{C_{mm} + C_{ms}}\right) = 1\ \text{V}\left(\frac{1.75\ \text{fF}}{1.75\ \text{fF} + 0.25\ \text{fF}}\right) = 0.875\ \text{V}$$

This voltage will turn on N1, since $V_{tn} = 0.35$ V. The circuit function will be correct if the gate driving N1 in the fault-free circuit also makes a low-to-high transition, but an erroneous logic behavior would occur if N1 was off in the fault-free circuit.

(b) The N1 input to ground capacitance is $C_{ms} = 280$ μm × 0.05 fF/μm = 14 fF, and $C_{mm} = 10$ μm × 0.35 fF/μm = 3.5 fF, since both metal lines run parallel for 10 μm. The N1 gate voltage after the N2 transition is

$$V_{GN1} = 1\ \text{V}\left(\frac{3.5\ \text{fF}}{3.5\ \text{fF} + 14\ \text{fF}}\right) = 0.2\ \text{V}$$

In this case, N1 will remain off and the gate output will stay high. This will lead to a logic error if N1 input is supposed to switch in the fault-free circuit.

In general, the lower the interconnect level and the farther away the defect from the gate input, the more unperturbed the value of the floating node. ■

8.2.3 Influence of the Charge from MOSFETs

In many cases, the floating node drives the gate of one or more transistors. Renovell and Cambon [8], Champac et al. [1], and Johnson [4] analyzed the device terminal's charge influence on the gate voltage. The charge stored at the gate terminal of a MOSFET has a strong dependence on the coupling from the drain and channel terminals. Hence, the drain voltage plays an important role on the final gate voltage value. Neglecting the effects from other nodes, the floating gate voltage can be expressed as [4]

$$V_{FG} = \frac{Q_{FG}}{C_G} + \alpha V_{DS} \quad (8.13)$$

where Q_{FG} is the charge at the floating gate, C_G its capacitance, and α ranges between zero and one. A detailed expression for Q_{FG} can be found in [1].

Experimental results [1, 4] report floating voltage values up to 3 V for a 5 V technology, demonstrating that the device can conduct significant drain current while its gate is floating.

8.2.4 Tunneling Effects

Narrow cracks can appear in metal lines or in a contact or via. Figure 8.5 shows such a defect whose electrical behavior is another unusual feature of open defects. The crack is narrow enough to support quantum mechanical electron tunneling when an applied voltage creates an electric field [3]. Barriers must be on the order of less than 100 Å to support a probability of significant tunneling. Another study proposed tunneling conduction mechanisms caused by incomplete etching or native (room temperature grown) oxidation after etching [5].

Tunneling is a quantum effect by which a small particle can cross through a finite potential barrier as a consequence of the wave–particle duality. There are different tunneling mechanisms, one of which is electric-field dominated. This field-dependent tunneling is called Fowler–Nordheim tunneling, and the current density J_e associated with this mechanism can be quantitatively described by

$$J_e = \alpha \cdot \mathcal{E}^2 \cdot e^{-\beta/\mathcal{E}} \quad (8.14)$$

where J_e is the current density in A/m², and α and β are constants depending on the physical properties of the structures through which tunneling takes place. The electric field in the metal void depends on the size of the open, the applied voltage, and the morphology.

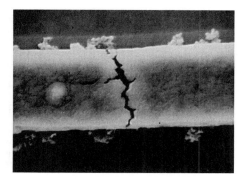

Figure 8.5. SEM photo of metal tunneling open [3].

As the crack narrows, direct tunneling increases. The direct tunneling current relation to the electric field is more complex than Equation (8.14). An important observation is that metal has a high thermal coefficient of expansion (TCE) so that tunneling efficiency increases as the metal is heated and expands to close the crack. An IC will show faster operation at higher temperatures when metal cracks are present. No defect-free circuit ever shows this behavior.

Electron tunneling across opens enables a logic gate to function at low frequencies but fail at high frequencies, depending on the size of the open. Therefore, detection of these defects strongly depends on the careful application of speed testing discussed later in Chapter 10.

■ EXAMPLE 8.2

A metal-1 line of width 0.3 μm has a crack of $d = 25$ Å located 10 μm away from the transistor gate that the line is driving. The node line has been at 0 V for a long time, thus completely discharging the transistor gate node. Determine the final voltage at the transistor gate if the node driver connects its output at V_{DD} for 1 μs. Assume that the polysilicon capacitance is negligible and that for this technology $V_{DD} = 3.5$ V, metal-1 height t = 5500 Å, $V_{tn} = 0.35$ V, $\alpha = 1$ μA/V², and $\beta = 3 \times 10^8$ V/cm.

The initial voltage difference across the crack is 3.5 V, while the crack size is 25 Å indicating that the dominant conduction mechanism will be Fowler–Nordheim tunneling [Equation (8.14)]. Figure 8.6 illustrates the problem.

Initially, C is discharged and when the node driver switches to a logic 1, the voltage across the crack is $V = V_{DD}$. This creates an initial electric field of value $\mathscr{E} = V/d = 3.5$ V/25 Å = 14 MV/cm, which creates a tunneling current that charges the capacitor and reduces the electric field. Hypothetically, the charging mechanism would continue until the capacitor is charged at V_{DD}. In practice, when the voltage across the metal crack goes below around 3 V, direct tunneling also takes place and Equation (8.14) is no longer valid.

The current through the crack is given by

$$I = J_e \cdot s = s \cdot \alpha \cdot \mathscr{E}^2 \cdot e^{-\beta/\mathscr{E}} = s \cdot \alpha \cdot \frac{V^2}{d^2} \cdot e^{-\beta \cdot d/V}$$

where s is the crack section area given by

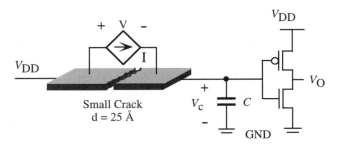

Figure 8.6. Tunneling through a metal crack.

8.2 MODELING FLOATING NODES IN ICs

$$s = w \cdot t = 0.3 \ \mu m \cdot 5500 \ \text{Å} = 165 \ \text{fm}^2$$

The current–voltage equation at the parasitic capacitor is

$$I = C\left(\frac{dV_C}{dt}\right)$$

since

$$V_{DD} = V + V_C \Rightarrow V_C = V_{DD} - V$$

then

$$I = -C\left(\frac{dV}{dt}\right)$$

Equating this current to the current through the crack, the equation describing the problem is

$$-C\left(\frac{dV}{dt}\right) = s \cdot \alpha \cdot \frac{V^2}{d^2} \cdot e^{-(\beta \cdot d)/V}$$

that can be rewritten as

$$\left(\frac{1}{V^2}\right) e^{(\beta \cdot d)/V} dV = -\left(\frac{\alpha \cdot s}{C \cdot d^2}\right) dt$$

To solve this equation we define

$$y = e^{(\beta \cdot d)/V}$$

and rewrite

$$dy = \left(\frac{\beta \cdot s \cdot \alpha}{C \cdot d}\right) dt$$

Integrating

$$\int_{y(t=0)}^{y(t=T)} dy = \frac{\beta \cdot s \cdot \alpha}{C \cdot d} \int_{t=0}^{t=T} dt \Rightarrow e^{(\beta \cdot d)/V_f} - e^{(\beta \cdot d)/V_{DD}} = \left(\frac{\beta \cdot \alpha \cdot s}{C \cdot d}\right) T$$

where V_f is the final voltage at $t = T = 1 \ \mu s$ in this case. The solution is:

$$V_f = \frac{\beta \cdot d}{\ln\left[e^{(\beta \cdot d)/V_{DD}} + \left(\frac{\beta \cdot \alpha \cdot s}{C \cdot d}\right) T \right]}$$

Note that the solution is consistent since for $T = 0$, $V_f = V_{DD}$, whereas if $T \to \infty$, then V_f goes to zero. Substituting the interconnect and technology values, and making $T = 1 \ \mu s$, $V_f = 3.3$ V. Since the voltage across the crack is beyond 3 V, the equation is valid during the charging process. The voltage at the gate capacitor is $V_C = 0.17$ V, which is not enough to turn on the nMOS device. ■

8.2.5 Other Effects

Additional charge components may be present when a floating node due to an open defect affects a transistor gate. These components come from the charge induced during the fabrication processes such as plasma etching or ion implantation. Although this charge can be removed or masked during subsequent annealing steps, the oxide traps created by these reactive process steps may manifest later due to hot-carrier injection during circuit operation [14]. The final trapped charge due to this mechanism depends on the amount of charge injected during the reactivation process step, the amount of charge removed during successive annealing, and, finally, the amount of charge trapped in the oxide during circuit operation. This makes the contribution of these effects on the total charge in the floating node difficult to compute if an open defect occurs, thus further complicating the exact modeling of opens in ICs.

8.3 OPEN DEFECT CLASSES

This section organizes the diverse behavior of CMOS open circuit defects into six classes or models. Each class is described with supporting evidence. The outward effect of an open defect in an IC can give immediate clues to the type of defect. Importantly, the clues also eliminate what the possible defect is not. The impact of an open on circuit behavior depends on the transistors driven by such a floating gate and the gate topology to which they belong.

Six general classes of opens are identified from failure analysis and test research:

1. Transistor on
2. Transistor pair on
3. Transistor pair on/off
4. Delay
5. Memory (transistor off)
6. Sequential

Although the names are awkward, they roughly describe open defect class behavior. The first five open categories appear in combinational logic circuits, and in certain instances in sequential circuit open defect behavior. We will look at each type and the supporting behavioral data.

8.3.1 Transistor-On Open Defect

An open gate to a single transistor is called a transistor-on defect class, and it has an unusual response [1, 2, 7]. Figure 8.7(a) shows a CMOS test inverter whose polysilicon gate to metal was removed at the mask level. Its static transfer curve is shown in Figure 8.7(b). Surprisingly, the curve shows functionality despite the loss in noise margin caused by the shift of the voltage curve to the right [2].

When $V_{in} - V_{DD} \ll V_{tp}$, the pMOS transistor in Figure 8.7(a) turns on strongly into the

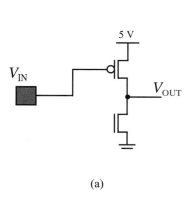
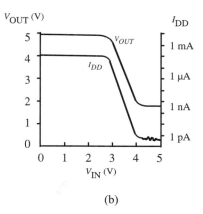

Figure 8.7. Transistor-on open defect class. (a) Open transistor in inverter. (b) Transfer characteristic [2].

nonsaturated state and 5 V is fed to the output drain. The nMOS transistor initially has no gate drive and is off. When the drain voltage rises and approaches V_{DD}, the DC capacitive coupling to the nMOS transistor from drain to gate and from gate to source creates a capacitive voltage divider to the gate. When the capacitively induced voltage at V_{GSn} is larger than V_{tn} the nMOS transistor turns on, drawing current from the pMOS transistor.

The static transfer curve in Figure 8.7(b) shows V_{out} = 4.96 V for V_{in} < 2.5 V in a technology with $V_{tn} \approx |V_{tp}| \approx 0.8$ V. Above V_{in} = 2.5 V, the pMOS transistor begins to turn off, but charge continues to pass from the output node to the nMOS transistor, as shown by the I_{DD} curve. V_{out} drops until the attenuated signal V_{GSn} is below threshold. The nMOS transistor turns off when $V_{out} \approx 1.8$ V, and the output node is in a high-impedance state with slow discharge leakage through the reverse-bias junctions of the MOSFETs. The flat portion of the curve at the lower-right-hand side reflects this slow discharge. Power supply current is strongly elevated at 96 μA for $0 < V_{in} < \approx 2.8$ V. It is elevated in only one logic state, which is typical for many defects that elevate I_{DDQ}. It is important to observe that contrary to electrical intuition, open circuit defects often elevate power supply current.

■ **EXAMPLE 8.3**

The circuit in Figure 8.7(a) approximates the DC conditions of a transistor gate open. If V_{DD} = 5 V, V_{tn} = 0.8 V, V_{tp} = −0.8 V, K_n = 200 μA/V², K_p = 125 μA/V², W/L = 1, C_{gd} = 20 fF, and C_{gs} = 9 fF, what is V_{out} when the nMOS transistor just turns off?

$$V_G = V_D \left(\frac{C_{gd}}{C_{gs} + C_{gd}} \right)$$

$$V_{out} = V_D = V_G \left(\frac{C_{gs} + C_{gd}}{C_{gd}} \right)$$

$$V_{out} = (29/20) \times 0.8 \text{ V} = 1.160 \text{ V}$$

■

234   CHAPTER 8 OPEN DEFECTS

A dynamic transfer curve would include the effects of the C_{gd} capacitance from the *p*MOS transistor. C_{gd} would allow charge injection quickly out of the output drain node, causing the voltage on that node to drop quicker. This would tend to shift the voltage transfer curve to the left of the static curve shown in Figure 8.7, providing some assistance to the logic change.

8.3.2   Transistor Pair-On and Transistor Pair-On/Off

A wide open-circuit interconnect defect on a logic gate input affects two transistors—the *p*MOS transistor and its complement *n*MOS transistor. Figure 8.8(a) shows such a defect situation. The affected node floats in a high-impedance state, searching for a voltage that satisfies its environment. The node electrically lies between V_{DD} and ground coupled by parasitic capacitances. The local topology will settle the node somewhere between V_{DD} and 0 V. It is typically not predictable where a floating node finds a steady state value, therefore we expect a range, and that range has three distinct regions. Figure 8.8(b) illustrates the variable region where the floating node can settle. If the gate floats to a value in region B, then both transistors turn on, V_O is at a weak stuck voltage, and quiescent power supply current I_{DDQ} is elevated.

If the node voltage is near the power rails at a value outside region B, then only one

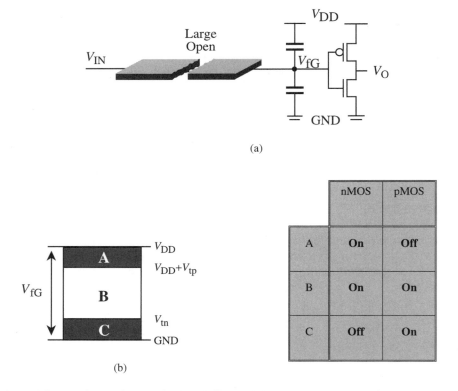

Figure 8.8. Transistor pair on and pair on/off open defects. (a) Large open defect and (b) its floating gate voltage range.

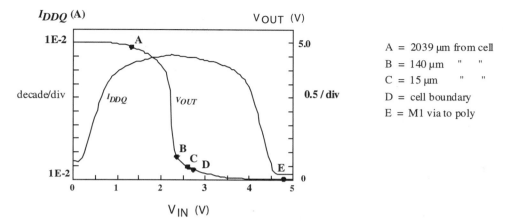

Figure 8.9. Floating node response to decreasing gates capacitance (A–E) [2].

transistor is on and the other off (regions A and C). V_0 is then a strong stuck voltage, and I_{DDQ} is not elevated in regions A and C. Thus, a wide open defect to a logic gate can cause (1) a weak stuck voltage and I_{DDQ} elevation (transistor pair on), or (2) a strong stuck voltage and no I_{DDQ} elevation (transistor pair on/off). The table in Figure 8.8 summarizes this.

Supporting data for these two open defect classes are shown in Figure 8.9 [2]. Points A–E are results from five CMOS inverter test structures in which the length of the metal interconnect floating node to the inverter was varied from 2039 μm (A) down to an open polysilicon gate to metal missing contact (E). The longer floating metal structures had a larger capacitance to ground, pulling the floating gate voltage lower (an effect similar to the one analyzed in Example 8.1).

The measured output voltages and I_{DDQ} were superimposed on a normal inverter transfer curve from the same die to estimate the floating gate voltage on the x-axis. Experimental points A–D are measured from transistor pair on open defects (region B in Figure 8.8) and Point E is a transistor pair on/off defect (region A in Figure 8.8). Circuits A–D showed weak stuck output voltage behavior and significant elevated I_{DDQ}. Circuit E showed strong stuck output voltage behavior and no I_{DDQ} elevation. The floating gate voltage for structure E was $V_{Gfl} \approx 4.8$ V. This turns off the pMOS transistor clamping V_{out} at 0 V. These distinctions are important when understanding symptoms in failure analysis or devising test detection methods.

■ EXAMPLE 8.4

Assume that the two open defects in Figure 8.10 lie (a) 50 μm to the right and (b) 50 μm to the left of the lower-left node. What electronic behavior differences would you expect?

The defect in Figure 8.10(a) affects one transistor. It is a transistor-on defect whose response is linked to capacitive coupling between its drain and source. It will probably pass a functional test, but have elevated I_{DDQ} in one logic state (AB = 10). Figure 8.10(b) shows a transistor pair on or pair on/off defect. If the transistors connected to node B float between the threshold voltages, then I_{DDQ} will

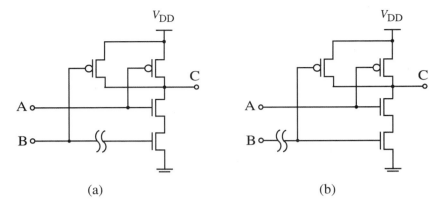

Figure 8.10. Open defect difference. (a) Open defect to transistor gate. (b) Open defect to logic gate.

elevate, and a weak stuck V_C appears for the AB = 10, 11 states. If the floating nodes clamp hard to one of the rails, then I_{DDQ} is not elevated and the output is a strong stuck voltage. ∎

8.3.3 The Open Delay Defect

This effect was discussed earlier in the chapter where tunneling was the main conduction mechanism through the open [3]. Figure 8.11 shows an equivalent circuit with a tunneling current density J_e crossing the crack. ICs with these defects can operate into the hundreds of MHz, depending on the small dimension of the crack.

This metal defect has an interesting temperature property that relates to test and failure

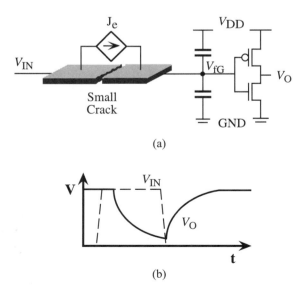

Figure 8.11. (a) Tunneling open and (b) response to electron tunneling.

analysis. When metal is cooled, it contracts and the crack widens. Fewer tunneling electrons are supported, and the maximum operating frequency drops. This is an unusual property for an IC since, normally, ICs increase their maximum operating frequency as the circuit is cooled. It is a clue to the presence of incomplete metal arising, perhaps, from stress voiding, electromigration, or a flaw in fabrication to a via or contact. These defects are often referred to as tunneling defects. They can be difficult to detect in test or locate in failure analysis. Resistive interconnect opens (specifically, high-resistive vias) are discussed further in Chapter 9, as they are also classed as parametric delay defects.

8.3.4 CMOS Memory Open Defect

Figure 8.12 shows a 2NAND with an open defect in a *p*MOS transistor source—no current can pass through this transistor. The truth table in Figure 8.12 shows correct logic results for the first two rows. These are expected. When AB = 00, both *n*-channel pull-down transistors are off, and the good *p*-channel transistor pulls the output to a correct logic one. When AB = 01, the pull-down transistor is again blocked, but the good *p*-channel transistor pulls node C to a correct logic one. However, row 3 gives a correct result that is subtle. The AB = 10 puts node A in a noncontrolling logic state, but the result seems to depend on conduction through the defective *p*MOS transistor. The reality is that the AB = 10 vector puts the output node in a floating or high-impedance state, and there is no outlet for rapid discharge of its high logic voltage charge from the previous logic state. The previous logic high state was stored in the load capacitance C_L and a correct value was read. The fourth vector AB = 11 pulls node C down to its expected 0 V.

This is a dramatic defect causing no error even though the logic gate was stepped through the whole truth table, and one transistor was incapable of conduction. The same result would have occurred if the defective transistor were removed from the circuit. The clue to potential error is the sequence of vectors. If the AB = 11 state was followed by the AB = 10 state, then an error of 0 V is measured on the second vector (last two rows in table).

We could examine an IC for this defect by testing each transistor with a sequence of

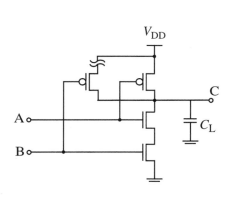

A	B	C
0	0	1
0	1	1
1	0	1
1	1	0
⋮	⋮	⋮
1	1	0
1	0	0

Figure 8.12. Stuck-open defect.

two-vectors [15]. This test may become numerically intractable for large circuits not designed for easy testing. Design for testability (DfT) circuits such as the scan circuits described in Chapter 10 partition the larger circuit into smaller combinational logic blocks. In this case, two-vector-sequenced patterns may be feasible.

Figure 8.13 shows a logic gate output voltage and I_{DDQ} timing response when a memory defect was placed in its high-impedance state at room temperature [13]. The circuit was a 2NOR driving a ROM row decoder inverter whose V_0 node was probed. At $t = 12$ s, the circuit was put in the high-impedance state of the memory defect. The output voltage shows a drift and change in logic state over about 6–8 s. This altered logic state under normal operating conditions might occur many millions of clock cycles after setting the high-impedance state.

I_{DDQ} responds with about a 200 ms step followed by instability to about 30 s and then a slow decline. $I_{DD}(t)$ reflects the biasing of one or more transistors as the floating node drifts through a range of values. In practice this defect is usually caught by an accidental sequence of voltage-based test patterns or by I_{DDQ} as shown here. The memory defect is difficult for test detection and failure analysis.

8.3.5 Sequential Circuit Opens

Figure 8.14 shows several wide open circuit defect locations in a sequential circuit. Open-1 (O_1) affects the *n*MOS and *p*MOS transistor gates of the two transmission gates. O_1 will cause those nodes to float to the same voltage. If the node floats to an intermediate voltage that turns on the *n*- and *p*-channel transistors, then the latch loses its signal isolation. Transistor contention or conflict occurs since both transmission gates are always on. The incoming signal conflicts with the stored feedback signal $\overline{Q}$. The latch will functionally fail, and I_{DDQ} will elevate during contention.

If the O_1 node floats to near V_{DD}, then the *n*-channel pass transistor is permanently on, and the *p*-channel transistor is permanently off. The feedback transmission gate still works with a single *n*-channel pass transistor, but conflict again occurs when the input transmission gate is supposed to be off. A similar situation occurs when the O_1 node floats near ground. The *p*-channel transistor of the feedback transmission gate is always on, and the *n*-channel transistor of the input transmission gate always off. Signal contention oc-

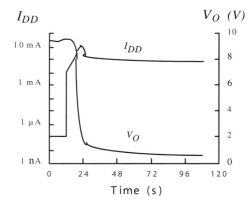

Figure 8.13. Timing response of 2NOR memory open defect [13].

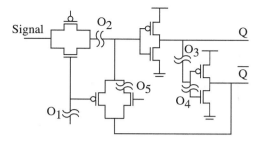

Figure 8.14. Open circuit defects in a sequential circuit.

curs when the feedback transmission gate is supposed to be off. I_{DDQ} elevation and signal error occur during the contention. The transistor contention is between one of the transistors of the second inverter ($\overline{Q}$) through the transmissions gates to the transistors driving the signal input.

Defect O_2 will float the input to the inverter when the feedback transmission gate is off and the response will be a transistor pair on or pair on/off. The returned signal in the latch will alter the floating node voltage and try to drive it to a strong voltage when the CLK signal turns on the latch T-gates. When the latch T-gate is off, then the floating node will again seek its steady-state level. The latch is dysfunctional with O_2 and an error is detected when applying a functional voltage test. I_{DD} may or may not be elevated. Defect O_3 also gives a response similar to that described for transistor pair on or pair on/off; the latch is again dysfunctional.

Defect O_4 exhibits the transistor-on open response of Figure 8.14 that should elevate I_{DDQ}, and it may or may not fail a functional test. O_5 opens the pMOS transistor so the T-gate will pass weak logic low voltages and unattenuated logic high voltages. Overall, the open circuit responses in the flip-flop are similar to the open behaviors described in Section 8.3. The failing symptom can be either that of a functional (voltage-based) test or an I_{DDQ} failure, or both simultaneously. All these defects affect the critical setup and hold times.

8.4 SUMMARY

The variety of open circuit responses is due to the sensitivity to (1) open defect location (drain, source, gate), (2) open to logic gate input, (3) sequential circuit, or (4) narrow cracks in flat metal or vias. Local topography determines capacitive coupling, and temperature changes will affect the dimensions of some opens. Although these responses are more complex than bridging defects, they are electronically understood so that later we can intelligently apply test methods that target each form of open circuit behavior.

REFERENCES

1. V. Champac, A. Rubio, and J. Figueras, "Electrical model of the floating gate defect in CMOS IC's: Implications on I_{DDQ} Testing," *IEEE Transactions on Computer-Aided Design of Integrated Circuits and Systems, 13.* 3, 359–369, March 1994.

2. C. Hawkins, J. Soden, A. Righter, and J. Ferguson, "Defect classes—An overdue paradigm for testing CMOS ICs," in *IEEE International Test Conference (ITC)*, pp. 413–424, October 1994.
3. C. Henderson, J. Soden, and C. Hawkins, "The behavior and testing implications of CMOS IC Open Circuits," in *IEEE International Test Conference (ITC)*, pp. 302–310, October 1991.
4. S. Johnson, "Residual charge on the faulty floating gate MOS transistor," in *IEEE International Test Conference (ITC)*, pp. 555–560, October 1994.
5. J. Li and E. McCluskey, "Testing for tunneling opens," in *IEEE International Test Conference (ITC)*, pp. 85–94, October 2000.
6. T. Miller, J. Soden, and C. Hawkins, "Diagnosis, analysis, and comparison of 80386EX I_{DDQ} and functional failures," *IEEE I_{DDQ} Workshop*, Washington D.C., October 1995.
7. W. Maly, P. Nag, and P. Nigh, "Testing oriented analysis of CMOS ICs with opens," in *International Conference on Computer Aided Design (ICCAD)*, pp. 344–347, 1988.
8. M. Renovell and G. Cambon, "Electrical analysis and modeling of floating-gate fault," *IEEE Transactions on Computer-Aided Design of Integrated Circuits and Systems, 11*, 1450–1458, Nov. 1992.
9. M. Renovell, A. Ivanov, Y. Bertrand, F. Azais, and S. Rafiq, "Optimal conditions for Boolean and current detection of floating gates," in *IEEE International Test Conference (ITC)*, pp. 477–486, October 1999.
10. W. Riordan, R. Miller, J. Sherman, and J. Hicks, "Microprocessor performance as function of die location for a 0.25 μm five layer metal CMOS logic process," in *International Reliability Physics Symposium (IRPS)*, pp. 1–11, April 1999.
11. R. Rodriquez-Montanes, J. Segura, V. Champac, J. Figueras, and A. Rubio, "Current vs. logic testing of gate oxide shorts, floating gates, and bridging failures in CMOS," in *IEEE International Test Conference (ITC)*, pp. 510–519, October 1991.
12. A. Singh, H. Rasheed, and W. Weber, "I_{DDQ} testing of CMOS opens: An experimental study," in *IEEE International Test Conference (ITC)*, pp. 479–489, October 1995.
13. J. Soden, R. Treece, M. Taylor, and C. Hawkins, "CMOS IC stuck-open fault electrical effects and design considerations," pp. 423–430, in *IEEE International Test Conference (ITC)*, pp. 302–310, August 1989.
14. R. Tu, J. King, H. Shin, and C. Hu, "Simulating process-induced gate oxide damage in circuits," *IEEE Transactions on Electron Devices, 44,* 9, 1393–1400, September 1997.
15. R. Wadsack, "Fault modeling and logic simulation of CMOS and MOS integrated circuits," *Bell Systems Technical Journal,* 1449–1488, May-June 1978.

EXERCISES

8.1. Observe the schematic and transfer curve in Figure 8.7. Would the shape of the curve change if the input were swept from $V_{in} = 5 - 0$ V instead of being swept from $V_{in} = 0 - 5$ V as shown in Figure 8.7(b)?

8.2. A 2NOR gate has an open defect in one of the pull-down source leads (Figure 8.15). Verify whether this defective circuit will
   (a) Correctly pass an ordered truth table beginning with 00
   (b) Correctly pass an ordered truth table beginning with 11

8.3. The stuck-open fault behavior was described as a behavior related to the parallel transistors in a logic gate. In theory, a stuck-open behavior can occur in a logic gate

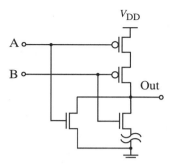

Figure 8.15.

series stack of transistors initiated by events occurring during power-up of the circuit. Figure 8.16 illustrates this with a single series stack (an inverter) in which an open defect appears in the drain of the *n*-channel transistor. Describe how a stuck-open behavior could occur in this series stack circuit.

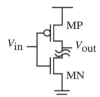

Figure 8.16.

8.4. An integrated circuit has the following failure symptoms: It fails a functional (Boolean) test and I_{DDQ} is elevated. If you suspect an open circuit defect is causing this, what type of open defect class would it be? Explain.

8.5. An integrated circuit has the following failure symptoms: The maximum clock frequency of the IC declines as the temperature is increased. There is no Boolean failure at slow clock frequency, nor is there I_{DDQ} elevation. If you suspect an open circuit defect is causing this, what type of open defect class would it be? Explain.

8.6. An integrated circuit has the following failure symptoms: It fails a functional (Boolean) test and I_{DDQ} is not elevated. If you suspect an open circuit defect is causing this, what type of open defect class would it be? Explain.

8.7. A metal-6 interconnect that is 3 μm wide has a 27 Å crack located 50 μm away from its end where it goes down to a gate node. Assuming that the floating portion of the metal is at 0 V, compute the final node voltage when the portion of the line connected to its driver is settled to V_{DD} during 1 μs. For this technology, the metal-6 height is $t = 1.04$ μm, its capacitance to the substrate is 0.008 fF/μm, and $V_{DD} = 3.5$ V. Determine the dominant conduction mechanism and the validity of the equation used.

8.8. The dynamic NAND shown in Figure 8.17 has been unpowered for a long time. Determine the voltage at the gate of N1 2 μs after the signal lines driving the gates of N1 and N2 simultaneously make a 0 to 1 transition. Assume the technology values given in Example 8.2 and a crack of 20 Å.

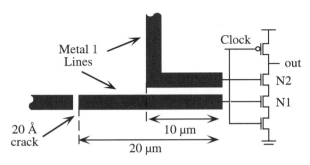

Figure 8.17.

8.9. Two different failure symptoms were found upon speed testing a microprocessor at 30°C and at 100°C. If you suspect an open defect, describe which it might be if:
(a) The parts run faster at 100°C than at 30°C.
(b) The parts run slower at 100°C than at 30°C.

8.10. An ASIC showed peculiar failure symptoms. The failure was isolated to a subcircuit that failed certain verification tests and not others. It did not seem to make sense. What type of an open defect might it be?

CHAPTER 9

PARAMETRIC FAILURES

9.1 INTRODUCTION

Parametric failures are the third and most difficult class of IC failures. They have elusive properties that may let a CMOS circuit function under specific conditions, but not under all environmental possibilities. A part may function at certain power supply voltages, but not over its whole specified V_{DD} range. There can be unusual temperature properties; the IC may pass a high-speed test at a hot temperature, but fail at a colder temperature. It may have windows of pass/fail. There are even regions on the Earth where the IC can show different pass/fail properties, such as high-latitude, mountainous regions having high cosmic radiation.

Parametric failures can be defined as failures due to the variation in one or a set of circuit parameters such that their specific distribution in a circuit makes it fall out of specification. Parametric variation has always been present in digital ICs, but their variance grew spectacularly in deep-submicron technologies. Parametric failures show two general forms: one failure form is caused by *defect-free* (intrinsic) parameter shifts; a second failure form affects functionality through environmentally sensitive *defects* on the die (extrinsic). The intrinsic failures can be due to physical factors that come from variations in the electrical and physical parameters of transistors and interconnects, and the failures can arise from an unfortunate statistical distribution of these parameters. Intrinsic failures can be also due to environmental factors such as variations in supply voltage and/or temperature, or due to crosstalk and switching noise influence. These broad failure classes are called *parametric* because the intrinsic or extrinsic failures are either due to variation of IC process parameters, or due to sensitivity to environmental parameters such as power supply, temperature, clock frequency, and/or radiation.

Individual transistor and interconnect parameters vary widely within a die, from die to die, wafer to wafer, and from lot to lot, making exact transistor drive (speed), and interconnect properties (resistance, capacitance, and inductance) prediction difficult [4, 7, 8, 9, 15, 41, 49]. These variations come from optical effects during lithography patterning processes, resulting in wafer images that can be very different from those drawn on the layout. In particular, optical proximity effects such as pitch-dependent *critical dimension* (CD) variation and line shortening can degrade transistor parameters or even lead to catastrophic defects (shorts or opens) when occurring in the polysilicon layer. The loss of pattern fidelity may happen during mask making, wafer imaging, and/or etch steps. Many researchers have studied CD variation in polysilicon and interconnect metal lines, and different techniques such as optical proximity correction (OPC) or phase shift mask (PSM) have been proposed. These methods require modification of the physical design layout on the photo mask to compensate for the proximity effects [24], and automated solutions are devised [3]. Although some of the variations can be corrected using OPC, significant discrepancies between measurements and models even after corrections still exist [35].

Another source of variability in the metal interconnect system comes from chemical-mechanical polishing (CMP) that has emerged as the primary technique for planarizing interlayer dielectrics (IDL) [54]. Its present applications include important process steps like shallow-trench isolation (STI) and multilevel inlaid copper interconnection [24]. However, it has been observed that the post-CMP topography shows an important variation that is strongly dependent on the layout pattern. This causes certain regions on a chip to have differences in dielectric layers thickness, depending on the underlying topography [54]. A method to reduce this layout-pattern-dependent dielectric thickness variation is to fill large metal-free areas with dummy metal. This extra dummy metal structure changes the parasitic capacitance of the interconnect system, and may complicate RLC extraction and impact timing and noise, as explained later in this chapter.

Both sources of variation impact circuit performance and may lead to parametric failures. Table 9.1 lists several forms of intrinsic and extrinsic parametric failure mechanisms and the physical effects that may significantly alter circuit speed.

Table 9.1. Parametric Failure Sources and Effects

	Failure Mechanism	Physical Effect
Intrinsic	Process Fabrication Parameter Variation	V_t shift ILD variation Interconnect Ω shift, metal width, spacing, thickness, granularity L_{eff} shift W_{eff} shift nMOS-to-pMOS length ratio shift Diffusion resistance
Extrinsic or intrinsic	*Metal:* Via-interconnect defect; Electromigration Stress Void	Resistive metal
	Oxide: Defect or wearout; Hot Carriers	Gate oxide short Hot carrier injection

Parametric failures are typically insensitive to many test methods such as I_{DDQ}, stuck-at fault, delay fault, or functional tests (these methods will be covered in Chapter 10), and can emerge as failures in the field [13, 18, 39, 49]. Tim Turner of Keithley Instrument's yield test structures group estimates that among the several hundred CMOS process steps, there are on the order of 10,000 variables that can affect the final performance of an IC [59]. He refers to interactive effects when two or more variables act in synergy to cause an IC to go out of specification. These interactive effects are one form of the parametric failures discussed here.

We emphasize that most of these parametric failures are speed-related, and are not conditions causing DC parametric failures such as I/O pin leakage or continuity. Since parametric defects are mainly speed-related failures, they usually require expensive test equipment and fixturing for detection and characterization. The IC maximum operating frequency (F_{max}) or propagation delay times are timing measurements that can indicate the presence of a parametric failure.

Process-induced parameter variation in IC manufacturing has mainly impacted die-to-die shifts in the past, and worst-case analysis was enough to predict its impact [37]. In deep-submicron and nanometer technologies, integrated circuits are large enough that within-die variations have become as important as interdie ones. These variations have two components; one is systematic and the other is random. A substantial portion of within-die variations are layout-dependent and, therefore, deterministic. These parameter variations may not be treated deterministically in many environments due to several reasons [37]:

- The models describing the dependence of a particular parameter on the design may not exist, be inaccurate, or be too expensive to evaluate.
- In the early design cycles, the circuit is described in high levels of abstraction, causing any interaction with physical implementations to be only estimates.
- A reusable design (like an IP core) cannot predetermine the physical design environment in which it will operate.

In these cases, the parameter can be effectively treated as random. This characteristic makes intrinsic parameter variations nonsystematic in circuits fabricated in the same lot, or even in the same wafer [32, 49]. This leads to inaccuracies that impact circuit quality, and can provoke erroneous behaviors that occur at very specific circuit states or environmental conditions. Present design technology is unable to characterize the whole complexity of parameter combinations, so that present strategies usually characterize only the corner parameters. Therefore, unfortunate parameter combinations can be very difficult to detect and screen. For example, an IC that is normally on the fast edge of the distribution could have a delay defect that puts it at the slow edge, but in an acceptable range, so the part passes. Defective parts that pass function tests are an increased reliability risk. Riordan et al. measured over one million ICs and reported F_{max} as a function of wafer location and yield [44]. This is an example of the uncontrollable parameter variations that affect performance.

Another source of uncertainty arises from the rapid picoseconds of signal rise and fall times of intrinsic nanotechnology ICs. This can cause specific noise mechanisms such as crosstalk or switching noise (these mechanisms are discussed later) that may lead to timing errors with associated Boolean upsets that may not be systematic for circuits manufactured from the same design. Although crosstalk and switching noise are traditionally categorized as design-related problems, technology-scaling complexity has diffused the boundaries between design and functionality, so that designers must take statistical variation into account in their designs [33].

Production ICs are characterized by measuring speed performance at different environmental conditions. Since there are many sets of conditions with which to do this, one efficient practice is to characterize performance at two sets of power supply voltage and temperature maximum and minimum points. These two sets are done at high (H) and low (L) V_{DD} and T, and are designated $V_L T_H$, and $V_H T_L$ [9]. The speed performance of the IC will vary from slowest at $V_L T_H$ to fastest at $V_H T_L$. Process variation may affect corner values through parameters involving transistors and interconnects, so that some circuits from the same fabrication lot fail, while others do not.

If parametric failures sound difficult, you are correct. When the failure analysis lab characterizes these failures, they often find that we lack the technology to economically detect them at the test and failure analysis stages. Several failure analysis experiences with parametric delay defects showed that one to three months of effort may be necessary to locate one defect in an IC. This is intolerable, and research efforts are currently directed at more efficient location techniques [10, 18].

There is presently no unified view of what must be considered within the umbrella of parametric failures. The material collected in this chapter does not pretend to define any new effects under a given defect class, but intends to organize and structure a number of "deviations" that have not previously been systematically categorized.

We will next detail three distinct origins of parametric failures. The first describes these origins as they relate to fabrication process variations that affect IC performance. Data are shown to illustrate how these parametric failures impact the speed capability of individual transistors and ICs. The next section describes parametric failures due to inherent noise generation within the chip. The final section describes how certain defects cause parametric failures.

9.2 INTRINSIC PARAMETRIC FAILURES

Two factors cause intrinsic parameter variation: environmental and physical [15]. Environmental factors include variation of the power supply levels within the die, or on the board, or switching activity and temperature variation across the circuit. Physical variation comes from the inherent weaknesses in circuit manufacturing that allow transistor and interconnect structural variations. These deviations from targeted values are limitations or imperfections in process and mask steps.

9.2.1 Transistor Parameter Variation

Individual transistor parameters vary widely within a die, die to die, and lot to lot, making transistor drive (speed) prediction difficult. The main parameters that determine the transistor drive properties are [7, 15]:

- Channel length variation
- Channel width variation
- nMOS-to-pMOS length ratio variation

- Effective gate oxide thickness variation
- Doping variation—threshold voltage and diffusion resistance

These parameters directly modulate two of the most important speed parameters of the device: the drain saturation current (I_{Dsat}) and the threshold voltage (V_t). V_t and I_{Dsat} are interrelated, and have a significant impact on the delay of the logic gates in the IC. Subthreshold leakage current (I_{off}) is another electrical parameter that is affected, and is now significantly elevating static power dissipation levels in modern submicron technologies.

Channel Length Variation. Channel length has a first-order influence on transistor delay, and thus on circuit performance. The MOSFET model equations given in Chapter 3 show that effective channel length (L_{eff}) lies in the denominator of the drain current expression. As L_{eff} gets smaller, drain current rises, and load capacitances can be charged and discharged faster. Channel length variations also directly impact the threshold voltage, reducing V_t as L_{eff} gets smaller. Channel length variations are due to a combination of photolithography, gate etching, ion implant, spacer formation, and thermal processing effects. Present limitations in optical photolithography make the variance worse as technologies scale to 90 nm and below. Bernstein et al. presented data showing that variation in L_{eff} has the greatest affect on IC performance compared to other process parameters [7].

Channel Width Variation. Channel width variation has a second-order impact on device switching speed, but is important in minimum-width devices due to narrow-channel effects. The main narrow-channel effect is that the transistor effective threshold voltage varies across the transistor width, changing from the nominal value at the center region to an altered value toward the device edge. The shift in threshold voltage may be positive or negative depending on the technique used to isolate the device. The main techniques are "local oxidation of silicon" (LOCOS) and "shallow trench isolation" (STI). LOCOS patterns silicon dioxide areas that isolate adjacent devices, and was the dominant technique until shallow-trench isolation was adopted. In general, LOCOS shifts the threshold voltage positive, whereas STI tends to decrease the effective V_t at the edges.

The narrow-width effect is typically seen for channel widths ≤ 0.4 μm. This means that the effect will influence the 180 nm technology when transistor W/L ratios are ≤ 2. The effect on V_t occurs more often for 130 nm, 90 nm, and below since more transistors have widths below 0.4 μm. It is difficult to compensate for this narrow-width modulation on V_t since it implies that different doping implants be done for transistors that are scattered across the die.

nMOS to pMOS Length Ratio Variation. This variation is important in ICs that use ratioed logic styles. The width-to-length aspect ratio between *n*MOS and *p*MOS transistors determines the noise margins and switching point of the logic gates. The channel length variation attributed to the two types of devices may come from implant dose, energy, and diffusion tolerances of the dopants associated with the different MOSFET types [7]. Bernstein et al. offered guidelines that *n*MOS-to-*p*MOS transistor tracking tolerance is about 10% of measured length and that *n*MOS-to-*p*MOS I_{Dsat} ratios can vary by ± 10% [7]. This statistically modulates the individual gate-voltage transfer curves, affecting pull-up and pull-down speed and noise margins.

Effective Gate Oxide Thickness Variation. Gate oxide thickness has a first-order impact on device performance, directly affecting transconductance, threshold voltage, and device drive current. Gate oxide physical thickness variance is related to the tolerance of the thermal growing process. Analysis of the oxide thickness must distinguish between physical and electrical equivalent insulator thickness [7]. The electrically equivalent insulator thickness of the oxide takes the polysilicon gate depletion into account, since its depth is no longer negligible in front of the oxide physical thickness. Gate oxide variation occurs across wafer lots over an extended period of time, whereas little variation is typically observed across a given wafer. In modern processes, this variation is kept below 0.5% of the effective oxide thickness. However, for the ultrathin oxides, ≤ 25 Å, the SiO_2 molecules show surface monolayers about 3.5 Å thick. The absence or presence of a portion of a monolayer can cause T_{ox} to discretely vary by 15–20% in local oxide regions [11], as shown in Figure 9.1. This figure shows the crystalline silicon, the amorphous SiO_2, and the polycrystalline gate material.

Random Doping Fluctuations. Doping variations may differ even for devices in the same die and are due to variations in the implant dose, angle, or energy. These variations change the junction depth and doping profiles that impact the effective channel length and threshold voltage. The angled halo doping implant technologies used in modern submicron technologies to reduce short-channel effects further increase parameter variations since the angle of implant is critical.

Another performance noise source that impacts the threshold voltage is related to the distribution of dopant atoms. A variation in doping density beneath the gate makes local threshold voltage uneven under the gate. Also, short-channel effects are controlled if substrate doping density is larger near the drain. This suppresses *pn* junction depletion width. When variations occur here, the drain-induced barrier lowering (DIBL) reduces the transistor threshold voltage.

This effect is starting to be identified in SRAM circuits, and is expected to increase in importance [15]. For small devices on the order of 100 nm and less, it is possible to track the locations of silicon and dopant atoms, considering the impact of their distribution on

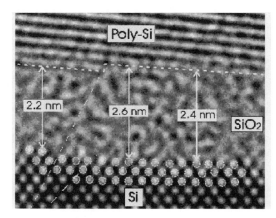

Figure 9.1. Gate oxide photo of a 27 Å oxide showing thickness variation. (Reproduced by permission of Doug Buchannan, IBM Corporation.)

the device behavior. A detailed study on 25 nm devices quantified the magnitude in threshold voltage variation, reporting a V_t uncertainty of $10/W^{1/2}$ (mV/μm$^{1/2}$) (W being the width of the device) [23].

9.2.2 Impact on Device Intrinsic Electrical Properties

The parameter variations listed earlier cause fluctuations of drain saturation current, threshold voltage, or both, in devices from the same circuit. The transistor effective length L_{eff} and threshold voltage V_t are the main parameters that determine current drive and, therefore, the circuit clock speed. These parameters relate to each other in deep-submicron technologies such that the effective transistor length and width impact the V_t value. The threshold variation modulates off-state current, impacting the static power consumption that is increasingly important in high-performance applications.

Transistor Threshold Voltage (V_t) Variation. Figure 9.2 shows threshold voltages measured for transistors designed with two different target channel lengths; the inset illustrates the threshold voltage dependence with the effective transistor length. The spread in effective transistor length is typical of parameter variations in submicron technologies. Ideally, since all devices are designed to have one of two target transistor lengths, only two points should appear in the graph. In practice, the spread in transistor effective length translates to a spread in device threshold voltages and circuit speed performance.

Figure 9.3 shows threshold voltages measured in 4,000 devices with different sizes and V_t adjustments using superretrograde halo implants to control V_t for these devices. Square devices were designed with large, equal L and W values, whereas the *small* devices target smallest length and width for that technology. *Narrow* devices were designed for minimum width and large length, whereas *short* devices refer to wide devices with minimum transistor length.

Short devices show a mean transistor target threshold voltage of about 350 mV, but width reduction (when moving from short to small devices) provides a mean V_t of about

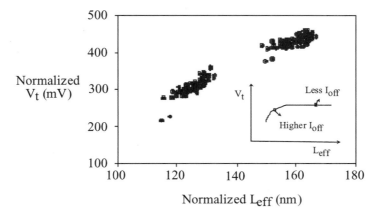

Figure 9.2. Measured transistor threshold voltages for two set of *n*MOS devices with differently designed channel lengths. The inset plot shows the V_t roll-off dependence with the effective transistor length [31].

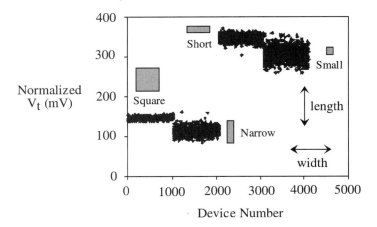

Figure 9.3. Measured threshold voltages for four sets of nMOS transistors. Square refers to transistors designed with large length and width, whereas small refers to devices with minimum length and width. Short devices are wide minimum length transistors, whereas narrow transistors have large L and small W [49].

300 mV due to STI. This widens threshold voltage variations, since L_{eff} and W_{eff} variations are worsened for minimum dimensions. Square test devices have a small spread in V_t so that the impact of process variation in large-sized devices (both length and width) is smaller. The lowered V_t value of the square and narrow transistors is due to the superretrograde halo implant effect on a large-dimension transistor.

The short and small transistors are the typical geometries expected in digital ICs, for which the W/L ratio varies from unit size to integer values from 2–20 (and larger for buffer drivers). The short devices are typically the targeted threshold voltages for the transistors of the IC. As technology nodes go from 130 nm to 90 nm and below, the number of transistors affected by the narrow width V_t reduction effect increases since this effect typically occurs for transistor widths below 0.4 μm. We emphasize the significant variation when the channel lengths or widths are taken to their minimum dimension.

Figure 9.4 shows the 3σ variation of transistor threshold voltage at three technology nodes. Variation ranges from 12% at the 360 nm die to more than 29% for the 180 nm die [36]. Figure 9.5 shows the percentage of V_t variation within the die for three technologies, showing that at 130 nm, the variation in V_t is markedly increased. This occurs while the absolute value of threshold is reduced from about 0.45–0.4 V at the 0.23 μm and 0.18 μm nodes, respectively, to about 0.3 V at the 130 nm node [36].

Impact on I_{Dsat}. Transistor L_{eff} intrinsic variation is the primary variable that influences F_{max} [7]. Keshavarzi et. al., also presented L_{eff} variation data and its impact on I_{Dsat}, which is the transistor speed parameter [31]. Figure 9.6 plots transistor drain saturation current versus device threshold voltage [31]. I_{Dsat} measures the current drive of a transistor in the saturation region, reflecting its ability to drive load capacitance during logic transitions. This graph illustrates how threshold voltage fluctuations translate to the transistor saturation current and circuit operation speed.

Transistor length variations impact circuit operation speed not only through V_t variations, but also directly, as was shown in the device model equations discussed in Chapter

9.2 INTRINSIC PARAMETRIC FAILURES   251

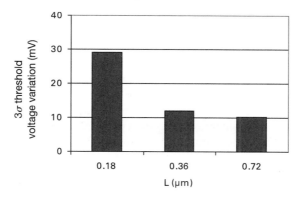

Figure 9.4. 3σ variation of transistor threshold voltage at three technology nodes [36].

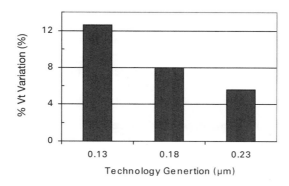

Figure 9.5. Threshold voltage variation within the die [36].

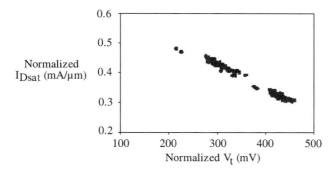

Figure 9.6. Relationship between transistor threshold voltage and drain saturation current I_{Dsat}[31].

3. Figure 9.7 shows this length effect in the saturated drain current versus the transistor effective length plot. Again, the spread in I_{Dsat} is an indication of device-to-device variation in speed capability.

Impact on I_{off}. Leakage or off-state current is another important transistor parameter, since it directly relates to the total static dissipation power at the IC level. Ideally, nonconducting transistors should pass negligible current. However, a characteristic of submicron technologies is an increase in subthreshold current due to the reduced transistor threshold voltages required to maintain performance when the supply voltage is lowered (see Chapter 3). Transistor length has a first-order impact on the device subthreshold current, threshold voltage modulation, and IC speed performance.

Figure 9.8 plots transistor intrinsic leakage current versus the inverse of the linear drain current ($1/I_{Dlin}$, an indicator of L_{eff}) for two groups of devices designed with different target effective lengths. It shows the dependence of the subthreshold current on the transistor effective length. The spread of the effective transistor length induces changes in the leakage current of about one order of magnitude between target or nominal L_{eff} (L_{nom}) and worst-case L_{eff} (L_{wc}). The impact of subthreshold current variation on main circuit parameters and its implication for test technology are described in detail in the next chapter.

These data illustrate the manufacturing variation that we expect in modern transistor parameters. The variations are larger than in the past, so that overall IC performance also has wider variation. This increased variance property of modern ICs profoundly affects testing, since testing traditionally assumes a single test limit. We might say that "modern dies won't hold still."

9.2.3 Line Interconnect Intrinsic Parameter Variation

Traditionally, device and design engineers target transistor drive strength and reduced input capacitance as the most effective ways to improve IC performance. Transistor parameter variance was the dominant statistical influence on speed in older technologies, but a worse statistical influence on speed failures now comes from the interconnections. The smaller interconnect sizes and increased packing density of deep-submicron technologies

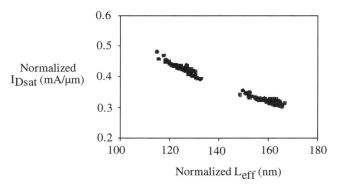

Figure 9.7. Measured saturation current values for two groups of *n*MOS transistors with two separately designed transistor lengths [31].

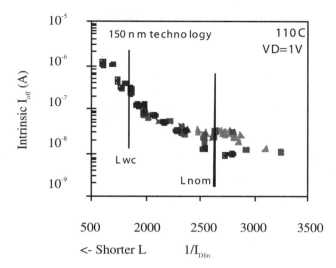

Figure 9.8. Transistor leakage current versus $1/I_{Dlin}$ for a 150 nm technology showing worst-case (L_{wc}) and nominal (L_{nom}) channel length values. (Reproduced by permission of Intel Corporation.)

add line resistance and cross-couple capacitance that make metal statistical distribution equally if not more important than transistors [56]. Additionally, the increasing clock speed and length of selected global interconnect lines (clock distribution and buses) make inductive effects important not only at V_{DD} and GND, but also in signal lines.

Parasitic capacitance, resistance, and inductance effects in IC interconnections depend on the driver and line characteristics that determine the signal properties traveling the line. The statistical fluctuation of device and interconnect characteristic parameters combine to impact delay variation and noise fluctuation that can lead to incorrect, unpredictable behaviors.

Interconnection coupling is the source of undesired noise called crosstalk and/or switching noise. These mechanisms are discussed later. Analysis of crosstalk and supply/ground bounce and their effects on ICs for half-micron and larger technologies are understood. There are even CAD software tools that take these effects into account and assist in preventing circuit malfunctions at the design stage. The crosstalk and supply/ground noise that affect deep-submicron technologies also relate to parameter fluctuation and design, similar to that of the transistors. The intrinsic variation of the fundamental parameters that determine these noise-related effects at the interconnection (as parasitic resistance, capacitance, and inductance) and device level (as driver strength) may cause well-designed circuits to fail due to parameters lying at process corners.

Figure 9.9 shows two cross sections of ICs with six and eight levels of metal. Lower metal levels are used for local interconnect and power distribution of custom logic cells that require high density to minimize area and achieve high clock frequency. The upper metal layers are typically used for the global V_{DD}, GND, clock, and control lines. The lower-level metal lines and vias lie closer to each other since they have minimum width. Notice the different aspect ratios (H/W) between 2–3 and the pitch of the vias and contacts. Each technology node has a minimum metal width of about that technology node number. A 90 nm node has a minimum metal width of about 90 nm [28]. The dielectric spacing be-

 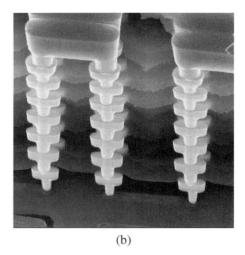

(a)   (b)

Figure 9.9. Cross section of an IC with (a) six [60], and (b) eight metal levels. (Reproduced by permission of Bob Madge, LSI Logic.)

tween minimum metal geometry is also about 90 nm. The vias and contacts have diameters on the order of the metal widths, and since the intermetal level dielectric vertical spacing scales more slowly, the via and contact aspect ratios tend to get larger. Via aspect ratios between 5–10 are used on modern ICs that have hundreds of millions or billions of these structures, and defect-free vias cannot be guaranteed. Vias and contacts are a difficult challenge for fabrication, test engineering, and failure analysis.

Figure 9.10 shows the lower two metal layers of a large microprocessor die. The upper layers of metal interconnect were etched off. Metal–1 is tungsten, and metal-2 is copper. The narrow, high-aspect-ratio lines on metal-1 and metal-2 are for local signal connection of drains, sources, and gates. The small cross sections result in line resistances that can range from 1000 Ω/cm to 3000 Ω/cm even for higher-conductivity copper metallization [22]. The higher-aspect signal lines show the trend that lateral area capacitance is larger

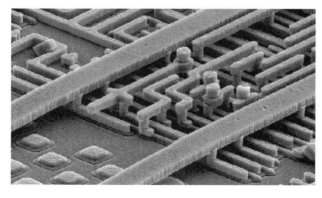

Figure 9.10. First two metal layers in a microprocessor IC. (Reproduced by permission of Chipworks Inc.)

than vertical area capacitance. This increases crosstalk while still providing a load capacitance. Lateral signal coupling capacitance depends on the design spacing, the metal length and height, the dielectric material, and the statistical variation. The longer flat lines in Figure 9.10 are power and GND interconnections, and it is here that power line inductance can cause problems. Signal-induced current surges raise the GND voltage, and lower the V_{DD} rails due to induced Ldi/dt changes. The 90 nm and 60 nm nodes are increasingly at risk if crosstalk from narrow spacing is not characterized and controlled [22]. The power lines have more horizontal surface area so their major capacitance is at top and bottom. Higher capacitance in V_{DD} and GND lines is usually a good feature, since it helps buffer the power lines against transient voltage changes.

The 10–30 ps rise times of modern ICs aggravate these inductive and capacitive conditions, causing increased noise and transient errors. Induced timing delays and speedups can result from these phenomena as well as from large surges induced by the simultaneous switching of large output buffers (ΔI noise). Figure 9.10 illustrates the R, L, and C parametric complexity of modern IC interconnections and via/contact metals. This complexity is aggravated by the dummy metal structures added to meet density requirements.

Process variations can induce worst-case sheet resistance fluctuation ranging from 10% in metal lines to about 38% in polysilicon, as reported for a 0.8 µm CMOS process [38]. Line width metal variations of about 10% and area capacitance variations from 17% to more than 25% were measured for intermetal lines. These parameter fluctuations were measured in a 0.8 µm process, but deep-submicron processes have larger process-induced variation, especially below 180 nm.

In summary, we saw that significant variations occur in V_t, L_{eff}, W_{eff}, I_{Dsat}, I_{off}, and interconnect capacitance and resistance. These variations translate directly to performance and potential failures of the ICs. This is the new reality of dealing with deep-submicron circuits and that is what this chapter is about.

9.2.4 Temperature Effect

Temperature impacts the electrical properties of both devices and transistor interconnects, although devices are more sensitive to this parameter. Carrier mobility and threshold voltage are the two transistor parameters most impacted by temperature, and their dependencies have compensating effects. Mobility decreases with increasing temperature, as moving carriers experience more collisions with the Si-crystal lattice whose vibration thermal energy gets higher with temperature. The effective mobility in the inversion layer can be expressed as

$$\mu(T) = \mu(T_0)\left(\frac{T}{T_0}\right)^{-m} \quad (9.1)$$

where T_0 is room temperature ($T_0 = 300$ K) and m is the mobility temperature exponent that depends on the channel doping, typically taking values between 1.5 and 2.

The temperature dependence of the threshold voltage is of the form

$$V_t(T) = V_t(T_0) - \alpha(T - T_0) \quad (9.2)$$

where α is a constant for each scaled technology, and has a typical value of about –0.8 mV/K [31].

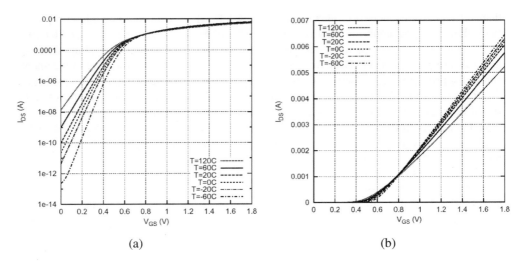

Figure 9.11. Drain current versus gate voltage for a nMOS transistor at different temperatures. (a) Shows the drain current in logarithmic scale for the I_{off} component; (b) shows the linear dependence for the saturation current.

The threshold voltage temperature variation mainly impacts the leakage current since I_{off} has an exponential dependence with V_t. The mobility temperature variation has an impact on the transistor drain saturation current. Figure 9.11 shows simulations of the drain current of a nMOS transistor from a 0.18 μm technology in logarithmic and linear scales to show the variation of the off-state and saturation currents, respectively. Figure 9.11(a) shows an I_{off} reduction of about three orders of magnitude when lowering the temperature from 120°C to 0°C. Figure 9.11(b) shows that drain current increases when temperature is lowered, although its variation is not as high as that of I_{off} (there is about 15% improvement in I_D when cooling from 120°C to 0°C). The reduced leakage current and increased drain saturation current gained at lower temperatures makes active cooling a promising technique for future scaled technologies if power-efficient cooling technologies can be developed.

Figure 9.11(b) shows an interesting common intersection of the lines when temperature is varied. If the transistor were biased at $V_{GS} \approx 0.8$ V for the given V_{DS}, then drain current would not change with temperature. Analog designers have long known of this property. The temperature-invariant intersection is a result of decreased mobility as temperature rises versus the decrease in V_t. The mobility drops, but the gate overdrive voltage $(V_{DD} - V_t)$ increases in the drain current equation.

Interconnect resistance rises with temperature, thus increasing circuit delay. It has been reported that in modern fabrication processes, copper has a temperature coefficient of resistivity of 3.6% to 3.8%/10°C [1]. This implies a change of the load resistance with temperature and an impact on delay due to interconnect variation.

9.3 INTRINSIC PARAMETRIC FAILURE IMPACT ON IC BEHAVIOR

Capacitive, resistive, and inductive noise were always present in previous IC technologies, but were considered design problems due to their systematic and well-characterized low-impact behavior. As technology scales, more noise sources appear as ICs lose noise margins. Moreover, the increasing process parameter variations cause larger fluctuations in the predicted design parameters with technology scaling. Noise is now less predictive, causing more potential "nonsystematic" circuit malfunctions. This nonsystematic behavior and the growing percentage of circuit malfunctions in current technologies due to noise-related issues motivated us to categorize these interconnect and/or signal routing noise sources as parametric failures, as well as design issues.

The impact of device and interconnect parameter variations on noise and delay mechanisms requires knowledge of electrical models and parameters that characterize these elements. MOSFET behavior and its modeling were covered in detail in Chapter 3. The electrical models that describe the interconnect and related coupling mechanisms are briefly described in the first part of this section.

9.3.1 Interconnect Models

We introduce the digital IC interconnect models that depend on the characteristics of the logic gates and the interconnect line parameters. These models will be required later to understand the effects of signal coupling and noise mechanisms.

Line Capacitance Model. The line capacitance model is a first-order model of the effects of gate-to-gate delay. The near-infinite resistive component of the MOSFET transistor gate allows us to model the load effect of a CMOS logic gate as a capacitor. One inverter driving another inverter [Figure 9.12(a)] can be modeled with the first inverter driving the equivalent input impedance of the second one, i.e., a lumped capacitor [Figure 9.12(b)]. This has inaccuracies since the input capacitance of a CMOS gate changes with the applied bias (Chapter 3). However, the lumped capacitor load can be computed from the various components of the gate input capacitance, as explained in Chapter 3. This

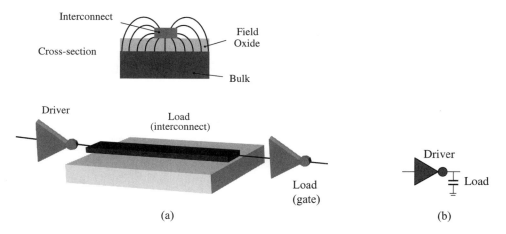

Figure 9.12. (a) CMOS gate driving another CMOS gate through a short interconnect. The field lines from the interconnect to the substrate (shown in the figure inset) result in parasitic capacitance. (b) Simplest equivalent electrical model in which the electrical effect of the interconnect and gate load gate are modeled with a single capacitor.

first-order lumped capacitor is a pragmatic choice when metal interconnect properties can be neglected.

A more accurate interconnect line model would include capacitive, resistive, and inductive effects. However, if the interconnected gates are close to each other, and the interconnect length is small, then none of these effects are important and the electrical effects of the interconnect can be neglected.

When the distance between logic gates increases, the metal line lengthens, and the first effect that becomes important is the line parasitic capacitance. Metal lines form small capacitor plates with the substrate isolated by the field oxide (as illustrated in Figure 9.12), thus forming a capacitance with one terminal grounded. Line capacitance is measured as (F/mm) so that a longer line has a larger capacitance. When the interconnect capacitance is of the same order of magnitude or larger than the load gate input capacitance, then its value is simply added to that input capacitance. IC local connections around the compactly designed standard cells have maximum lengths of about 10–100 μm, and they can be described as lumped capacitive loads, since their resistive components are negligible [20].

Line RC Model. A long interconnect has considerable effective resistance that must be taken into account. For frequencies up to several hundred megahertz, the interconnect is well modeled by a distributed RC line segmenting the line into several RC sections [Figure 9.13(a)] [4]. The effect is that the resistance component [Figure 9.12(a)] now shields the gate driver from the capacitance at the distal end of the line. In other words, the *effective capacitance* seen by the driver is smaller [42]. Therefore, the transition of the signal at the output of the gate will now be faster, since the effective capacitance that the driver must charge is smaller [46]. This does not imply that the total signal delay is smaller for longer interconnects, only that the signal transition at the gate output is faster, and that the transition at the input of the load gate will occur later.

A good circuit-level load model for signal delay requires only two capacitors and one resistor [Figure 9.13(b)], and is referred to as the π-model [40]. A detailed description about RC models can be found in [4].

Line RLC Model. The upper interconnect levels of an IC are typically used for global wiring to drive signals among blocks far away in the IC. They are mainly data buses, control signals, and clock distribution networks. These high-speed signals operate synchronously, and seek the minimum path delay over long distances. Power supply distribution trees are also used at higher interconnect levels, since they must connect to all the blocks in the IC.

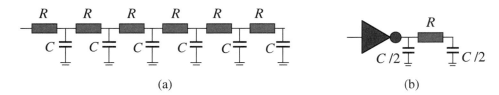

Figure 9.13. (a) Distributed RC model of an interconnect line. (b) Equivalent model of the circuit in Figure 9.12(a) when the line resistance plays a role.

Transmission line effects become important when the interconnect line is very long and transistor signal rise/fall times are fast. The RC model is no longer adequate. Deutsch showed that RC models can underestimate delay by more than 20% compared to models that account for transmission line effects [21]. Although global interconnects are a small fraction of the total wiring in an IC, they are usually part of critical paths that limit the system maximum speed. These global wiring interconnects are expected to grow in number and complexity from 10k–20k lines to 100k lines in the future [21].

A straightforward single-line model that takes these inductances into account connects a lumped inductance in series with each lumped RC segment, forming an RLC model (Figure 9.14). The line inductance effects in ICs are complex, since not only is the inductive component important, but also its inductive coupling to other lines [6, 22, 27]. An additional challenge to these models is related to parameter extraction, i.e., the determination of the R, L, and C values. In a complex design, these components are distributed and depend on the surrounding structures to the line. Moreover, the adoption of dummy metal structures to alleviate CMP induced IDL thickness complicates the extraction problem.

9.3.2 Noise

Diminishing noise immunity challenges deep-submicron technologies. Noise is now viewed as a design metric comparable to timing, area, and power [51]. The aggressive scaling of power supply voltage from 5 V to 1 V or less in current nanotechnologies severely reduces the noise margin budget. The values of modern supply voltages were the noise margins for older technologies. Threshold voltages are scaled accordingly as V_{DD} scales to maintain circuit performance, contributing even more to the noise margin reduction. The noise sources come from many factors with the constant increase in IC operational speed, rise and fall times, leakage currents, capacitive and inductive crosstalk effects, and power supply noise. All are predicted to further aggravate noise immunity of future technologies.

Noise can induce transient voltage perturbations on nodes that are at quiescent logic voltage states. If this noise propagates through the circuit and arrives at a memory element, it may cause a permanent logic error. In other cases, noise acts on nodes in transition, impacting circuit delay and, thus, operation speed. The former type of noise is known as static noise, whereas the second one is referred to as noise on delay [50].

Signal coupling and switching are two of the most important noise sources in modern digital CMOS ICs. Coupling noise is a form of crosstalk, impacting signal delay and amplitude. Noise and delay are related parameters [15]. Noise due to switching occurs at the supply lines (V_{DD} and GND), and comes from the parasitic resistance and inductance of the supply/ground distribution tree. The frequency components of this noise identify IR drop noise (resistive mid–low frequency range noise) or delta-I noise. Delta-I noise (inductive high-frequency noise) occurs when large currents are drawn during simultaneous switching of many logic gates, particularly output buffers.

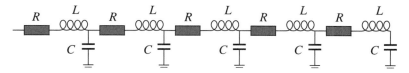

Figure 9.14. Distributed RLC model for a single line.

260   CHAPTER 9 PARAMETRIC FAILURES

Coupling Noise. Crosstalk or coupling noise is the most common effect of signal coupling between closely spaced interconnect lines. It produces a noise spike in one static line (called the victim line) when a neighboring line (aggressor line) makes a transition. Rubio et al. analyzed crosstalk noise due to the capacitive coupling between interconnect lines [47]. Figure 9.15(a) illustrates this type of coupling: two metal lines running parallel within the circuit are coupled through their parasitic capacitor. The capacitive coupling between the signal lines provokes an undesired voltage spike at the victim line when the aggressor transitions. We saw in Chapter 8 that if the victim line is floating, then the induced voltage change can be significant. Figure 9.15(a) shows a victim line driven by a gate (an inverter) that maintains the static logic levels at this node. The noise amplitude and duration are dependent on many factors such as the aggressor and victim driver's relative strength, capacitive coupling, resistive losses, and the receiver circuit bandwidth [21]. Figure 9.15(b) shows simulation results of noise spikes induced at the victim line. The victim line (V_{outv}) has a positive induced 0.2 V peak voltage that is prolonged for

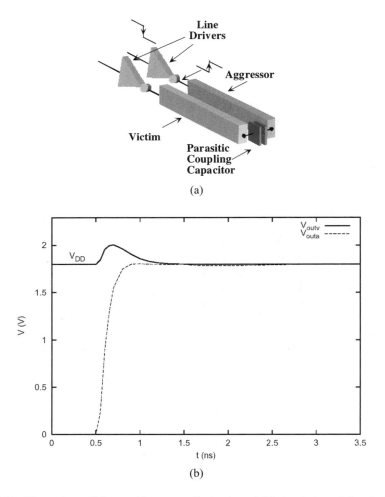

Figure 9.15. Illustration of (a) capacitive crosstalk circuit, and (b) simulation of electrical behavior.

about 500 ps. That bump in signal voltage would speed up a circuit transition. Speeding up a signal is not inherently good since timing synchronous success depends on predictable arrival times.

Crosstalk noise in modern technologies is quite complex since the length of metal lines and clock operation speed may require transmission line models that take inductive components into account. If the inductive component is important, but only the RC components are considered, then the noise levels will be underestimated. This is an important issue for circuit failures since functional errors can occur if noise margins are unexpectedly violated.

Modeling IC interconnect capacitive coupling is relatively simple, since it occurs only between signal lines that are physically close. Capacitors are conductive plates separated by insulators, and capacitance increases with decreasing plate separation. Modern IC local logic gate metal spacing can be less than 130 nm.

Inductive coupling originates when an electromagnetic field created by a time-varying current in one line (aggressor) induces a current in the metal lines influenced by such a field (victims) (Figure 9.16). These currents are known as return currents, since they close the current path. The victim line has a resistance component to the ground node large enough to build a potential drop caused by the return current. This potential opposes the return current. As a result, the victim line close to the aggressor can sustain only a fraction of the return current. Other victim lines further away in physical distance must sustain the remaining return current. The inductive coupling extends across large distances in the IC if nearby low-resistance return paths to ground are not present (Figure 9.16) [15]. Inductance-noise modeling is difficult. The identification of the return paths in the ICs is not straightforward, complicating the whole modeling process [22].

Power Supply Noise. This noise appears at the supply and ground distribution networks with two components: one is associated with the resistance of the distribution network and the other with its inductance. The former is referred to as *IR drop noise*, whereas the second is named *delta I-noise* (or ΔI-noise) [16, 51], although the term "IR drop" is used sometimes to refer to both cases.

IR drop noise causes variations in the DC level of the supply/ground lines due to the localized current demand of the various gates in the circuit. This current causes a voltage

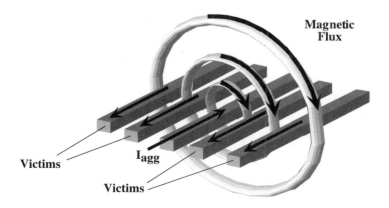

Figure 9.16. Inductive coupling showing the electromagnetic field extending over a large area.

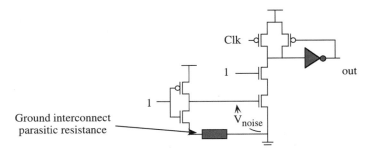

Figure 9.17. Illustration of the IR drop effect on a ground line and possible logic error if for a given time $V_{\text{noise}} \geq V_{tn}$.

drop between different parts of the circuit due to the line resistance, such that different gates may be at different supply/ground levels. The V_{DD} level drops while the V_{SS} (GND) level rises. Figure 9.17 shows this effect for a ground line and its possible impact on circuit behavior. If the first logic gate ground terminal has a voltage above zero because of IR drop, then the logic zero at the output of the first inverter will not lead to zero voltage at the input of the second inverter. If the second gate is a static inverter, then the IR drop noise will probably not impact its logic value unless the ground reference difference goes significantly beyond the transistor threshold voltage. A different situation might appear if the second gate is a dynamic logic gate. The IR noise could trigger an *n*MOS device to the on-state when the gate is the evaluation phase, and lead to a permanent logic error. In practice, a well-designed robust distribution power grid, with appropriate packaging technologies, makes the effect of IR drop negligible when compared to delta-I noise.

Delta-I noise occurs when several gates switch simultaneously, and they are coupled to the parasitic inductive components of the supply/ground lines. This effect is illustrated in Figure 9.18(a), where the supply and ground nodes of an inverter are connected through inductors (representing the parasitic inductance of the supply lines) to the main supply nodes. When the input changes, a transient current is demanded by the logic gate to charge its output capacitor. This transient current causes an inductive voltage drop at the V_{DD} terminals of magnitude $V_{\text{drop}} = L(di/dt)$ so that the supply and ground values seen by the gate are degraded. The parasitic inductance causes the GND terminal voltage to temporarily rise while the V_{DD} line voltage falls. The voltage degradation increases with the inductance value and the rate of change of the current. This last parameter is related to the fast rise and fall times at the gate input. Rise and fall times are decreasing to 10–30 ps in leading-edge technology. These effects are illustrated in Figure 9.18(b), where we show simulation results of the circuit in Figure 9.18(a).

■ EXAMPLE 9.1:

Calculate $v_{\text{drop}} = L di/dt$ if $\Delta i = 2$ A, $\Delta t = 20$ ps, line inductance = 5 pH.

$$\Delta i / \Delta t = 2 \text{ A}/20 \text{ ps} = 10^{11} \text{ (A/s)}$$

$$v_{\text{drop}} = (5 \text{ pH}) \cdot (10^{11} \text{ A/s}) = 500 \text{ mV}$$

■

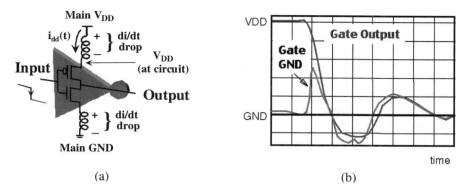

Figure 9.18. Illustration of supply/ground bounce, (a) circuit representation, and (b) electrical behavior.

Power line inductance is minimized in flip-chip packaging that has many closely spaced I/O leads. Microprocessors may use two out of every three leads for power and ground connections. This reduces line inductance, since the power lines are in parallel.

Figure 9.19 shows noise spikes simulated on a 2,601 node circuit power grid versus planar *ij* position using a 1 ns clock period [34]. We will see that these many mVs of V_{DD} noise have a significant impact on delay. Design margins try to incorporate these noise effects; however, designs are now too complex to accurately predict all temperature, ground bounce, and cross-coupled effects.

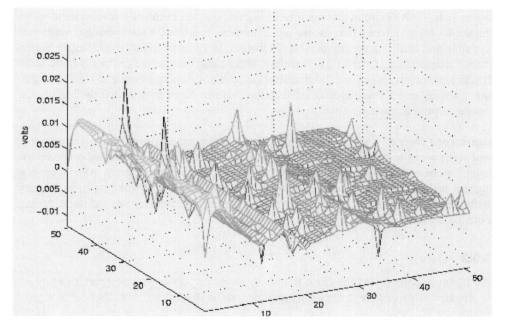

Figure 9.19. V_{DD} versus *ij* plot on chip [34].

The main problem is that the power supply variation effects on delay cannot be fully analyzed in detail since it is not feasible to simulate the entire power grid together with the circuit [14]. The voltage and current at a given point in the grid are dependent on the currents at all points in the grid, thus requiring a whole simulation. A typical IC power grid may have between 1M–100M nodes, and the analysis required for a 10M node grid might take about 5 months [14]. Some simplification techniques can analyze power grids by treating the effects of devices separately from those of the power grid model. This may lead to a number of effects combining at a given gate or IC region that provoke a malfunction of the circuit. These effects were not characterized together.

The impact on timing is due to the effective supply voltage reduction seen by the affected gates; they become weaker, and the propagation delay through them increases. Chapter 4 discussed fundamental aspects of gate delay in CMOS circuits from the analysis of inverter gates. IR drop is critical in the clock distribution networks since the clock activity is large. Delta-I noise also affects line delay since the transient change in the supply/ground levels change the driving capabilities of the gates, thus impacting speed. These transient power supply reductions are sometimes referred to as brownout.

Delta-I noise was traditionally identified with I/O driver switching activity. Recently, the simultaneous switching noise from internal circuitry was found to have a similar delta-I impact in high-performance ICs. Delta-I noise voltage was reported as high as 0.35 V (30%) at gigahertz frequencies and a 1.2 V supply technology [58]. Delta-I noise is minimized by using decoupling capacitors within the IC and at the package level. Decoupling capacitors are high-frequency capacitors connected between V_{DD} and ground, close to the gates. These capacitors are charged during the circuit power-up and static operation periods, and support the sudden transient current demands of the switching gates. The closer the capacitor to the gate supply node, the lower the impedance of the power line and the smaller the induced noise.

Noise spikes from interconnect coupling or supply/ground noise can cause circuit failure in static CMOS circuits if they travel a signal path and eventually reach a memory element. Dynamic circuits have worse noise margins than CMOS static designs since both dynamic and static stages are unbalanced toward the evaluate-signal direction for performance purposes (see Chapter 5). Therefore, static noise in dynamic circuits is more sensitive to logic errors, even for combinational gates. A noise spike arriving at a dynamic gate input whose output was not discharged in the evaluate phase may switch the logic value, causing a permanent logic error.

Substrate Noise. CMOS circuits are fabricated on conductive substrates. The substrate and wells serve to polarize the bulk terminals of devices, but they are also a conductive layer for noise. Although substrate noise injection is not as significant as other sources in current technologies for standard digital CMOS applications, it was shown to increase with substrate doping [2]. Future technologies are expected to increase substrate doping, potentiality increasing this noise.

9.3.3 Delay

All ICs eventually fail as the clock frequency increases. If the same test pattern sequence is repeated at successive decreasing clock periods, a Boolean error will occur at a given frequency. This frequency is called F_{max} (i.e., the maximum operating frequency), and it is usually related to the minimum time required for a signal to travel between internal memory elements (generally flip-flops) within the circuit. If the signal arrives later than the next clock period, then a timing error can occur.

Figure 9.20 illustrates this concept. The circuit state is temporarily stored in flip-flop memory cells. If the circuit is static and the clock is stopped, then the present state is unaltered as long as power is maintained. The flip-flops at the left in Figure 9.20 hold the outputs of the previous combinational block (not shown in the figure), and these data are the input to the combinational logic block shown. This block processes its inputs from the flip-flops, and places the result temporarily at the inputs to the subsequent flip-flops. When the next clock edge arrives, these combinational block outputs are evaluated, and stored by the flip-flops on the right side in Figure 9.20. The longest time required for the combinational blocks to process the inputs and deliver the output (for all the blocks in the circuit) limits the maximum operating frequency F_{max}. The period of the clock that drives the memory elements must be larger than the delay of the longest signal path in the combinational circuits.

Not all signal paths in the circuit have the same propagation time between memory elements. The signal paths imposing the strongest limitations to signal propagation (i.e., the slower ones) are called *critical paths*. A design challenge is to identify critical paths in the circuit, and redesign their topology, allowing faster signal travel and overall higher speed.

Process parameter variations alter signal delays for the same given path signal when changing from chip to chip within a wafer, or die from different wafers in a lot, or different lots. This causes equivalently designed circuits to operate at different maximum speeds. Dies are tested for F_{max} and ICs are then separated, or binned, into different speed groups. If the required F_{max} is fixed by the design specification, such as in ASICS, then faster chips have more margin, whereas slow ones are yield loss. In other applications, like microprocessors, they are then sold at a price depending on their maximum operating frequency. This availability of parts at different clock rates and prices is not intentional, but a result of not controlling process variations.

When a defect in the signal path slows signal propagation, no Boolean error may be detected or generated at that defect site, but, rather, the signal waveform is simply shifted by a few nanoseconds or less. The actual Boolean failure may occur at a flip-flop whose setup, hold, or clock pulse width time requirements are violated (see Chapter 5). The Q output of the flip-flop then incorrectly reads the data at the D input.

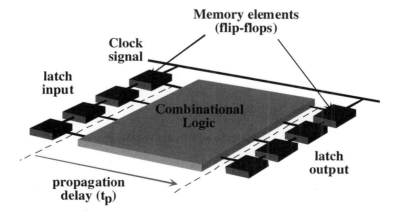

Figure 9.20. Propagation delay of a combinational block between state memory elements.

Figure 9.21 illustrates the statistical timing variation measured on 25 wafers in a 0.25 μm technology with cumulative propagation delays measured on 910 dies in each wafer [5]. The initial rise in the plots shows a near-straight-line relation to the normal distribution. The difference in propagation delay in Figure 9.21(a) between fastest and slowest dies at $V_{DD} = 2.5$ V is about 4 ns against a mean of about 8 ns. When the power supply was reduced to $V_{DD} = 1.2$ V [Figure 9.21(b)], the propagation delays increased almost three times. The break in the curve at the 95% cumulative point is more distinct. Test limits could be set tightly at the 95% cumulative point to reject the outliers for high-reliability products, or set somewhat higher, trading yield loss against product reliability expectations; however, tight limit setting is not practical in a production test environment.

There is a physical difference between parts in the normal and outlier distributions. The outliers have defects causing delay in addition to the part's otherwise normal variation. An important point is that real tests that target delay faults do not have the capability and resolution to measure to this fine degree for each delay path. Each timing test pattern does not have an individual test limit, so that delay fault testing typically uses the period of the system clock, and adds an amount to account for tester noise and parameter variance.

Similar results were obtained in [38] from process variation data measured in a 0.8 μm process. A test limit must take into account the slowest parameter measurement. Slower outliers from faster dies overlap normal data in the slower die. This is part of the challenge in detecting these statistical variance failures. We must look for properties that distinguish normal from outlier data.

Parametric failures identified in the preceding sections impact the circuit delay in a different fundamental manner. We now discuss four forms of parametric-induced delay: interconnect, crosstalk, supply noise, and temperature-induced delay.

Interconnect Delay. Path-line resistance is influenced by the statistical variation in metal granularity, width, height, length, and via quality. Variations in metal separation and dielectric uniformity affect line capacitance, as well as the height and length variation. The interactive effects of these parameters contribute to the signal path variation in propagation delay shown in Figure 9.21.

More complex problems frequently arise when parallel signal paths converge at a common gate. We may also see abnormalities when V_{DD} is increased, such as reduced F_{max}, thus contradicting the principle that circuits run faster if V_{DD} is increased. V_{DD} changes can affect the timing failures describe above. When V_{DD} increases, the transistor gate drive increases, allowing the device to switch faster. If the data path has a defect delay insensitive to V_{DD} increase, but the clock path reduces its propagation delays due to the V_{DD} increase, then the clock pulse can arrive too early and violate data setup time. A similar effect of a V_{DD}-insensitive delay defect in the clock line can cause hold time violations.

Crosstalk Delay. Crosstalk can significantly impact circuit delay [12]. It has a unique property of increasing or reducing signal delay depending on the swing direction of the aggressor and victim nodes. Although delay reduction might initially seem an advantage, it is not, especially if the node being speeded up is the clock node. Clock speedup is equivalent to data-path slowdown (all the paths in the combinational block of Figure 9.20), thus leading to violation of setup and hold times and logic malfunction.

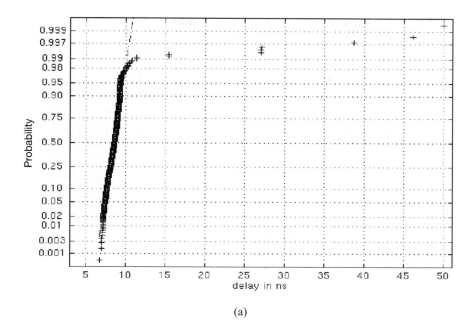

(a)

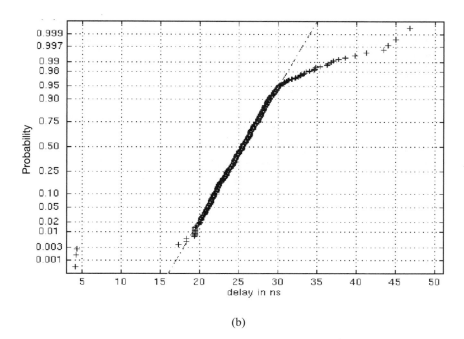

(b)

Figure 9.21. Cumulative distributions for propagation delay from 25 wafers with data including 910 dies per wafer. (a) $V_{DD} = 2.5$ V. (b) $V_{DD} = 1.2$ V [5].

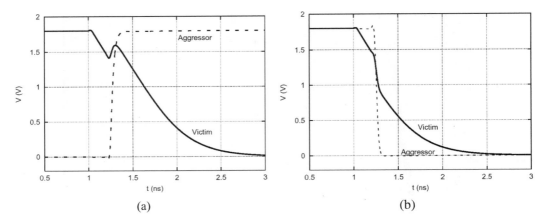

Figure 9.22. Illustration of crosstalk-induced (a) delay, and (b) speedup.

Figure 9.22 illustrates both types of crosstalk-induced delay, showing signal speedup and slowdown. The crosstalk-induced delay happens when the victim node and the aggressor node make simultaneous transitions in opposite directions [Figure 9.22(a)]. The crosstalk-induced delay can be quantified as the additional time required for the victim line to reach its 90% signal swing with respect to the case in which the aggressor does not switch [12]. Signal speedup [Figure 9.22(b)] occurs when a node makes a transition in a given direction (say, low to high), and a neighboring coupled node also makes a transition in the same direction.

Supply Voltage Variation Induced Delay. Chapter 4 showed that the delay of a CMOS gate is determined by the supply voltage, the output capacitance, and the transistor's drain saturation current. Once the design is fixed for a given circuit, its speed depends mainly on the supply voltage.

Figure 9.23 shows the V_{DD} sensitivity of an Intel 1 GHz, 6.6 million transistor, router chip built with 150 nm technology. The chip has an approximate 1.8 MHz change per mV of power supply voltage at $V_{DD} = 1.3$ V. Figure 9.23 shows about a 14% increase in F_{max}

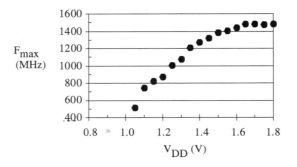

Figure 9.23. F_{max} versus V_{DD} for a 150 nm IC [49].

when V_{DD} increases 10% from V_{DD}=1.3 V. The sensitivity is higher at the low end V_{DD} and saturates above V_{DD} = 1.6 V. Bernstein et al. reported a change in F_{max} with V_{DD} of 200 kHz/mV for a 180 nm microprocessor, and gave a performance rule of thumb that chips vary by 7–9% when the power supply varies by about 10% [7]. Vangal et al. showed F_{max} versus V_{DD} plots for a 130 nm dual-V_t 5 GHz technology with a sensitivity of 11.3 MHz/mV at V_{DD} = 1.0 V [61]. Figure 9.23 emphasizes that noise-induced millivolt changes in V_{DD} induced on a critical signal node in normal operation can measurably affect IC speed. A related system problem occurs when board power supplies have a ±5% accuracy. This could move a test-verified 1 GHz IC in Figure 9.23 to an operational range of about 883 MHz to 1.18 GHz on the board. One protection used by manufacturers is to guard-band the shipped product, anticipating power supply influences, but this is wasteful of IC capability. Parts at the low end of a particular bin are moved to the next lower bin as a safety precaution, but it is a yield and financial loss for ICs at the higher bins.

Figure 9.19 gave simulation results showing many millivolts of V_{DD} noise spikes over the die. These intrinsic noise disturbances can cause large MHz changes in F_{max}, especially if a critical path is affected. The timing of the clock distribution network is one of the most critical control signals in the circuit. IR drop impacts clock skew, which is the maximum difference between the arrival times of the clock signal to the memory elements in the circuit. It was reported that an X% IR drop in the supply voltage of a gate can cause from X/2% to X% increase in gate delay [17, 48]. Saleh et al. simulated an industrial design, and found that nodal IR drops caused a 30% change in the skew value [48]. They also found an approximate rule of thumb that a 10% drop in V_{DD} caused a 5–10% change in clock timing. The complexity of today's large IC power supply grid does not allow a full simulation of the entire circuit to check for the impact of supply noise spikes on localized signal paths. Designers usually check for maximum and minimum temperature and V_{DD} conditions (called process corners), using oversimplified models that can miss these combined effects [9].

Temperature-Induced Delay. Figure 9.24 shows temperature data taken from a small, 20,000 transistor test circuit [49]. The temperature sensitivity in the regression equations in Figure 9.24 is –13.1 kHz/°C and –12.3 kHz/°C for the fast and slow ICs. We

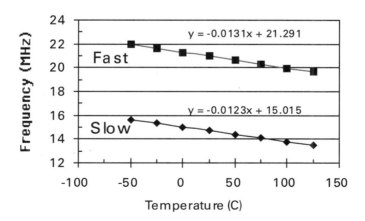

Figure 9.24. Speed versus temperature dependence on a 20k transistor circuit [49].

say that the major effect that slows an IC with temperature rise is the decrease in carrier mobility. A compensating speed effect is that the absolute values of the n- and p-channel transistor threshold voltages decrease as temperature rises. For $V_{DD} = 0.5$ V and $V_t = 200$ mV technologies, the compensation of the threshold behavior can even lead to ICs with a positive temperature performance [30]. There may be thermal instability issues, but temperature characterization must be known for each technology to allow recognition of anomalous parametric behavior.

Interconnect properties also vary with temperature, as explained previously. Since the total line resistance change with temperature depends on the line total length [22], different interconnect paths may change their resistance in different amounts due to temperature and local thermal impedance variation. This may change the relative delay of the different signal and clock paths on the critical path signal distribution.

The temperature distribution within the circuit is difficult to predict since it is strongly related to the activity of each block and the local thermal impedance. Figure 9.25 shows a temperature map corresponding to IR emission from the back side of the die, illustrating that different regions of the circuit can have more than 40–50°C differences. There is no circuit simulator to date that can link temperature as an intrinsic parameter to current in a way that predicts the intrinsic interaction between temperature-induced current variation and current-induced temperature changes. As a result, temperature spreads out the distribution of circuit fundamental parameters with a corresponding impact on prediction of its behavior.

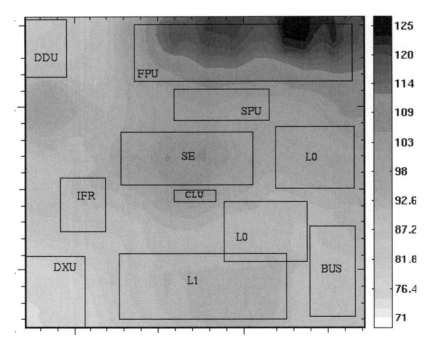

Figure 9.25. Within-die temperature gradients due to different activity zones. (Reproduced by permission of Intel Corporation.)

9.4 EXTRINSIC PARAMETRIC FAILURE

We will describe five extrinsic IC mechanisms associated with parametric failures: (1) weak interconnect opens, (2) resistive vias and contacts, (3) metal mousebites, (4) metal slivers, and (5) gate oxide shorts in ultrathin technologies. The weak opens and resistive vias are major defect-related parametric failure mechanisms. Mousebites occur when sections of metal are missing from an interconnect line. Slivers are common defects in which a metal particle lies between two metal conductors and barely contacts the signal lines. Another failure that shows a variety of responses is the gate oxide short. Some gate shorts show timing- and power-supply-dependent failures, and the recent ultrathin oxides have a unique failure mode that raises questions about the implied reliability risk for some gate oxide shorts.

9.4.1 Extrinsic Weak Interconnect Opens

Chapter 8 dealt with open defects. Weak opens are included here since they cause a relatively small increase in interconnect resistance, and do not prevent current in the line. These defects include a continuous range of resistances, and only those having a very large value (beyond 1 GΩ) are considered hard opens.

Figure 9.26 shows a photo of a weak open in which the line resistance is locally higher due to aluminum overetching caused by the presence of a crack. The Ti barrier provides a parallel current path, although its resistance is much higher than aluminum.

A recent study [45] characterized both interconnect and via resistance distributions for a six-metal-layer, aluminum-based CMOS, 0.18 μm process technology analyzing yield evaluation monitor structures. Table 9.2 summarizes their results, providing the resistance distribution for each metal level. Among the total number of opens, between 65%–70% of opens in metal-1 to metal-4 levels were hard opens, whereas for the two top interconnect levels, 100% of opens were hard opens. Weak opens in the four lower metal levels ranged from 15% to 25%. Weak opens are a significant portion of the open defect population.

9.4.2 Extrinsic Resistive Vias and Contacts

Vias are the short, vertical metal structures that connect the layers of horizontal metal interconnections. Contacts are the short, vertical metal structures that connect the transistor

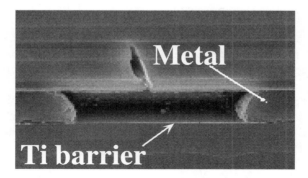

Figure 9.26. Weak open due to crack-induced metal overetching. The current goes through the Ti barrier with a much higher resistance [45].

Table 9.2. Resistance (R_M) Distribution for Weak Opens

Resistance [Ω]	Metal 1	Metal 2	Metal 3	Metal 4	Metal 5	Metal 6
3k< R_M < 100k	0.4%	10.9%	12.02%	6.90%	0%	0%
100k < R_M < 1M	5.2%	3.5%	6.25%	4.31%	0%	0%
1M < R_M < 10M	9.2%	4.9%	6.25%	5.17%	0%	0%
Partial total weak opens	14.8%	19.3%	24.52%	16.38%	0%	0%
10M < R_M < 100M	8.9%	5.7%	2.88%	9.48%	0%	0%
100M < R_M < 1G	7.0%	6.0%	1.92%	8.62%	0%	0%
R_M > 1G	69.77%	69.0%	70.67%	65.52%	100%	100%
Total	100%	100%	100%	100%	100%	100%

Note: From [45].

drain, source, or gate to the first level of metal interconnections. The physics and fabrication of contacts and vias are different, but their resistive failure modes can be similar. Modern ICs can have billions of transistors and perhaps ten times that number of vias, so that defective vias with elevated resistance are not surprising. Figure 9.9 showed vias and contacts with different sizes depending upon the metal level. Cracks in flat metal lines also show properties similar to resistive vias, but are a less common failure mechanism, particularly with the shunting barrier metals deposited around the Al or Cu metal interconnect (see Chapter 6).

Several fabrication-related mechanisms can cause resistive via failures [5, 18].

- Via pushups
- Etch ash polymers left in the via bottom after the reactive ion etch
- Incomplete etches that leave a thin dielectric layer across the bottom of the via
- Residual moisture that occupies a space in the via, creating voids when W is deposited and leading to increased resistance and possibly corrosion
- Insufficient metal filling of via
- Via misalignment that affects connecting areas and via size
- Small random particles that lie within the via, or large particles that can cause mask defects, even where the via is missing
- Chemical reaction between W and Fl

Figure 9.27 shows a defective via with poor etching [45]. The via makes contact only at the edges. This can and does cause temperature-dependent timing failures.

The same study that characterized weak interconnect opens also analyzed the resistance distribution of resistive vias and contacts [45]. Table 9.3 lists their results. Contact problems resulted in hard opens for more than 90% of the cases, whereas via problems caused between 52% to 81% of strong opens. Weak opens ranged from 11% to 36% in via contacts. Resistive vias and contacts are electromigration-sensitive, especially weak opens.

9.4.3 Extrinsic Metal Mousebites

Missing regions of interconnect metal are called mousebites. They can be due to particle defects, electromigration, or stress voids. Mousebites have a minor electrical effect, but

9.4 EXTRINSIC PARAMETRIC FAILURE **273**

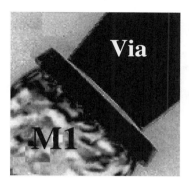

Figure 9.27. Defective via in a deep-submicron CMOS IC [45].

are a major reliability risk [49]. Figure 9.28(a) sketches a defect-free and a defective (mousebite) section of interconnect. If sheet resistance is $R_\square \approx 70$ mΩ/$\square$ and a 90% bite is taken out of the middle section, then that line resistance changes from 210 mΩ to 840 mΩ. This major defect in the line would not elevate resistance sufficiently to be detected by a speed test. Analysis shows that line resistance rises exponentially, but not significantly, until the mousebite exceeds about 95% [Figure 9.28(b)]. These conclusions extend to voided vias and contacts that also show this resistance dependency on volume voiding. The via or contact must be well voided to cause an RC delay failure sensitive to temperature.

There will be some increase in local temperature due to the resistance increase in the mousebite region, but this defect is virtually undetectable with present technology. The failure analysis lab may simply find an open circuit, or the defect would not open completely if protected by a barrier metal sheath.

Analysis is easier using the sheet resistance parameter. Resistance of a structure with resistivity ρ, length L, area A, height H, and width W is

$$R = \rho\left(\frac{L}{A}\right) = \rho\left(\frac{L}{HW}\right) \qquad (9.3)$$

Table 9.3. Distribution of Resistive (R_M) Open Contacts and Vias

Resistance [Ω]	Contacts	Via 1	Via 2	Via 3	Via 4	Via 5
$R_M < 10k$	0.8%	13.8%	1.45%	15.69%		7.33%
$10k < R_M < 100k$		6.9%	1.45%	1.96%	9.38%	6.00%
$100k < R_M < 1M$	0.8%	3.44%	4.35%	5.88%	3.12%	12.67%
$1M < R_M < 10M$	2.8%	17.24%	4.35%	11.76%	6.25%	10.00%
Weak opens	4.4%	41.38%	11.6%	35.29%	18.75%	36.00%
$10M < R_M < 100M$	4.4%			3.92%		12.00%
$R_M > 1G$	91.2%	58.62%	88.4%	60.78%	81.25%	52.00%
Total	100%	100%	100%	100%	100%	100%

Note: From [45].

274 CHAPTER 9 PARAMETRIC FAILURES

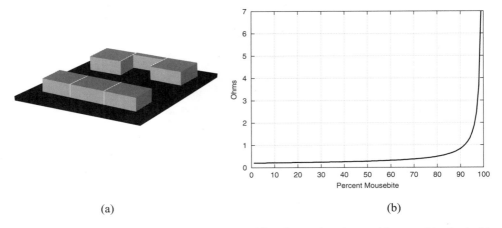

(a)   (b)

Figure 9.28. (a) Control metal line: normal metal line (bottom) and one with mousebite (top). (b) Voided metal resistance (mΩ) versus percent metal voiding using R = 70 mΩ/□ [49].

Since metal height is constant in an IC metal layer for each technology, we can divide ρ by H to get the sheet resistivity $R_\square$

$$R_\square = \frac{\rho}{H} \qquad (9.4)$$

This equation assumes that $L = W$ so that $R_\square$ is the symbol for the resistance of one square of metal no matter the size. We use $R_\square$ to quickly estimate the resistance of an interconnect line by counting the number of squares in the metal line. The number of squares in a line is the length L divided by the width W

$$R = R_\square \frac{L}{W} \qquad (9.5)$$

An example will show its use.

■ **EXAMPLE 9.2**

Let resistivity $\rho = 3 \, \mu\Omega \cdot cm$, $W = 0.5 \, \mu m$, $L = 1.5 \, \mu m$, and $H = 0.5 \, \mu m$. (a) If 90% of the middle segment of an Al line is removed [Figure 9.28(a)], what is the increase in resistance? (b) If the line is 100 μm long, what is the percent increase in resistance if, again, 90% of one square is removed?
   (a) The sheet resistance is

$$R_\square = \rho/H = 3 \, \mu\Omega \cdot cm / 0.5 \, \mu m = 60 \, m\Omega/\square$$

The resistance of the defect-free line in Figure 9.28 (3 squares) is

$$R = (60 \, m\Omega/\square)(3 \, \square) = 180 \, m\Omega$$

If 90% of the middle square is missing [Figure 9.28(a)], then the middle square becomes 0.5 μm / 0.05 μm, or it has 10 squares. The middle resistance is

$$R_\square = (60\ \text{m}\Omega/\square)(0.5\ \mu\text{m}/0.05\ \mu\text{m}) = 600\ \text{m}\Omega$$

The resistance of the original segment of one square was 60 mΩ and it has now increased to 600 mΩ. The resistance of the original three squares was 180 mΩ and that has increased to 720 mΩ. A speed test could not detect such a small change in RC time constant despite this gross defect.

(b) If the line was initially 100 μm long, then the resistance would increase from 12 to 12.54 Ω, an increase of 4.5%. Again, a speed test could not come close to detecting this small resistance change for such a flagrant defect. ■

9.4.4 Extrinsic Metal Slivers

Metal slivers gained importance with the advent of CMP (chemical mechanical polishing). A small metal sliver lies between two interconnect lines, barely or not even touching them (Figure 9.29). When the temperature rises, the metal will expand, and the sliver will now touch the signal lines. Higher voltages at burn-in can promote the rupture of the high-resistance oxide surface of the metals, bonding the three metal elements [43]. When the temperature returns to lower values, the metals are now tightly bonded. The bridge resistance is now low enough to reduce noise margins or even cause functional failure.

9.5 CONCLUSION

This chapter brought together a diverse failure set different from the study of hard open and bridging defects. Intrinsic parametric failures arise from skewed statistical distribution of circuit-speed-related parameters, and from extrinsic subtle defects causing the same type of failure modes. The mechanisms for these failures were described in figures and plots. Cross-coupled noise, ground bounce, IR drop, and delta-I inductive noise are intrinsic failure modes that span design, test, reliability, and failure analysis issues. The extrinsic failure modes from resistive vias, mousebites, and metal slivers were described. Their detection depends on variation of circuit environmental parameters. Parametric failures are expected to worsen in occurrence and diversity as circuit dimensions shrink with technology.

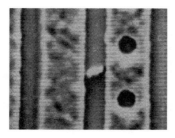

Figure 9.29. Metal sliver [43].

REFERENCES

1. I. Aller, K. Bernstein, U. Ghoshal, H. Schettler, S. Shuster, Y. Taur, and O. Rorreiter, "CMOS circuit technology for sub-ambient temperature operation," in *IEEE International Solid-State Circuits Conference,* pp. 214–215, February 2000.
2. X. Aragonés and A. Rubio, "Experimental comparison of substrate noise coupling using different wafer types," *IEEE Journal of Solid State Circuits, 34,* 10, 1405–1409, October 1999.
3. V. Axelard, et. al. "Efficient full-chip yield anaysis methodology for OPC-corrected VLSI designs," in *IEEE International Symposium on Quality Electronic Design,* pp. 461–466, 2000.
4. H. Bakoglu, *Circuits, Component, and Packaging for VLSI,* Addison-Wesley, 1990.
5. K. Baker, G. Grounthoud, M. Lousberg, I. Schanstra, and C. Hawkins, "Defect-based delay testing of resistive vias-contacts—A critical evaluation," in *IEEE International Test Conference (ITC),* pp. 467–476, October 1999.
6. M. Beattie and L. Pileggi, "IC analyses including extracted inductance models," in *Design Automation Conference (DAC),* pp. 915–920, June 1999.
7. K. Bernstein, K. Carrig, C. Durham, P. Hansen, D. Hogenmiller, E. Nowak, and N. Rohrer, *High Speed CMOS Design Styles,* Kluwer Academic Publishers, 1998.
8. K. Bowman, X. Tang, J. Eble, and J. Meindl, "Impact of extrinsic and intrinsic parameter fluctuations on CMOS circuit performance," *IEEE Journal of Solid State Circuits, 35,* 8, 1186–1193, August 2000.
9. M. Breuer and S. Gupta, "Process aggravated noise (PAN): New validation and test problems," in *IEEE International Test Conference (ITC),* pp. 914–923, October 1996.
10. M. Bruce, V. Bruce. D. Eppes, J. Wilcox, E. Cole, P. Tangyungyong, and C. Hawkins, "Soft defect localization (SDL) on ICs," in *International Symposium on Test & Failure Analysis (ISTFA),* pp. 21–28, Nov. 2002.
11. D. Buchanan, "Scaling the gate dielectric: Materials, integration and reliability," *IBM Journal of Research and Development, 43,* 3, 245–264, 1999.
12. F. Caignet, S. Delmas-Bendhia, and E. Sicard, "The challenge of signal integrity in deep-submicrometer CMOS technology," *Proceedings of the IEEE, 89,* 4, April 2001.
13. A. Campbell, E. Cole, C. Henderson, and M. Taylor, "Case history: Failure analysis of a CMOS SRAM with an intermittent open circuit," in *International Symposium on Test & Failure Analysis (ISTFA),* pp. 261–269, Nov. 1991.
14. R. Chaudhry, D. Blaauw, R. Panda, and T. Edwards, "Current signature compression for IR-drop analysis," in *Design Automation Conference (DAC),* pp. 162–167, June 2000.
15. A. Chandrakasan, W. Bowhill, and F. Fox (Eds.), *Design of High-Performance Microprocessor Circuits,* IEEE Press, 2001.
16. H. Chen and J. Neely, "Interconnect and circuit modeling techniques for full-chip power supply noise analysis," *IEEE Transactions on Components, Packaging, and Manufacturing Technology-Part, 21,* 3, 209–215, August 1998.
17. D. Cho, K. Lee, G. Jang, T. Kim, and J. Kong, "Efficient modeling technique for IR drop analysis in ASIC designs," in *IEEE International ASIC/SOC Conference,* pp. 64–68, 1999.
18. E. Cole, P. Tangyunyong, C. Hawkins, M. Bruce, V. Bruce, R. Ring, and W.L. Chong, "Resistive interconnection localization," in *International Symposium on Test and Failure Analysis (ISTFA),* pp. 43–50, November 2001.
19. R. Degraeve, B. Kaczer, A. De Keersgieter, and G. Groeseneken, "Relation between breakdown mode and breakdown location in short channel NMOSFETs," in *IEEE International Reliability Physics Symposium (IRPS),* pp. 360–366, May 2001.

20. A. Deutsch et al., "When are transmission-line effect important for on-chip interconnections?," *IEEE Transactions on Microwave Theory and Techniques, 45,* 10, October 1997.
21. A. Deutsch et al., "Frequency-dependent crosstalk simulation for on-chip interconnections," *IEEE Transactions on Advanced Packaging, 22,* 3, August 1999.
22. A. Deutsch et al., "On-chip wiring design challenges for gigahertz operation," *Proceeding of the IEEE, 89,* 4, April 2001.
23. D. Frank, Y. Taur, M. Ieong, and H. Wong, "Monte Carlo modeling of threshold variation due to dopant fluctuations," in *VLSI Technology Symposium,* pp. 169–170, June 1999.
24. W. Grobman et. al., "Reticle enhancement technology: Implications and challenges for physical design," in *Design Automation Conference,* June 2001.
25. D. Halliday and R. Resnick, *Fundamentals of Physics,* 3rd ed., Wiley, 1988.
26. W. Henson, N. Yang, and J. Wortman, "Observation of oxide breakdown and its effects on the characteristics of ultra-thin nMOSFET's," *IEEE Electron Device Letters, 20,* 12, 605–607, December 1999.
27. Y. Ismail and E. Friedman, *"On-Chip Inductance in High Speed Integrated Circuits,"* Kluwer Academic Publishers, 2001.
28. International Technology Roadmap for Semiconductors 2001 (http://public.itrs.net/).
29. B. Kaczer, R. Degraeve, G. Groseneken, M. Rasras, S. Kubieck, E. Vandamme, and G. Badenes, "Impact of MOSFET oxide breakdown on digital circuit operation and reliability," *IEEE International Electron Device Meeting,* 553–556, December 2000.
30. K. Kana, K. Nose, and T. Sakurai, "Design impact of positive temperature dependence on drain current in sub–1V CMOS VLSIs," *IEEE Journal of Solid State Circuits,* 1559–1564, October 2001.
31. A. Keshavarzi, K. Roy, and C. Hawkins, "Intrinsic leakage in low power deep submicron CMOS ICs," in *IEEE International Test Conference (ITC),* pp. 146–155, October 1997.
32. S. Kundu, S. Zachariah, S. Sangupta, and R. Galivanche, "Test challenges in nanometer technologies," *Journal of Electronic Testing: Theory and Applications (JETTA), 17,* 209–218, 2001.
33. F. Lakhoni, Sematech, Lecture Series on Microelectronic Failure Analysis at University of New Mexico, November 20, 2001.
34. Y. Lee and C. Chen, "Power grid transient simulation in linear time based on transmission-line-modeling alternating-direction-implicit method," in *International Conference on Computer Aided Design (ICCAD),* pp. 75–80, November 2001.
35. V. Mehrotra and D. Boning, "Technology scaling impact of variation on clock skew and interconnect delay," in *IEEE International Interconect Technology Conference,* pp. 122–124, 2001.
36. S. Narendra, D. Antoniadis, and V. De, "Impact of using adaptive body bias to compensate die-to-die V_t variation on within-die V_t variation," in *IEEE International Symposium on Low Power Electronics and Design,* pp. 229–232, 1999.
37. S. Nassif, "Design for variability in DSM technologies," in *IEEE International Symposium on Quality Electronic Design,* 2000.
38. S. Natarajan, M. Breuer, and S. Gupta, "Process variations and their impact on circuit operation," in *IEEE International Workshop on Defect and Fault Tolerance in VLSI Systems,* pp. 73–81, November 1998.
39. W. Needham, C. Prunty, and E.H. Yeoh, "High volume microprocessor test escapes, an analysis of defects our tests are missing," in *IEEE International Test Conference (ITC),* pp. 25–34, October 1998.
40. P. O'Brien and T. Savarino, "Modeling the driving-point characteristic of resistive interconnect for accurate delay estimation," in *IEEE International Conference on Computer-Aided Design,* pp. 512–515, November 1989.

41. M. Orshansky, L. Millor, P. Chen, K. Keutzer, and C. Hu, "Impact of spatial intrachip gate length variability on the performance of high-speed digital circuits," *IEEE Transactions on Computer-Aided Design, 21,* 5, May 2002.
42. J. Qian, S. Pullela, and L. Pillage, "Modeling the effective capacitance for the RC interconnect of CMOS gates," *IEEE Transactions on Computer-Aided Design, 12,* 12, 1526–1535, December 1994.
43. W. Righter, C. Hawkins, J. Soden, and P. Maxwell, "CMOS IC reliability indicators and burn-in economics," in *IEEE International Test Conference (ITC),* Washington DC, November 1998.
44. W. Riordan, R. Miller, J. Sherman, and J. Hicks, "Microprocessor reliability performance as a function of die location for a 0.25 μm, five layer metal CMOS logic process," in *IEEE International Reliability Physics Symposium (IRPS),* pp. 1–9, April 1999.
45. R. Rodriguez Montanes, J. Pineda de Gyvez, and P. Volf, "Resistance characterization for weak open defects," *IEEE Design and Test of Computers,* pp. 18–25, September–October 2002.
46. J. Rosselló and J. Segura, "Power-delay modeling of dynamic CMOS gates for circuit optimization," in *IEEE International Conference on Computer Aided Design (ICCAD),* pp. 494–499, 2002.
47. A. Rubio, N. Itazaki, X. Xu, and K. Kinoshita, "An approach to the analysis and detection of crosstalk faults in digital VLSI circuits," *IEEE Transactions on Computer-Aided Design of Integrated Circuits and Systems, 13,* 3, 387–395, March 1994.
48. R. Saleh, S. A. Hussain, S. Rochel, and D. Overhauser, "Clock skew verification in the presence of IR-drop in the power distribution network," *IEEE Transactions on Computer-Aided Design of Integrated Circuits and Systems, 19,* 6, 635–644, June 2000.
49. J. Segura, A. Keshavarzi, J. Soden, and C. Hawkins, "Parametric failures in CMOS ICs—A defect-based analysis," in *IEEE International Test Conference (ITC),* October 2002.
50. [50. K. Shepard and V. Narayanan, "Conquering noise in deep-submicron digital ICs," *IEEE Design and Test of Computers,* 51–62, January–March 1998.
51. K. Shepard, V. Narayanan, and R. Rose, "Harmony: static noise analysis of deep submicron digital integrated circuits," *IEEE Transactions on Computer-Aided Design of Integrated Circuits and Systems, 18,* 8, 1132–1150, August 1999.
52. H. Shibata, T. Matsuno, and K. Hashimoto, "Via hole related simultaneous stress-induced extrusion and void formation in Al interconnects," in *IEEE International Reliability Physics Symposium (IRPS),* pp. 340–344, April 1993.
53. A. Singh, H. Rasheed, and W. Weber, "I_{DDQ} testing of CMOS opens: An experimental study," in *IEEE International Test Conference (ITC),* pp. 479–489, October 1995.
54. B. Stine, et. al. "The physical and electrical effects of metal-fill patterning practices for oxide chemical-mechanical polishing processes," *IEEE Transactions on Electron Devices, 45,* 3, 665–679, March 1998.
55. J. Suehle, "Ultrathin gate oxide reliability: Physical models, statistics, and characterization," *IEEE Transactions on Electron Devices, 49,* 6, 958–971, June 2002.
56. D. Sylvester and K. Keutzer, "Impact of small process geometries on microarchitectures in systems on a chip," *Proceedings of the IEEE, 89,* 4, 467–489, April 2001.
57. K. Takagi and Y. Kohno, "Failure analysis of tungsten stud defects from the CMP process," in *International Symposium on Test and Failure Analysis (ISTFA),* pp. 315–321, Nov. 2000.
58. K. Tang and E. Friedman, "On-chip ΔI noise in the power distribution networks of high speed CMOS integrated circuits," in *IEEE ASIC/SOC Conference,* pp. 53–57, 2000.
59. T. Turner, "A step-by-step method for elimination of burn-in as a necessary screen," in *96IRM Final Report,* pp. 82–86, 1997.
60. S. Tyagi et. al., "A 130 nm generation logic technology featuring 70 nm transistors, dual V_t tran-

sistors and 6 layers of Cu interconnects," in *IEEE Electron Devices Meeting (IEDM)*, pp. 567–570, 2000.
61. S. Vangal et al., "A 2.5 GHz 32b integer-execution core in 130 nm dual-V_t CMOS," *IEEE Journal of Solid-State Circuits, 37,* 11, 1421–32, Nov. 2002.
62. J. Walls, "Stress-voiding in tungsten-plug interconnect systems induced by high-temperature processing," *IEEE Electron Devices Letters, 16,* 10. October 1995.
63. M. Walker, "Modeling the wiring of deep submicron ICs," *IEEE Spectrum,* 65–71, March 2000.
64. B. Weir et al., "Ultra-thin gate dielectrics: they break down, but do they fail?," in *IEEE International Electron Device Meeting (IEDM),* pp. 73–76, 1997.

9.7 EXERCISES

9.1. Parametric failures are basically insensitive to our present test methods, such as stuck-at fault testing, I_{DDQ} testing, delay fault testing, and functional tests. Therefore, why do we care about parametric failures?

9.2. Discuss why I_{Dsat} is an indicator of circuit speed performance.

9.3. Transistor gate width dimensions have a statistical variation as does channel length. Discuss two effects that Δw has on circuit speed performance.

9.4. Figure 9.1 shows a monolayer step change in oxide thickness T_{ox}. What would this effect have on the thin oxide reliability? Why would it be a negligible effect on older technologies?

9.5. Use the model equation for a saturated long-channel transistor, and explain why variation in L_{eff} has a larger effect on I_{Dsat} than a similar variation in W_{eff}.

9.6. I_{off} is the small drain leakage current when the transistor is in the off-state. Why do we worry about this small current?

9.7. The reliability and test issues of vias and contacts were not a great concern in older technologies as they are now. Explain.

9.8. Temperature affects the transistor threshold voltage and carrier mobility. Typically, V_t drops in magnitude as temperature increases (increases gate drive voltage), and carrier mobility degrades (reduces I_{Dsat}). Can you locate the transistor bias point in Figure 9.11(b) where the device parameters are invariant with temperature? What does this imply?

9.9. What circumstances determine whether you would use the line RC model or the line RLC model in taking logic gate loading effects into account.

9.10. Crosstalk noise is an increasing source of parametric failures in modern ICs. Explain the variables that worsen crosstalk noise.

9.11. Describe the variables that worsen ground bounce noise.

9.12. All ICs have an upper maximum frequency of operation called F_{max}. Explain why F_{max} is not a constant value for identically designed parts from the same wafer.

9.13. Figure 9.23 shows F_{max} variation with V_{DD}. How might a customer experience negative effects of this property?

CHAPTER 10

DEFECT-BASED TESTING

10.1 INTRODUCTION

The life cycle of an IC from conception to market to obsolescence requires many engineering man-hours plus complex equipment. The goal is to ensure that the product meets its specifications and performs as predicted. Two main verification processes are involved: (1) *design verification* is the process that ensures that the design (not a physical entity) meets specifications before the IC is manufactured, and (2) *testing* verifies that the ICs were manufactured correctly, assuming a good design. A perfect design can lead to ICs that do not meet specifications because of errors or imperfections during the fabrication process. This chapter is about testing, not design verification.

Testing evolved from simple practices to complex procedures with the growth of IC technology. Testing can now be more than 50% of the overall development/fabrication cost, and it starts with the design. Testing must ensure that the circuit was manufactured correctly, and that it will function during the lifetime of the product. This involves reliability and test escape issues. We start with a historical perspective, and then define the various types of IC testing.

Test practices began with simple ICs, and we will describe its evolution, introducing the concepts of *automatic test pattern generation* (ATPG) and illustrate these concepts with examples. We will describe *design for test* (DfT) practices since test is an integral part of design in today's technologies. We follow with a detailed look at *defect-based testing* (DBT) since this is the strongest foundation on which to build a test plan.

DBT develops a detection strategy that matches defect-class electronic properties to particular test methods. We describe those test methods, and also special designs for test structures now necessary to understand test practices and provide test success. We then link that test knowledge to the defect properties of Chapters 7–9. There are many test details, but our goal is to provide introductory concepts of testing, and focus on the physical strengths and weaknesses of the various tests. The chapter ends with a description of the serious negative impact of nanometer technologies on all test methods.

The practice of DBT dates back in electronics history to when printed wire boards were visually inspected for solder blobs (bridges), wire trace opens, or mousebites. The tests were visual, but they targeted specific defects. The formal study of CMOS IC defect properties grew rapidly in the 1980s. A paper presented at the International Test Conference in 1994, titled "Defect Classes—An Overdue Paradigm for Testing CMOS ICs," collated diverse data on bridges, opens, and parametric delay defects, and then proposed a unified test strategy based on these defect electronic properties. About the same time, Manual d'Abreu is believed to have first coined the phrase "defect-based testing" at an Intel Corp. test workshop.

10.2 DIGITAL IC TESTING: THE BASICS

Testing does several things. Digital testers must evaluate whether ICs meet their product specifications by looking for gross or random defects that cause failure. Testers also sort products into performance bins, such as maximum frequency performance. Customers pay considerable price differences for microprocessors or microcontrollers that run faster than others. Testers are also useful tools in failure analysis.

Testers are expensive computers with special I/O features and large memory capacity, and they are often referred to as automatic test equipment, or ATE. ATEs apply a logic signal to the IC input pins on each clock cycle. This signal is a *test vector*, and a sequence of test vectors is a *test pattern*. The measured result at the output pins is compared with an expected good result from a simulation. Digital ICs are also subjected to DC and AC parametric tests in addition to the logic testing. Testers measure analog parameters, such as input and output pin voltages and currents, or propagation delay times. Testing faces the complex challenge of finding the oftentimes obscure defects in the IC.

The major test costs are (1) test development to create and simulate the test vectors, (2) the initial ATE purchase (several million dollars per ATE), and (3) the recurring costs of the actual manufacturing tests. We know much about the test process, and use that knowledge to minimize defective ICs escaping to the customer. Our primary philosophy is *detection of defects*. That may seem obvious, but it is not. It is a critical decision made early in product planning. The goal of testing is to determine if a given IC will follow specifications or it will not because of a defect. The primary test purpose is not to find the defect and its location within the IC. Finding defects within ICs is referred to as failure analysis. Most ICs that do not pass testing are just discarded, but in some cases, certain ICs are diagnosed to monitor deviations in the fabrication process.

Digital IC testing uses voltage-based and current-based measurements. Voltage-based testing measures the potential at IC output pins. It examines Boolean correctness by applying logic voltages to the input pins and measuring the voltage levels at the output pins of the IC. The expected high and low logic output voltages are computed from expensive logic simulations and stored in computer memory. Voltage-based testing, in practice, is often done at slow application rates, using fault models, or at high clock rates that evaluate subsets of IC functionality over a range of temperatures and power supply voltages (V_{DD}). ATE testing is done at the wafer and package-part level.

Current-based testing analyzes the rich information content of the IC power supply current in either the DC quiescent logic state or in the transient current pulse. The quiescent current test, called the I_{DDQ} test, measures the power supply current when all logic states are settled at their stable, steady-state values. The I_{DDQ} test is by far the most sensitive of the CMOS digital IC tests [39, 44]. The transient current test takes several forms, depending upon when in the transient time period the current is measured, or it may use statistical techniques on the whole waveform. We will describe the various forms of voltage-based and current-based testing.

10.2.1 Voltage-Based Testing

The major voltage-based test techniques are (1) functional testing, (2) stuck-at fault testing, and (3) delay fault testing. These three sometimes overlap, but understanding their strengths and weaknesses is necessary for defect-based test practice.

Functional Testing. Functional testing literally means to test the IC for its intended use by the customer, but there are different interpretations. Functional testing may refer to applying all of the logic states of the truth table to the part. Or it can refer to the case when the input voltage signals are applied only to the IC primary input pins, as opposed to special test pins called scan pins (discussed later in this chapter). We will describe later the design for test structures that support this interpretation.

Functional testing to replicate customer function is appealing. We might say, "If the part passes a functional test for the customer, what could be better." However, the statement is flawed, as an example will illustrate.

■ EXAMPLE 10.1

Calculate the time to functionally test a 1K static RAM for all of the memory states that a customer might use. Assume a test clock rate of 100 MHz.

There are 1024 memory bits in the 1K SRAM and each bit has two possible states (zero or one). Therefore, the total number of logic states (test vectors) is

$$2^{1024} = 1.8 \times 10^{308} \text{ logic states}$$

The time for one test is $(100 \text{ MHz})^{-1} = 10$ ns. Therefore, the total time for the functional test is

$$\frac{10^{308}(\text{tests}) \times 10(\text{ns/test})}{31.5 \times 10^6 \text{ (s/year)}} \approx 10^{292} \text{ years}$$

The point is made. Even relatively simple ICs are numerically intractable to the task of replicating customer function. Microprocessors, microcontrollers, and DSP (digital signal processing) ICs are worse. ■

Functional testing has this profound weakness, but it has a pragmatic role in testing. When a large number of test vectors are cycled through an IC at fast clock rates, we find that defect detection is significant enough to warrant its use even though we cannot predict what defect classes are actually detected. Also, functional testing is the method of choice for testing at high clock rates to separate the product into its various speed performance bins. These test patterns usually originate with the designer's design verification simulations. Functional testing is useful, but beware of its limitations.

Stuck-At-Fault (SAF) Testing. Test engineers in the early days of digital IC production quickly found that the time to generate test vectors became prohibitive as die complexity rose. Test development costs and lost time to market forced new ways of thinking. Why not let a computer generate the test if it is simply fed the logic design netlist? Since computers do not understand defects, the notion of a *fault model* arose. A fault model is a hypothesis of how a circuit will fail. It is imaginary, and the fault model test process does not care what caused the failure. A wonderfully simple fault model is the *stuck-at fault (SAF)*, mentioned in Chapter 7, whose hypothesis is that the IC failed its Boolean logic exercise. A signal node is assumed to be stuck, or tied to one of the power rails. The SAF test method does not concern itself with bridge defects and their critical resistance, or even open defects. It just tests for a Boolean upset at a signal node. Faults are not real. No one has ever photographed a fault, but defects are real and have been photographed.

Figure 10.1(a) shows the SAF concept applied to inverter. The output node C is clamped at zero ohms to the strong power supply V_{DD} so node C is stuck at logic-1 (SA-1). This is one electrical equivalent of how a SAF might occur. Another SAF behavior arises when an open defect causes a floating gate that goes to one of the rail voltages. If it floated to V_{DD}, then the *n*MOS transistor would be permanently on, and the *p*MOS transistor would be permanently off, causing a permanent output stuck at zero (SA-0).

The number of SAFs is twice the number of signal nodes, or four SAFs for the inverter. The 2NAND in Figure 10.1(b) has three signal nodes and six SAFs. Input node B on the 2NAND is shown with a symbolic stuck-at-zero (SA-0) fault. Since computers are the perfect machine for Boolean algebra, automatic test pattern generation (ATPG) for SAF testing developed rapidly in the 1960s, and it is now a convenient adjunct tool to the design process [5]. We will look next at how SAF test patterns are derived.

Stuck-At-Fault Test Pattern Generation. Computers read a logic circuit netlist, and then compute a list of test vectors that test for a particular failure mode. There is broad value in understanding how SAF test patterns are generated. The stuck-at fault is the earliest and simplest of fault models, and despite its many weaknesses, it is still a commonly used fault model [20].

A test for a SAF would try to drive the node under test to a logic value opposite to the one being tested. If we were testing the 2NAND gate in Figure 10.1(b) for a SA0 at node B, then we would attempt to drive node B to a logic-1. We could not test node B for a SA0 by driving it to a logic-0, since the error would not be exposed. This is the

Figure 10.1. SAFs shown for (a) an inverter, (b) a 2NAND.

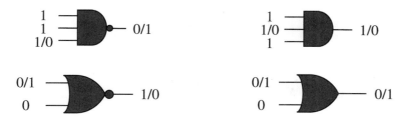

Figure 10.2. Logic gates showing path sensitization signals for a NAND, AND, NOR, and OR gate.

first rule of fault testing: always drive the node under test to the opposite logic state of the suspected error.

Figure 10.2 illustrates how signal information from a node under test can be passed from one logic gate to another until finally reaching a primary output pin where it is evaluated by a tester. When a logic gate has multiple inputs, there is a way to disable the influence of those input nodes not under investigation using the concept of the *noncontrolling logic state* (Chapter 4). For example, if we want the 3NAND gate in Figure 10.2 to pass the lower input node signal through the gate, then we set the other inputs to a logic-1. When the top two inputs are at logic-1, then the output signal is the complement of the signal at the bottom input terminal. Construction of a 3-bit truth table will show this. The logic-1 has the effect of turning off the parallel pMOS transistor tied to that node, and turning on the series nMOS transistor in a NAND or AND gate. The two complementary transistors then have no effect on the output if the other logic inputs change. The process of using noncontrolling input states to pass a signal from the inner core logic to a primary output is called *path sensitization*. Path sensitization can be confirmed for the other three logic gates in Figure 10.2, leading to two rules to memorize: (1) the noncontrolling logic state for an AND and a NAND gate is logic-1, and (2) the noncontrolling logic state for a NOR and an OR gate is logic-0.

There are four basic steps in generating a SAF vector for a selected node:

1. Select a node for testing and select one of the two SAFs for that node.
2. The logic value of that node must be set to the logic state opposite to the SAF condition under test.
3. A sensitive path must be constructed between the node under test and a primary output, using the principle of noncontrolling logic states.
4. Do a backward trace of the logic to set primary input nodes at values that will achieve steps 2 and 3 above.

■ **EXAMPLE 10.2**

Use the circuit in Figure 10.3 and derive the SAF test vector to test the designated node for a SA-0.

Repeating the step procedure:

1. Select a node to test for a SA0 (see designated node in Figure 10.3).
2. We set that node to the opposite logic state and pencil in a logic-1 in the figure.

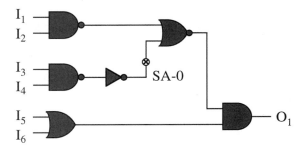

Figure 10.3. A circuit showing SAF test vector generation for the designated SA0 node.

3. The path sensitization to a primary output would pencil-in a logic-0 at the 2NOR gate upper terminal and a logic-1 at the 2AND lower terminal that drives primary output O_1.
4. The backward trace would set inputs I_3 and I_4 to the 11 state, forcing the node under test to a logic-1. I_1 and I_2 are set at logic-1 to sensitize the 2NOR gate at logic-0. I_5 and I_6 are set to any combination that forces a logic-1 out of the 2OR gate. That could be 11, 10, or 01.

Therefore, three possible test vectors will test the designated node in Figure 10.3 for a SA0: 11 11 11, 11 11 10, or 11 11 01. Notice that the same test vectors would test the inverter input node for a SA1. Another fault-equivalency example is that the output node O_1 is driven to logic-0 in a good circuit. If we measured a logic-1 state at O_1, then that is equivalent to detecting a SA1 error on O_1. These fault equivalence properties reduce the number of vectors to test the whole circuit, since one vector detects more than one SAF.

The number of signal nodes bound the number of SAFs in a given circuit. When the SAF test generation software is completed, the percentage of detected SAFs can be calculated. Fault models have the ability to provide a test coverage metric in contrast to functional testing. ■

■ EXAMPLE 10.3

Determine the test vector that detects node-m SA0 (Figure 10.4) (a) when the primary output is Node A, and (b) when the primary output is Node B.
Follow the four-step procedure and get

(a) Node-m is selected
Set to logic-1
Path sensitization for the 2-OR gates sets $I_1 = 0$, $I_2 = 0$
The backward trace assigns $I_3 = 1$ and $I_4 = 1$ to set node m at logic-1.
The test vector 0011 tests for node m stuck at zero.
(b) Node-m is selected
Set to logic-1
Path sensitization for the 2-OR gates sets $I_1 = 0$, $I_2 = 0$.

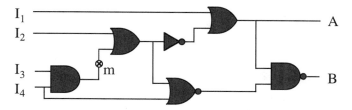

Figure 10.4.

The backward trace has a problem. If I_3 and I_4 are set to logic-1 to force node-m to logic-1, then the path sensitization on the 2NAND gate (or the 2NOR gate) feeding node B has a conflict. We cannot simultaneously set a logic-1 at node-m and the noncontrolling logic state of the 2NAND. Therefore, a SAF vector cannot detect the node-m SA0. Node-m is said to be *redundant* when tested through output node B.

Redundant nodes appear in circuitry, and a typical SAF coverage statement might be, "The circuit has 98.7% SAF coverage; the 1.3% undetected are redundant nodes." Redundant nodes often occur in designs to achieve timing constraints or as leftovers resulting from design reuse. ∎

Self-Exercise 10.1

Derive SAF test vectors to test node-c SA1 and node m SA0 in Figure 10.5.

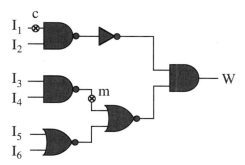

Figure 10.5.

The SAF test is the most popular of the structural test methods. However, it has weaknesses. SAF vectors applied through special scan test structures (described later) are typically clocked in at about 50 MHz–400 MHz limited by test clock layout and thermal heating of the die. The SAF test targets no bridge defects. It will detect a signal-to-rail bridge only if the defect is below the critical resistance. It will detect signal-node-to-signal-node bridges if, again, the bridge is less than R_{crit}, and also if a sensitive path exists by chance for that test vector. The SAF test method does target two of the five combinational logic open-circuit defect classes, but targets none of the parametric failure classes.

Despite these shortcomings, there are diverse applications for the SAF test pattern generation algorithm. We will show in later sections how it forms the basis for delay fault and I_{DDQ} test pattern generation.

10.2.2 Speed Testing

Speed testing refers to any voltage-based test that examines the IC at its user-specified clock frequency or clock period. This test is also used to classify and separate parts depending on their operating frequency. This procedure is called *speed binning*. SAF testing is done at slower clock rates, but ICs must be tested at their rated speeds and also need other forms of testing. We will describe the F_{max} and the delay fault test.

The F_{max} Test. The F_{max} test applies several thousands to millions of functional test vectors to the IC in a certain clock period. If the part passes, then the clock period is shortened until failure occurs. F_{max} is equal to the reciprocal of the minimum period of functionality. It measures how fast an individual IC can be clocked. Typically, the F_{max} test is done on parts that passed all other tests, so that F_{max} is often considered a speed binning procedure on good parts rather than a test. However, the F_{max} measurement has a significant role in testing resistive vias and contacts. If F_{max} is compared at two temperatures such as room temperature and hot (100°C), then resistive vias and contacts have a definite signature—an IC with resistive vias and contacts will typically run faster at the hot temperature. This defies normal circuit behavior and is explained later.

The F_{max} test is expensive in tester time since it replicates many test patterns until failure occurs, but it is the only test that is known to detect these forms of parametric defects. The F_{max} test has little knowledge of what paths and logic gates are being speed tested, but it is essential for speed binning and resistive via detection.

Delay Fault Test. Delay fault testing seeks to quantify speed test coverage of the IC. Delay fault tests evaluate individual timing paths, or evaluate the signal nodes of each logic gate, testing for their rise- and fall-time capability. The delay fault test uses a two-vector sequence. The first vector sets the initial conditions of the test, and the second vector activates a signal that typically tests for transition time of a gate or a path propagation delay. When the second vector is activated, *only one primary input should change on the clock command.* Multiple input changes on the same clock "confuse" the circuit and may present a noisy result. The resulting signal at the designated logic gate must be carefully steered through a sensitive path in the logic between that gate and a primary output where the tester makes a pass–fail timing measurement. The number of delay paths in most ICs is usually numerically intractable so that targeting gate node transitions is more practical. However, targeting only gate delays may result in a bad delay test for many parts since gate level test usually target shorter timing paths, leaving long ones untested.

The delay fault test presently cannot resolve timing defects at other than a gross level. Die-to-die statistical variation and inability to test propagation delay in different signal paths forces delay fault testing to gross test limits. Typically, the period of a test is set at the nominal clock period of the product plus a guard band for tester noise and product variances, and adjustment for "yield considerations." This frustrates the ability to detect finer-resolution defects that may cause customer failures.

10.2 DIGITAL IC TESTING: THE BASICS

Delay Fault Test Pattern Generation. Delay faults (DF) require a two-sequence pattern of test vectors, and both patterns use the SAF vector algorithm. The first DF pattern sets a node to a logic state, and is not concerned with reading a result (i.e., path sensitization). The SAF algorithm can do that by defining all logic gate outputs as primary outputs, so that path sensitization is disabled. The second DF test vector forces a change in the designated node, and develops a sensitized path to a primary output. The second vector should change only one node input to force the transition. This is a pure SAF vector, but used for timing in this case. An example will illustrate this procedure.

■ **EXAMPLE 10.4**

Generate the two-pattern delay fault test vectors for the node slow to rise in Figure 10.6.

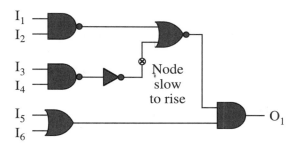

Figure 10.6. Delay fault patterns for a node slow-to-rise test.

Steps in solution:
1. Set the node to logic-0 (no path sensitization is required).
2. Set up path sensitization for the second vector. $I_1 \ldots I_6 = 11\ 01\ 11$.
3. Change I_3 from $I_3 = 0$ to $I_3 = 1$ with I_4 fixed at logic-1.
   First vector = 11 01 11
   Second vector = 11 11 11
   (You could interchange the roles of I_3 and I_4.)
4. Measure the time between the clock edge on the second vector and signal detection at the output. This is typically done with a timing limit, although the time value itself can be measured by test pattern iteration as the clock period is decreased.

■

10.2.3 Current-Based Testing

Data taken during the late 1980s and early 1990s showed that many CMOS IC defects caused intermediate voltages at internal nodes (i.e., voltage values out of the established ranges for logic 0 and logic 1). This caused circuit behaviors that could not be described using voltage-based approaches since internal mode intermediate voltages are restored by successive gates, and nonintermediate voltages are generated at the circuit outputs. These defect-induced behaviors were better described by circuit parameters other than the logic response. Current-based testing techniques were shown to be effective parametric test methods for CMOS ICs. Technology scaling pushed current-based testing methods from single-threshold I_{DDQ} to more elaborate quiescent current techniques. These techniques are covered at the end of the chapter, whereas the basic single-threshold I_{DDQ} is covered next.

I_{DDQ} Test. The quiescent power supply current test grew up with the CMOS IC industry in the 1970s and 1980s [30, 33, 38, 43], and later became known as the I_{DDQ} test [59]. Test engineers quickly concluded that since CMOS circuits had virtually zero supply current in the quiescent state, then an elevation in this current indicated the presence of a defect, design error, or fabrication problem. The tester measures I_{DDQ} when all transient currents have settled. The I_{DDQ} signature contains sensitive defect information.

Figure 10.7 illustrates the I_{DDQ} test applied to a pair of inverters. The signal timing waveforms show V_1, V_2, and the current through the power supply $i_{DD}(t)$. During the signal transition, both n and pMOS transistors are on, making a conductive path from V_{DD} to ground, and a current pulse appears (Chapter 4). When the signal reaches its post-transition levels, the current goes to near zero in a defect-free circuit since one of the complementary transistors is off. The I_{DDQ} test measures current in this quiescent state. A bridge defect shown between the N_2 gate and its source (or drain) elevates I_{DDQ} when V_1 is low and V_2 is high. The $i_{DD}(t)$ waveform in Figure 10.7(b) shows this. The I_{DDQ} measurement must be taken after the transient current settles, and this time is found by electrical characterization. Most defects elevate I_{DDQ}, but there are certain defect classes that do not, as we will review later.

The I_{DDQ} test has many positive features described in the bridge and open defect Chapters 7 and 8. The weaknesses of the I_{DDQ} test relate to certain open circuit defects, lack of direct detectability of parametric delay defects, the slow speed of the measurements, and the noisy background current of advanced technology ICs. Five open circuit defect classes for combinational logic were described in Chapter 8, and the I_{DDQ} test quantifies detection of two. I_{DDQ} has good nontarget detection of the memory or stuck-open defect [58, 68], but fails to detect the transistor pair-on/off and the narrow open defect. The open transistor pair on/off defect occurs when a floating gate acquires a voltage near V_{DD} or GND, thus turning off one of the complementary transistors. The I_{DDQ} measurement

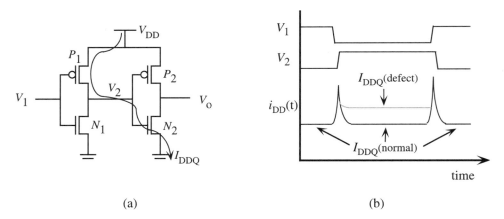

Figure 10.7. (a) One 2-inverter circuit with gate-source defects at N_2 and, (b) waveform showing I_{DDQ} elevation.

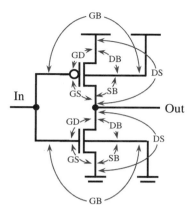

Figure 10.8. Inverter showing 12 possible transistor bridges.

speed can be solved with the Wallquist Quic-Mon load board circuit capable of submicroampere measurements above 100 kHz [67]. Although this rate is slow compared to voltage-based testing, considerably fewer I_{DDQ} vectors are needed for defect coverage. Measurement environments vary with different manufacturers, making instrumentation still not universally solved. Despite these weaknesses, the I_{DDQ} test is still the most powerful of the several test techniques [39, 44].

I_{DDQ} Test Pattern Generation. I_{DDQ} test patterns are typically generated for bridge defects across transistor nodes. Figure 10.8 shows an inverter for which I_{DDQ} transistor bridge vectors will be written. We count six possible transistor terminal bridge connections—GD, GS, GB, DS, DB, and SB—so that the inverter has 12 possible transistor bridge defect locations to test.

Inverters have only two logic states, and if I_{DDQ} is measured during the application of logic-1 and logic-0, we find the transistor node detection shown in Table 10.1. The I_{DDQ} coverage across a transistor node is defined as the presence of opposite voltage polarity across the bridged nodes with DC paths to V_{DD} and GND. Twelve transistor bridges are detected by the two logic states (Figure 10.8). Exceptions occur when terminals are intentionally connected, such as test-redundant source-to-substrate and well connections. The two inverter logic states are also the test vectors that give 100% SAF coverage to the circuit. An extension of this observation asks whether I_{DDQ} vectors will detect all transistor bridges in any circuit if the 100% SAF test vectors are applied to every logic gate. The answer is yes, and these patterns are called pseudo-stuck-at fault (PSAF) patterns [6, 39].

The three test vectors for 100% SAF coverage of a 2NAND gate are 01, 10, and 11 and 01, 10, and 00 for the 2NOR gate. If we apply these three test vectors to the 2NAND or

Table 10.1. I_{DDQ} Test on Inverter

Vector	Bridges detected with I_{DDQ}
In = logic-1	GSn, GDn, GBn, GDp, DBp, and DSp
In = logic-0	GDn, DBn, DSn, GSp, GBp, and GDp

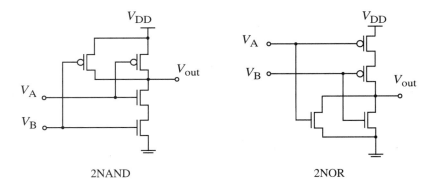

Figure 10.9. 2NAND and 2NOR circuits for I_{DDQ} test pattern verification.

2NOR gates and measure I_{DDQ} at each vector, then we attain 100% I_{DDQ} transistor node bridge defect coverage. You can manually verify the SAF and I_{DDQ} coverage using these vectors on the 2NAND and 2NOR circuits shown in Figure 10.9. There is no path sensitization since the power supply current pin monitors the test results, and not the primary signal output pins. This simplifies ATPG, and a SAF test pattern generator can generate the I_{DDQ} vectors by simply defining all logic gate outputs as primary outputs. The SAF patterns are applied to each gate, but the Boolean result is irrelevant to the test.

Transient Current Tests. Some defects, such as certain opens, do not induce intermediate voltages into the circuit, and, therefore, cannot be detected with the I_{DDQ} test method. In addition, I_{DDQ} testing has limitations when applied to deep-submicron ICs, as will be explained in this chapter. To solve some of these limitations, a number of techniques based on analyzing the transient portions of the supply current have been studied [10, 27, 37, 51, 53, 57].

The transient portion of the supply current contains information about the parts of the circuit registering activity, and the intent is to relate this information to the presence of defects in the circuit. There are many approaches to transient current testing, but a solution for production testing has not been demonstrated. However, some of the proposed techniques have been applied experimentally to commercial circuits [7, 31, 47].

One challenge of transient current testing relates to instrumentation since the transient portion of the current waveform contains very high frequency components that are difficult to measure. Additional challenges relate to noise on the supply pins when the circuit is switching. There is also possible information masking due to the internal and external decoupling capacitors that supply part of the transient current.

10.2.4 Comparative Test Methods Studies

Independent studies by Hewlett-Packard and IBM compared the relative defect detection efficiency of I_{DDQ}, delay fault, stuck-at fault (SAF), and at-speed functional tests. Figure 10.10(a) shows the Hewlett-Packard results; 758 total defects were found in a population of 26,000 ICs [39]. The uniquely detected defects are shown in bold, and the mutual detections are shown in smaller, regular type. Figure 10.10(b) shows the IBM results, a total of 665 defects from about 20,000 ICs in a joint study with the Sematech organization [44]. One of the IBM mutual detection numbers was omitted to disguise yield.

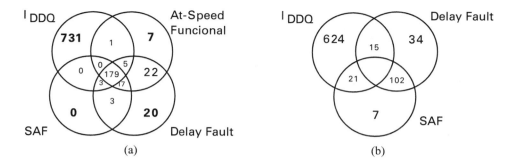

Figure 10.10. Test methods comparison of detection efficiency. (a) Hewlett-Packard [39], (b) IBM [44].

The data of Figure 10.10 are compared in Table 10.2. Since one set of overlapping data is missing in Figure 10.10(b), the unique detection percentages are compared. There were differences in the type of delay fault patterns by each company and other details, but the I_{DDQ}, delay fault, and stuck-at fault data match well in the two experiments. Defect electronic studies also correlate with the results of Figure 10.10, but do not provide comparative numbers. Bridging defects are dominant, and I_{DDQ} detects virtually all of these, whereas voltage-based tests detect only a small percentage. Open circuit defects are detected by voltage- and current-based tests. The lower detection rate of pure delay defects might correlate with the gross insensitivity of that technique.

10.3 DESIGN FOR TEST

Test and diagnosis are more efficient using design for test (DfT) structures. We will describe three major IC DfT techniques: (1) scan design, (2) built-in self-test (BIST), and (3) special test structures for the next level of assembly. The urgency in using DfT is that without it adequate testing is now virtually impossible given the complexity of present and future-generation designs. DfT is not an option, but required in nanometer ICs to meet overall product quality and reliability goals. Manufacturers and customers must accept DfT, or quality goals and time to market will be seriously jeopardized for most products.

DfT has several properties that slowed its acceptance. DfT circuits add active area to a die, reducing yield, increasing cost per die, and adding extra I/O pins. When DfT was introduced many years ago, ICs had small, fixed pin counts, such as 14, 24, 40, or 44 pins.

Table 10.2. Comparative Test Method Studies

Hewlett-Packard		IBM–Sematech	
Test	% Unique-Detect	Test	% Unique-Detect
I_{DDQ}	96.4	I_{DDQ}	93.8
Delay fault	2.6	Delay fault	5.1
Stuck-at fault	0	Stuck-at fault	1.1

Note: From [39, 44].

An extra four pins for DfT could not be added. Now, pin grid array packages allow hundreds to thousands of pins with more flexibility in their assignment. Extra pins for DfT are not usually an issue with modern packaging technologies. DfT structures must yield well or the whole die will fail. Increased die area also reduces the number of dies per wafer.

10.3.1 Scan Design

Circuit scan design is a popular design for test technique that dates back to the 1960s as a technique practiced on boards in Japan. Williams and Angel at Stanford University later described IC scan testing as a technique in which all circuit memory elements (flip-flops) had two modes: (1) normal circuit mode and (2) test mode [71]. The testability features of the flip-flops are transparent to normal operation. However in test mode, the flip-flops are transformed into a serial scan chain that could shift long test vectors deep inside the IC. Test vectors can now be efficiently delivered to deep logic portions of the IC, and the logic results can then be read and transported out of the IC by these same scan chains.

Figure 10.11(a) illustrates a single-phase clock design for a circuit with four combinational logic blocks separated by storage registers. Parallel storage registers hold information from each logic block until those data are clocked through to the next block. The timing is set so that all logic states are settled prior to the next clock.

In test mode, the flip-flop elements convert the parallel-in-to-parallel-out register into a serial-in-to-serial-out register [Figure 10.11(b)]. Four test pins are required: data-in, data-out, mode-select, and Clock. Test vectors serially enter the IC through the data-in pin, and are shifted a certain number of clock pulses. The shifting stops when bits in the chain are lined up with targeted combinational logic block input nodes. The bit pattern in the long serial register contains the segmented test vectors for each combinational logic block.

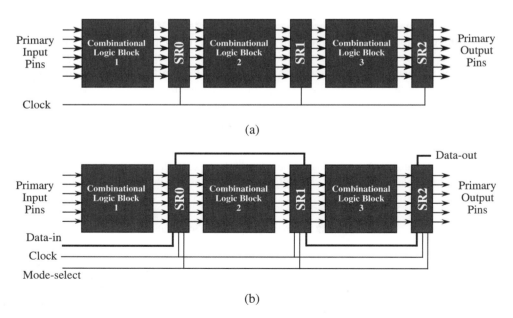

Figure 10.11. (a) Normal system mode. (b) Scan testing operation.

When the mode-select is activated for capture of the combinational logic output data, the results of each logic block are clocked into the scan chain registers at the output nodes of the block. This overrides any previous data in those storage cells. Once data are entered in the scan chain, the mode-select signal activates the scan chain, and the results are shifted out of the IC through the data-out pin to the tester. The scan clock controls the shifting operation. There are various clocking schemes and flip-flop designs, and we illustrate a simple one in Figure 10.11. The serial mode of testing is inefficient for test vector delivery, but it reduces the signal test pins to two. Scan circuit DfT is a brute force technique that succeeded.

Figure 10.12 is an expanded view of a typical scan flip-flop. The mode-select signal sets the 2-MUX so that it directs data from either the combinational logic block outputs or from a previous scan cell output path. A MUX mode-select signal of logic-0 directs the system signal data into a D flip-flop, and a logic-1 asserts the serial scan-in path.

A significant scan circuit property is that the combinational logic block nodes (Q_1 to Q_n) respond continuously to each pattern in the registers driving them. The combinational logic blocks move to a new state with each scan clock pulse. This causes excessive node toggling during the shifting process. It might make no difference, except that this is a source of unnecessary heat generation. The sum of the heat generated by the node toggling, the scan cells themselves, and the scan clock tree is large enough to melt test fixturing if clocked at high scan test rates [52]. This restricted typical scan clocks to the 5–50 MHz range, but thermal reduction techniques have increased the scan clocking to 50–400 MHz. These are described next.

Recent scan designs use techniques to isolate the Q-outputs of the scan registers to inhibit or defeat the high nodal activity in the combinational logic blocks during the scan phase. This can be done with a simple transmission gate inserted between the scan output nodes and the combinational logic block inputs. Saxena et al. found that the clock tree and scan cells themselves generated much more heat than the combinational logic blocks [52]. They redesigned their scan elements and their clock routing to reduce the heat generated by about a factor of two, allowing the scan clock frequency to double for the same heat generated. This thermal limitation has important effects, since virtually all stuck-at and delay fault testing is now done with scan chains.

Ideally, SAF tests should be applied at rates from 500 MHz to 2 GHz to keep pace with modern ICs, but a minor modification of the test allows some timing assessment. The last clock that shifts the scan test vector to its final position launches a test vector into the combinational logic block. If the capture clock driving the output scan registers is pulsed

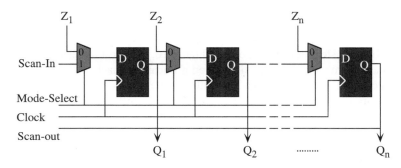

Figure 10.12. Scan flip-flop with MUX to convert between system and scan test mode.

a short interval later, then some measure of timing is achieved. This is sometimes called the AC scan technique [39].

The scan registers are tested first by putting them in the test mode. A test pattern of logic ones and zeros is shifted through the chain, and the tester compares the data-out pin result to the expected good result. When the scan chains are verified, then the combinational logic blocks are tested. This simple test procedure solved the numerically intractable problem of testing sequential elements as they normally exist in an IC. Flip-flops must be put in known logic states before they can drive combinational logic circuitry, and when controlled only from the IC primary inputs, it is a numerically intractable problem.

Delay fault test patterns require scan chains for test vector delivery and evaluation. The two-vector sequence makes the test more complex. The first vector is scanned in, setting the initial conditions in the combinational logic blocks. Since the second vector must immediately drive the nodes under test to allow for a pass–fail timing measurement, the second vector cannot be scanned in. A cleaver trick loads the second vector from the preceding combinational logic block, and then releases it with the mode-select signal into the scan register. The system clock releases the scan chain contents at the combinational logic block output. The delay between the launch and capture clocks sets the timing test limit that the circuit must pass, and this period is set by the system clock. This is called the "launch from capture" technique. When data are taken after the last data shift, the method is referred to as "launch from shift."

Scan circuits can deliver other types of test vectors, such as I_{DDQ} vectors; the ease of test pattern generation and node control make it ideal. Test patterns are generated rapidly only for the individual combinational logic blocks.

The development of scan-based test vectors can now be done in hours or days compared to months or years of functional test development. Short TTM (time to market) cannot tolerate long test development times. The benefits of scan design extend beyond testing. Designers can debug circuits much faster with efficient control and observation of nodes. Failure analysis can more easily narrow the area of the defect causing failure. Scan designs are efficient to further isolate the failure region to gates, transistors, and even interconnections. It is virtually impossible with ICs of any complexity to find an alternative to scan-based testing in order to meet TTM demands and good quality levels.

10.3.2 BIST

BIST (built-in self-test) refers to the process of letting the ICs test themselves [15]. BIST uses DfT circuits that are intended to reduce ATPG and ATE costs by letting the chip test itself. A totally self-testing chip has never been demonstrated, but many applications of partial BIST exist. BIST designs generate test stimuli internally on the IC and then can run them at system clock rates. It is a way to get high-frequency test patterns inside the IC. An early BIST system design is shown in Figure 10.13. A test pattern generator called a pseudorandom number generator (also called a linear feedback shift register—LFSR) [15] drives a circuit logic block with a few million patterns. The output of the logic block feeds an evaluation circuit called a MISR (multiple input shift register). The results of the output patterns are coded on-line into a single register result. The application of built-in self-test to combinational logic circuits is called LBIST.

One design area that systematically uses BIST is the embedded core memory, called MBIST. MBIST is relatively easy since x- and y-decoders can be stepped by a counter, and simple memory patterns can be generated internally bit by bit. The correct result of a

10.3 DESIGN FOR TEST **297**

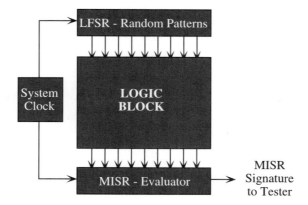

Figure 10.13. Built-in self-test block diagram.

read command is simple to generate and compare with the actual read result. Embedded core memory could not be adequately tested using the IC primary input pins, since access to the memory would require too many indirect control instructions.

Figure 10.14 illustrates a simple BIST structure that generates near-random patterns called pseudorandom patterns. The LFSR takes an ordinary serial shift register and adds feedback paths through exclusive-OR circuitry (XOR). The XOR gate delivers a logic-1 if a single high input is present. The all-zeros state (and several ones and zeros patterns) produces a logic-0 out of the XOR gate.

The LFSR example in Figure 10.14 first clears the register to the all-zeros state, and then a single logic-1 is placed on the input line. That logic-1 appears at the output of the XOR, and it is subsequently clocked into the first D flip-flop. Table 10.3 follows the logic states after each clock. If we pick up on the pattern of the second clock, we see a pseudorandom sequence of ones and zeros at the Q outputs of the three flip-flops. The ninth clock result is identical to the second clock, and the tenth clock is the same as the third clock result, etc. The sequence repeats because the total possible number of states for a 3-bit register is eight. The patterns show seven clock periods before repetition. The missing eighth state is the all-zero state. The circuit will not change state if the all-zero pattern is loaded and then clocked. The total number of logic states before repetition occurs are $2^n - 1$ where n is the number of flip-flop elements.

The test response of the logic block under test changes after each clock. These large

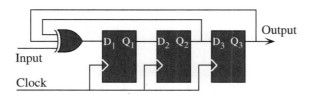

Figure 10.14. Clock-driven simple LFSR (linear feedback shift register) that generates pseudorandom vectors.

298 CHAPTER 10 DEFECT-BASED TESTING

Table 10.3. Logic States of BIST LFSR Circuit

Clock	XOR	Q_1	Q_2	Q_3
0	0	0	0	0
1	1	1	0	0
2	1	0	1	0
3	1	1	0	1
4	1	1	1	0
5	0	1	1	1
6	0	0	1	1
7	1	0	0	1
8	0	1	0	0
9	1	0	1	0
10	1	1	0	1

numbers of responses would be a problem to analyze except the LFSR technique processes the output data by compacting it. In effect, each response is "added" to the previous output, using a MISR (multiple input shift register) (Figure 10.13). The exact value or signature of the MISR bits is not important since each good circuit will have the same values in the bits of the MISR. When all LFSR input patterns are clocked through the circuit under test, then the actual output MISR bit signature is compared to the good result. There are large savings in data storage and analysis.

There are varieties of LBIST construction, particularly in the LFSR construction [15]. LBIST can drive parallel scan circuit chains by using one LFSR, where each Q-output stage drives one of the parallel scan chains. The output results are then scanned into a MISR. In Figure 10.14 this means that each Q-output stage drives a scan chain, and the pattern seen by the chain is the vertical pattern under each Q output in the table. Figure 10.15 shows this now popular BIST style. It is efficient and based on the IBM STUMPS system [5]. The STUMPS design merges scan DfT with BIST DfT.

The at-speed LBIST circuit has certain weaknesses. Although it can apply virtually the

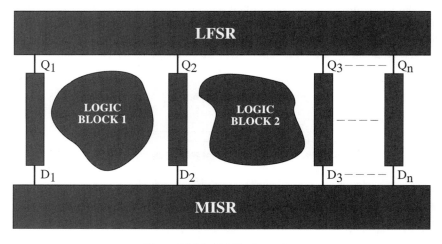

Figure 10.15. BIST schematics.

whole truth table to a circuit, this voltage-based test does not predict high success for bridging defects or certain opens. When driven at a maximum clock rate, a LBIST structure will detect some speed-related defects, but exactly which gate nodes or paths are examined is not known since none are targeted.

Another problem is the number of shift registers forming the LBIST. These registers are often modified from those in a scan circuit, and their number can be a problem. Since the number of possible test vectors is $(2^n - 1)$ where n is the number of storage elements, then registers with more than about 25 elements are of a size for which more vectors can be generated than are economically justified to deliver. A 30-bit register can generate $\approx 2^{30} = 1.07 \times 10^9$ vectors before repeating. This will take valuable test time and illustrates that LBIST patterns have a limit. Die overhead issues weighed against a virtually unknown defect coverage is another challenge to overcome in at-speed LBIST. A counterargument for LBIST in complex, modern ICs is that SAF test rates are relatively slow, expensive to generate, and require large ATE memory.

10.3.3 Special Test Structures for Next Levels of Assembly

Printed wire boards (PWB) and multchip modules (MCM) present special test challenges that can be partially or wholly solved by putting test structures on the IC I/O pad areas. The primary purpose of these test structures is to examine the interconnects at the higher level of circuit integration and, to a lesser extent, the ICs themselves. Electromechanical probers are too large, and data pathways through the module or board functional I/O pins are numerically impossible to achieve 100% testing. One solution designs special I/O DfT test circuits on the chip. We will briefly describe these techniques to complete the test description.

The IEEE organization exerted considerable good influence when volunteers from industry and the universities of the world wrote test method standards to solve these problems. The first one was the IEEE 1149.1 *Test Access Port and Boundary Scan Architecture* (BSCAN) standard, approved in 1990. The IEEE 1149.1 standard states that all nonmemory ICs have an input and output pin board test capability [1]. The board or MCM would have a serial test mode in which test vectors could be scanned in a chain through all I/O pins on the PWB. The principle is similar to the circuit scan used for testing ICs, but the primary purpose of the BSCAN is to test the interconnections between all ICs on the board. BSCAN does this by loading test vectors in the output ports of the ICs, and then capturing the result of those vectors as they travel to the input pin of another IC. BSCAN has the ability to shift data, retrieve data, and, secondarily, to drive input vectors into individual ICs. Virtually all microprocessors, microcontrollers, and digital signal processor chips now use BSCAN. More details can be found in [1, 12, 15].

Another IEEE Standard deals with testing mixed-signal components at the board level, where analog and digital ICs exist along with discrete components. The purpose of the IEEE 1149.4 *Mixed Signal Test Bus* Standard is "To define, document, and promote the use of a standard mixed-signal test bus that can be used at the device, sub-assembly, and system levels to improve the controllability and observability of mixed-signal designs, and to support mixed-signal built-in test structures in order to reduce test development time and costs, and improve test quality" [2]. Special analog boundary modules are inserted at the chip I/O pins. These modules are compatible with the 1149.1 BSCAN test modes, and have an analog mode that allows communication with analog source currents and voltages from a tester. More mixed-signal details can be found at the 1149.4 Web site [2], and in [48].

To deal with system test complexity, the IEEE 1149.5 *Module Test and Maintenance Bus* standard was approved in 1995 to provide an extensible and robust infrastructure for the communication of test and maintenance messages within a system [3]. Its bus protocol is designed for the communication of test, diagnostic, and maintenance information between a system test control module, or master, and up to 250 modules (slaves) within a system. This infrastructure was created to offer enhanced test solutions at the system level in terms of extensibility and error handling capabilities. Additional details can be found at [3] and in [25].

Digital networks are another important area in which testability needs special standardization efforts. The IEEE 1149.6 *Boundary Scan Testing of Advanced Digital Networks* draft proposal is an ongoing standard focussed on special features of advanced digital networks not adequately addressed by existing standards, especially those networks that are AC-coupled, differential, or both [4]. It is intended to work in parallel with the IEEE Std. 1149.1 testing of conventional digital networks and in conjunction with the IEEE Std. 1149.4 testing of conventional analog networks. The IEEE 1149.6 project seeks to complement IEEE 1149.4, specifically targeting parallel testing of advanced digital networks, whereas IEEE 1149.4 focuses on serial testing of more traditional analog networks.

The influence of these standards extends to the system on chip (SoC) test challenges. Here, embedded modules, such as microprocessors, digital signal processors (DSPs), RAMs, ASICs, mixed-signal and even RF subcircuits are incorporated into one IC design. The pin counts of the embedded cores may exceed that of the overall chip, illustrating the rapid explosion in test complexity. The IEEE P1500 *Embedded Core Test Interface* standard proposes common test structure approaches to deliver and retrieve test data from these complex circuits. More details can be found at the P1500 working group web site at [46].

Other test-related standards are IEEE 1450 *Standard Tester Interface Language (STIL) for Digital Test Vector Data,* IEEE 1532 *In-System Configuration,* and IEEE P1581 *Static Component Interconnection.* For additional information refer to [65].

These brief test standard descriptions are given here since they are an integral part of dealing with the complexity of testing in deep-submicron ICs. Our primary purpose is to understand failure mechanisms, but in so doing we will often be dealing with those high-level structures.

10.4 DEFECT-BASED TESTING (DBT)

DBT differs from fault model approaches. A fault model hypothesizes how a circuit will fail, and applies test patterns looking for failures of that hypothesis. Defect-based testing identifies the spectrum of defects that cause IC failure, and then develops test methods that are matched to the electronic properties of that particular defect class. We will briefly review the defect classes and mechanisms studied in the past chapters, and then assess their testability. We discuss each physical defect, analyzing when different failure mechanisms leading to the same defect type may cause different circuit malfunctions. Then we analyze which are the best options to detect these defects within the different test technology solutions.

An example of distinguishing faults from defects at the board level is a pattern of delay faults applied to a board. An IC is found to fail the delay fault test and removed from the board. Failure analysis shows the IC to be perfectly good, but failing due to an open connection from IC pin to board. The removal is costly considering the many, small pins involved. A defect-based approach would apply a test method targeted to open (or bridge) defects that is sensitive to that test. Defect-based testing at the integrated circuit level is the same. We identify the spectrum of defects that cause IC failure, and then develop test methods that are matched to the electronic properties of that particular defect class.

10.4.1 Bridge Defects

Physical defects resulting in bridges may be caused by electromigration, gate oxide shorts due to contamination or oxide wearout, and defect particles or fabrication-imperfection-caused shorts. Defect particle shorts exhibit low resistance when the mask defect causes a missing oxide region between two metal lines that is then filled with conducting material. If the bridging defect particle is a metal sliver between two lines then high ohmic resistances are more probable. Electromigration shorts may not have clean contacts and can cause higher-resistance bridges. Gate oxide shorts can have ohmic or nonlinear behaviors, as shown previously. In Chapter 7, we saw that bridges cause intermediate voltages, elevate the quiescent current, and may impact delay. We also saw that the critical resistance is an important concept when characterizing bridging defects, since it relates the defect resistance to the device properties and its impact on circuit logic behavior. A bridging defect will impact the quiescent current and/or delay depending on its critical resistance and may cause a voltage error. Each bridge-defect-induced behavior requires a different detection test strategy. Figure 10.16 shows microbridge connections for a 110 nm technology.

Only zero-ohm defects between a logic node and power/ground rails will guarantee a SAF without causing intermediate voltages when activated. A zero-ohm defect between two signal nodes will always cause an intermediate voltage when activated (different than V_{DD} or GND) due to the finite resistance and competition of the transistors driving the shorted nodes. Voltage errors will be observed depending on the relative strength of the shortened gates and the electrical properties of surrounding logic. Chapter 7 described experimental results that characterized bridging faults and gate oxide shorts, showing weak voltages and elevated I_{DDQ} for several defect resistances.

Bridge defects are predominantly sensitive to current-based testing. Voltage-based testing requires that the bridge resistance be less than the critical resistance for the nodes af-

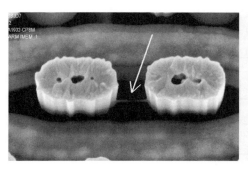
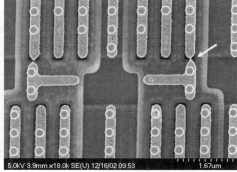

Figure 10.16. High-resistive shorts due to fabrication imperfections (reproduced by permission of Bob Madge, LSI Logic).

fected by the defect, and that a sensitive signal path exist between the SAF node and an output pin. We cannot predict when a voltage-based test will accidentally have a clear sensitive path to a primary output node, and we cannot predict defect resistance. Failure analysis results show that defect bridges have a wide ohmic range. We know that many gate oxide shorts are traditionally prone to escape detection by voltage-based tests [19, 60, 70], therefore, a strong risk of bridge defect escapes exists unless current-based test techniques are used.

The impact of bridging defects on circuit speed was also described. Comparative bridge defect simulations of static voltage testing, speed testing, and I_{DDQ} were done on many combinations of intratransistor bridging defects in a 2 μm technology. The results are repeated here for a 180 nm technology with near-identical results. Simulations corresponding to a bridge defect across the gate to drain of the upper n-channel transistor in the 2NAND are shown in Figure 10.17, and a sample result is shown in Figure 10.18. The resistance was varied over a wide range and the signal responses are shown. The DC voltage transfer curve [Figure 10.18(a)] starts to show a significant change in the drain output voltage when the bridge resistance went below 5 kΩ. At 5 kΩ, the bridge defect drops the high logic voltage by 17%. When the defect resistance was between 1.5 kΩ and 1 kΩ, the circuit failed. We would conclude that the particular bridge and its location in this gate results in a critical resistance of about 1.5 kΩ. The speed test [Figure 10.18(b)] was also sensitive to the bridging defect, showing functionally significant timing degradation at a resistance lower than about 3.5 kΩ. The definition of timing failure becomes smaller as circuits go beyond GHz frequencies, with clock periods less than 1 ns. When a bridge defect causes a pulse that is too narrow, then flip-flop setup and hold timing and pulse width requirements might initiate failure.

The I_{DDQ} test results in Figure 10.18(c) showed sensitive detection of the bridge up to resistances on the order of a megohm (out of the scale). It is the overwhelming choice to test bridging defects. Bridging defects are the dominant Pareto bin in CMOS manufacturing defects. The data indicate that companies that use only voltage-based testing should experience bridging defect escapes. Speed tests showed a reduced timing degradation sensitivity to the bridging defect, failing at a resistance lower than the DC critical resistance.

Bridge defects in sequential circuits have additional properties to those in combinational logic. The presence of bridge defects may impact the memory control loop, as explained in Chapter 7. Chapter 7 also showed that a bridge defect that connects one of the

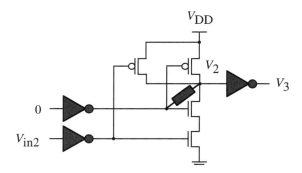

Figure 10.17. 2NAND gate with an input–output short.

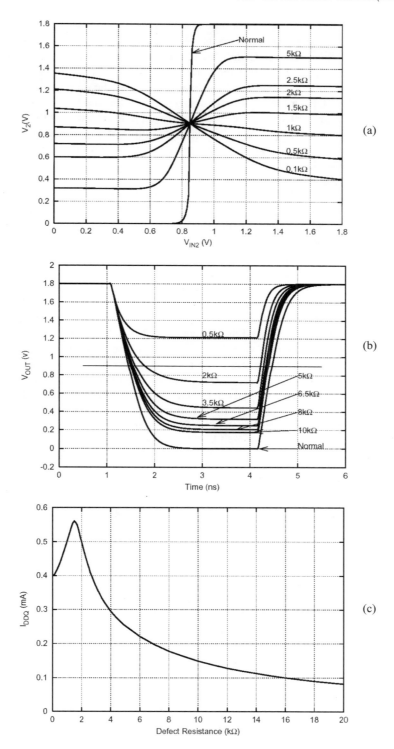

Figure 10.18. Test method comparison of (a) slow voltage (transfer curve), (b) timing voltage, and (c) I_{DDQ}.

nodes of a flip-flop to the output of a combinational logic gate has two responses depending on the size of the resistance. Assume that there are opposite polarity voltages across the bridge defect. If the resistance is high, then the combinational circuit has a weak effect on the flip-flop state, and the correct memory state is held. I_{DDQ} is elevated under this condition when opposite but weak logic states appear across the defect. If the resistance becomes small, then the combinational circuit can pull the flip-flop to a strong voltage error state, since memory transistors are typically designed to be small with little current drive strength. Both signal nodes are at full logic strength, and I_{DDQ} is not elevated. It is a paradox that if the resistance is smaller, then there is no I_{DDQ} elevation, but if the resistance is large, then we do see I_{DDQ} elevation [50].

Sequential circuits are now typically connected as scan circuits, and good testing would use both voltage-based and current-based tests. However, in practice, scan chains are tested by a voltage-based test that injects a simple sequence of logic-1 and logic-0 patterns. I_{DDQ} tests for bridges in the flip-flops are typically not done for economic reasons. Some of these bridge defects might be caught as nontarget detections when I_{DDQ} is applied to combinational logic portions of the circuit.

10.4.2 Opens

Opens have many causes, and may exhibit up to six circuit behavior classes. They may be due to electromigration, metal voiding, defect particles, fabrication imperfections, or missing or imperfect interconnect/vias. Partial opens are similar to bridges, exhibiting a spread of resistance values ranging from the nominal interconnect resistance (the resistance of the line when no defect is present) to virtually infinite resistance. Opens causing a relative small increase of the nominal interconnect line resistance are referred to as *weak opens,* whereas opens providing a significant increase in line resistance (say beyond 1 GΩ) or a total disconnect (Figure 10.19) are referred to as *hard or strong opens* since the current through these defects is negligible. The behavior induced by weak and strong opens and the test methods oriented to each of them is significantly different, so we treat them separately.

Hard Opens. Open defects require voltage, current, and F_{max} testing for best test coverage. Floating gates to single transistors are detected with I_{DDQ} tests, whereas floating logic gates affecting two complementary transistors are only detected by voltage testing

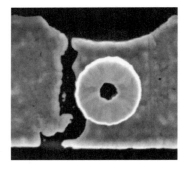

Figure 10.19. Hard open causing a total disconnect (reproduced by permission of Bob Madge, LSI Logic).

(SAF or delay fault) when the gate voltage floats to one of the rails. Logic gate opens whose node floats to an intermediate voltage between the rails are detected by either voltage or current-based tests. Sequential circuits require both forms of testing.

Tests for the CMOS stuck-open (memory) defect require a two-vector sequence to examine each of the parallel transistors in a logic gate and another two-vector sequence to test for the series stuck-open defect. Failure will occur if the drains or source are open, or if the transistor is otherwise prevented from passing drain current. A 100% stuck-open test also provides 100% delay fault coverage, but a 100% delay fault test only provides partial stuck-open-fault coverage. The delay fault test patterns are a subset of the complete stuck-open fault patterns.

In summary, open defects require a combination of voltage, current, and time-based tests. Delay fault testing using scan circuits and I_{DDQ} testing are the most efficient combination. Metal interconnect resistive vias are difficult to test. Metal interconnect defects are also a subset of the parametric delay class, and will be addressed in the next section.

Weak Opens and High-Resistive Vias. The dominant failure mode of resistive vias is a functional speed dependency on temperature. A typical IC with a resistive via may fail a speed test such as F_{max} or a delay fault test at a given clock frequency, but pass at that same clock frequency when the temperature is raised. This glaring failure signature means that the IC runs faster when hot than cold, defying all our experiences with circuits and temperature. Present technology ICs always slow down when temperature is raised because carrier mobility dominates performance, and it declines markedly with temperature. There may also be power supply amplitude dependencies (V_{DD}).

Why would a defective via allow better speed performance when heated? Figure 10.20(a) sketches a defective via at room temperature, showing metal voids. Figure 10.20(b) models how the voids shrink when temperature rises. Notice that the voided structure is a via whose volume is constrained by dielectric and interconnect metal. If the voided metal was unconstrained, then the void would enlarge as temperature increases. However, the passivation structures around the via outer surface force the metal at the void surface to grow inward with temperature, reducing the via resistance of the void at hotter temperatures. The result is a lower resistance via that imparts less delay to the signal through it. The signal has sufficient margin at this lower-resistance to pass a speed test that it failed at the cooler temperature [14]. One microprocessor test approach does speed testing at two temperatures for just this reason [42].

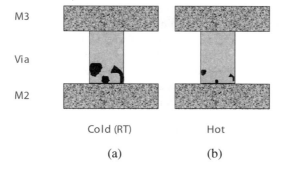

Figure 10.20. Sketches of (a) normal via at room temperature, and (b) defective via at higher temperature. Voids are shown as dark regions.

The typical intrinsic resistance of a via/contact may range from 2–20 Ω depending on material, liner dimensions and resistivity, annealing steps, aspect ratio, minimum diameters, and process variations. Resistive vias may range from a few ohms to MΩ, as shown in Chapter 9 [50]. SPICE simulations of via resistances show that logic gate timing failures occurred only when via resistance exceeded 150 kΩ to over 1 MΩ in two studies [8, 42]. We basically cannot detect most resistive vias at a single V_{DD} and temperature setting using a speed test (F_{max} or delay fault). Defect-based studies indicate that speed tests must be run with a multiparameter test method in order to detect these subtle defects. The relatively slower IC speed capability at lower temperature is the resistive defect indicator, not Boolean upset. Resistive vias are reliability risks if not detected and rejected at test [8, 14, 42, 69].

Weak opens impact circuit delay only as they increase the RC constant of the signal path in which the defect is located. Given the nature of the defect, it can only be detected using test strategies that verify timing. F_{max} and delay fault tests are the best candidates to detect these defects, although the percentage of delay must be significant for the defect to be detected. Simulation experiments on a 120 nm technology for an inverter driving an RC interconnect line showed that resistance must experience one order of magnitude increase beyond the critical value to increase delay by less than 4% [72]. Delay test methods practiced today have a small chance of detecting such weak opens due to parameter variations along with timing guard bands related to tester timing accuracy.

One might argue that if these weak opens have such a small impact on circuit timing and cannot be detected using delay fault testing methods, then why should we care about these defects. The problem of weak opens is that they are high-reliability risks. A weak open due to a missing portion of metal interconnect causes all the current through the line to go through a reduced area of via metal and the thin high-resistive Ti material. Current density increases, and as a result, increased local Joule heating appears and a weak open at a contact, metal flat line, or via is susceptible to electromigration. A better technique for our present technology is to burn-in the IC and displace the open resistance value from a weak to a complete open. Once a hard open is induced, detection probability increases significantly.

There are reports supporting these failure descriptions for vias. Campbell et al. reported failure analysis on an older technology for a part that failed cold-temperature testing in the field [14]. The failure was a drain contact in a memory cell. The contact had voids and eventually electromigrated to a complete open with continued testing in the failure analysis lab. Figure 10.21 shows a timing pass–fail histogram taken at test before the part was shipped. The test limit at 350 ns gave assurance that the part would function at the system level if its delay was less than 350 ns. Interestingly, the part that failed in the field was the outlier at 340 ns. A cardinal rule for a high-reliability product is that you do not ship parts whose test data indicate that they are outliers. This situation indicated in Figure 10.21 was a mistake, and root cause corrective action was taken. The difference between the normal population and the outliers in Figure 10.21 is that the outliers have a defect causing them to be slower. In this case, the defect was a resistive contact that electromigrated to failure. The risk of resistive vias in modern ICs is believed to be electromigration or stress voiding [69].

Needham et al. reported temperature-sensitive interconnect failures on three Pentium microprocessors [42]. All ICs failed at cold temperature and passed at hot. Failure analysis found a resistive via, a missing via, and a cracked polysilicon interconnect line. The study emphasized the inadequacy of best traditional test methods to detect these parametric defects that cause field failures. Resistive vias are a major reliability concern.

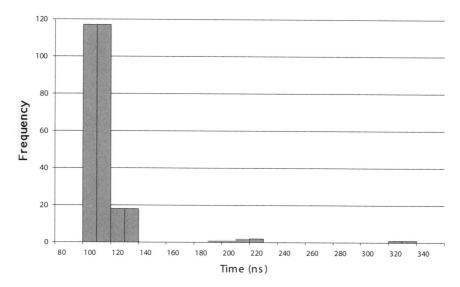

Figure 10.21. Pass–fail timing histogram of a sample of parts [14].

Resistive via detection strategies are clear, but there are complexities. ICs must be speed tested at two temperatures. The complexities lie in the two-temperature measurement procedure and in what will be compared in the speed measurements. Measurement temperatures should be as separated as much as economically feasible. The wider the separation, the more sensitive the speed tests. Wafer hot chucks can easily test from 125°C–140°C which is a good upper temperature. The cold-temperature tests can be done around room temperature, and the logical place to do this is at the package level. If the temperature is taken below the dew point, then water condenses, making the test more expensive. Testing at temperatures below 0°C is routinely done for military and other high-reliability parts, but requires an enclosure for the IC through which dry, cold nitrogen is blown. These more expensive, very cold temperature measurements are fortunately not necessary for resistive via testing.

Both temperature tests will not be done at the same time as the wafer probe or package tests. The thermal delay time to reach another temperature and the I/O pin thermal continuity issues with the tester suggest that one of the temperatures be done at wafer probe and one at package level. The tests at two temperatures may also be done at pre- and post-burn-in. That introduces the challenge of how to keep track of the die being tested. The second temperature test must compare its result with the other temperature measurement on the same die, and match the dies after they are cut and packaged from the wafer. Intel reported a technology that burns a die identification code into each die at the wafer probe test [49]. This identifies the die throughout its product life. Other companies use a similar strategy that has many uses in test and failure analysis. Two-temperature testing requires this or any equivalent technique that allows die identification throughout the test process.

10.4.3 DBT Summary

Table 10.4 summarizes many of the defect mechanisms that can appear in CMOS ICs. The information is taken from the past section and previous chapters. Defects are grouped into

Table 10.4. Physical Defects, with Failure Mechanism Source and Induced Circuit Malfunction

Source	Failure mechanism	Physical defect	Circuit malfunction							
			stuck-at	Intermediate Voltage	Elevated I_{DDQ}	Floating node	Memory effect	Delay impact	Crosstalk impact	V_{DD}/GND bounce
Interconnect reliability	Electromigration	Metal short	rarely	always	always	no	FTC	probable	no	no
		Broken metal	rarely	probable	very likely	MP	probable	if small cracks	no	no
	Metal stress	Metal void	rarely	probable	very likely	MP	probable	no	no	no
Oxide reliability	Oxide wearout	GOS	rarely	MP	always	no	no	very likely	no	no
Device reliability	Hot-carrier injection	GOS	rarely	MP	always	no	no	very likely	no	no
		Vt shift	no	no	no	no	no	MP	no	no
Process Fabrication	Defect particle/ fabrication imperfection	Short	probable	always	always	no	FTC	very likely	no	no
		Open	rarely	probable	very likely	MP	probable	no	no	no
	Via/interconnect problem	Open	rarely	probable	very likely	MP	probable	if small cracks	no	no
		High R line/via	no	no	no	no	no	yes	no	no
	Process control	GOS	rarely	MP	always	no	no	very likely	no	no
	Process variation	V_t shift	no	no	no	no	no	MP	no	no
		IDL variation	no	no	no	no	no	MP	very likely	rarely
		Interconnect ρ shift	no	no	no	no	no	MP	rarely	very likely
		Metal width shift	no	no	no	no	no	MP	very likely	very likely
		L_{eff} shift	no	no	no	no	no	MP	rarely	rarely
		W_{eff} shift	no	no	no	no	no	MP	rarely	rarely

Note: MP: most probable; FTC: few topological cases.

two large categories relating to reliability problems (interconnect, oxide, or device), or manufacturing imperfections/contamination (particle defects, via problems, process control, or parameter variation).

In each case, we identify the mechanism responsible for the defect, and the impact on the circuit behavior. In some cases the same defect type can have different failure mechanism roots. An open due to electromigration or a particle defect can have a different impact on the circuit behavior. A small electromigration open may lightly impact timing due to tunneling current, whereas a particle defect open may not cause that behavior. There are three basic types of defects: bridging defects (black background and white type in Table 10.4) as those inducing an undesired connection between two or more nodes; open defects (heavily shaded cells and white type)—those causing an interconnect to be broken or highly resistive; and parametric defects (lightly shaded background and black type)—those that cause a variation of an electric parameter either for devices or interconnects.

A defect-based analysis and testing plan depends upon the particular product and the company's test capability. However, once tests for gross defects have been applied, a general logic test sequence to optimize early detection of the most probable defects described to this point might be to do

- Scan chain testing with a sequence of logic ones and zeros.
- Wafer-level I_{DDQ} testing to detect bridge and certain open circuit defect classes.
- Wafer level delay fault testing for gross timing failures, and to quantify detection of two of the six open classes missed by the I_{DDQ} test: the transistor pair on/off defect and the stuck-open defect.
- F_{max} testing is difficult at the wafer level, but can be done for some ICs, especially if V_{DD} is reduced to a minimum value.
- At-speed package testing at cool temperature. Resistive vias and contacts have high probability of detection by comparing speed capability, using either F_{max} or delay fault tests.

The exact sequence of the tests and how they are economically implemented are company- and part-specific. However, the thought process of electrically evaluating the expected CMOS IC defects, and then applying a specific test to detect that targeted defect remains the core of defect-based test practice.

Faulty behaviors due to parameter variation are listed at the end of Table 10.4. These defects mainly impact delay, and/or aggravate noise and crosstalk problems. Parameter-variation-induced defects appear mainly in deep-submicron and nanometer technologies as technology nodes get to 90 nm and below. These effects are discussed in detail in the next section.

10.5 TESTING NANOMETER ICs

We will relate the nanometer effects of device and interconnect scaling to the IC parameters and higher speeds that impact the test and diagnosis functions. We then discuss the difficulties of the several test methods in dealing with defect detection in this environment. Most defect properties typically do not change as ICs are scaled, but the efficiency of the available test methods does. After analyzing these limitations, we will describe a number of test methods that were proposed in the few past years to deal with nanometer effects on testing.

The term deep-submicron technology refers to CMOS ICs whose transistor channel lengths are below about 0.5 μm, whereas nanometer ICs refer to dimensions approximately below 180 nm, at which noticeable changes appear in the physical characteristics of the transistors and in the chips themselves. Technology scaling follows Moore's law, in which the number of transistors per die doubles about every two years [26]. The goal of scaling for each technology node is to reduce gate delay by 30%, double transistor density, and reduce energy per transition by 30% to 60% [13, 18]. The industry has gone through generations of planned decrease in critical dimensions in the deep-submicron era from 350 nm, 250 nm, 180 nm, 130 nm, 90 nm, to 60 nm, and below. We call these technology nodes; each node usually demands new fabrication and test equipment and design tool capabilities. The nodes are defined by the minimum interconnect pitch of a DRAM divided by two [26]. Pitch is defined as the minimum metal width plus the minimum spacing between metal lines. DRAMS are the standard, since that technology has led the microprocessor and ASIC technology by about 2–3 years. The node geometric decrease derives from multiplying one generation by 0.7 to get the next size reduction (180 nm × 0.7 ≈ 130 nm, etc.). The 0.7 scale reduction on a side is significant since die area will halve. Each new technology node can place the same amount of transistors in half of the area. However, the true die size is actually growing by 14%–25% per technology node as more logic per chip is included [13].

At the smaller geometries, memory ICs have grown to billions of transistors, and ASICs and microprocessors to tens and hundreds of millions. The via counts are now in the billions, so that one effect on testing is simply the raw size of the ICs in the nanometer world. A staggering number of circuit structures must be examined in the test process. Test development must include a plan to generate test patterns quickly and cheaply. Suddenly, virtually all companies embrace scan DfT, whereas previously, individual companies might have viewed DfT as a cult practice. Historically, loud complaints have been that too much area was alloted for DfT. That extra area could be equilibrated to construction of another fabrication plant in some high-volume companies, just to make up for the lost space on the wafer. DfT also "slows the part down." However, the economics of test development and practice and diagnostics are overwhelming if primary pins are the sole entry and exit points for test patterns. Although scan DfT reduces the complexity of the task, unsolved challenges still remain in the testing and diagnosis of deep-submicron ICs.

A second challenge is the subnanosecond timing of these circuits. Testers must test leading edge performance ICs using the previous generation chip. ATE must place timing edges with picosecond resolution, overcoming the challenge of timing skew on several hundred signal pins. The applied test vector files, the measured response, and the expected response requires large data storage and retrieval capacity. The old standard of less than a second for an IC test has given way to minutes. Gigabit memories may take up to 30 minutes of test time, and a single wafer test may take an entire 8-hour work shift.

Another challenge is managing the test process for ICs with up to 1,000 I/O pins. Planarity must be exceedingly good for test probes to make assured electrical contact at the wafer probe. Open circuits at wafer- and package-level testing are a growing menace. The packages are typically flip-chips that put arrays of solder-interconnection balls on the backside of the chip.

Testers now deal with deep-submicron ICs generating from 20–200 W of power. Wafer probing and package sockets have a thermal differential interface to the chip on the order of 100°C, and the tester power current delivery can be 100 A or more. ATE companies are under pressure to create equipment to handle these environments and produce a tester product that is reliable, easy to operate, "affordable," and rapidly developed to keep pace with the technology nodes.

Another deep-submicron feature is the shortened time between technology nodes and the intense time-to-market pressure. Technology nodes have been achieved at a pace of about one every two years [26]. In this interval, the characterization and test development of the part must not lag behind the fabrication. The test engineering method must rapidly get it right the first time, or face large error costs. Defect-based testing practice forces methodical planning and best-test strategies.

Products have a wide range of cost, reliability, and performance needs. It is said that companies often pick one of these three as a priority and then try to manage the other two. Our discussion will deal mainly with the high-performance ICs, but we are not forgetting that the majority of products in the world created using nanometer technologies do not deal with GHz speeds or 200 Watt power supplies. Many high-performance ICs are high-volume products and this aggravates test problems to the extreme.

10.5.1 Nanometer Effects on Testing

Chapters 7 and 8 looked at bridge and open defect properties, mostly ignoring the parametric variance effects of modern circuits. This provided a clean look at how defects behave when undisturbed by the effects caused by the inability of manufacturing to achieve tight tolerances of key parameters in the deep-submicron era. It gave us a baseline of knowledge to assess the changes. Defects in deep-submicron and nanometer ICs do not lose their class distinctions. Bridges, opens, and parametric delay failures exist with no new categories; however, a shift in the defect spectrum has occurred. Bridges are still dominant in most technologies, but open circuits and parametric delay failures are growing in number, particularly as they relate to vias, contacts, and interconnect line noise. This means more possible defects for which we cannot guarantee detection. We know that parametric delay failure sites will escape testing. Bridge defect escapes will grow because we are losing sensitivity in the current-based tests. The electronics of defect responses have changed little, so we will examine the effects that deep-submicron technologies have on the test method.

Spread of Parameters. Deep-submicron structures do not affect testing and diagnosis just because they are necessarily small. They primarily impact testing because the manufacturing parameters are not tightly controlled. Wide variances exist in L_{eff}, W_{eff}, V_t, I_{off}, and I_{Dsat}, as shown in Chapter 9, and these variances give a wide distribution to the performance of each IC. As a first order, test limits must be set at the worst-case values for the intrinsic part performances. The background variance and amplitude leakage that increase with technology scaling limit the effectiveness of I_{DDQ} testing as introduced earlier. New, sophisticated I_{DDQ} and low-V_{DD}-based testing techniques are in use to extend these techniques to nanometer technologies.

Propagation delay and I_{DDQ} test limits must deal with the wide spread in values from intrinsic ICs described in Chapter 9. Figure 10.22 shows two normal plots with arbitrary units on the x-axis. The x-axis may represent propagation delay T_{pd}, memory access time, F_{max}, I_{DDQ}, or some other test parameter for intrinsic material. The narrow curve has a mean of 50 and a standard deviation of 4 units. This curve might represent older, long-channel technologies, for which parameters were more tightly controlled. The wider curve symbolizes the nanometer technologies. The mean in the wider curve is still 50, but the standard deviation is 10. If $\pm 3\sigma$ test limits are followed, then the nanometer ICs have a greater tolerance for pass–fail. Test limits must accommodate the weakest ICs in the distribution. This means that defective ICs can be buried at one end of the spectrum, but still not be detected as a defect. The better-performing ICs in modern technologies need a test limit adjusted to their level and not to the weakest level in the population.

The variance gets larger as transistor channel lengths and widths get smaller (Chapter

312 CHAPTER 10 DEFECT-BASED TESTING

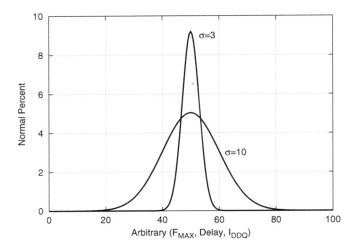

Figure 10.22. Effect of parameter variance on test limits for normal distributions.

9) . Optical lithography has lost its ability to maintain accurate dimensions for L_{eff} and W_{eff}. V_t is affected by changes in L_{eff}, and W_{eff}, and also by ion implant fluctuations [11]. These parameters and I_{Dsat} vary widely from die to die and even among transistors within the same die. The wide range of clock speeds available from a company's technology arise from these statistical variations at the transistor level, not from design intent. These variations weaken testing, and test engineers and failure analysts are left to deal with it. The major test methods affected are I_{DDQ}, delay fault testing, F_{max}, and other at-speed testing methods that use limits characterized from the ICs.

Background Current Increase. MOSFET miniaturization causes high electric fields, and the maximum supply voltage is limited by gate oxide wearout and power constraints. Voltage scaling reduces transistor gate drive ($V_{DD} - V_t$), which is partially compensated for by lowering V_t. The threshold voltage is scaled about 30% per technology node to compensate for V_{DD} reduction and to maintain a sufficiently large V_{DD}/V_t ratio. The final result of the scaling process is an I_{off} increase of about 5× per technology node [18], since the off-state current depends exponentially on V_t. Figure 10.23 shows the off-state current versus temperature, and its evolution per technology node as projected for deep-submicron ICs [13].

Leakage current can significantly contribute to the total power dissipated. Transistor off-state leakage is a primary design parameter for large deep-submicron ICs whose quiescent currents can reach several Amperes. Dynamic power (the power related to CMOS through current and charge and discharge of internal nodes) is the other major power component. The leakage power contribution is shortly expected to increase to about 50% of the total circuit power at high temperatures, as shown in Figure 10.24 [13]. The increase of leakage current variance and amplitude impacts the effectiveness of quiescent current testing since the intrinsic background may mask the current contribution from defects.

Radiation-Induced Soft Errors. Soft errors (SEs) are caused by alpha particles emitted by packaging materials, and high-energy neutrons resulting from cosmic rays

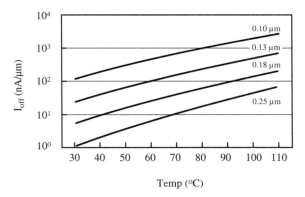

Figure 10.23. Estimated subthreshold leakage current of deep-submicron transistors [13].

colliding with particles in the atmosphere. These energetic particles traverse through the silicon bulk, creating minority carrier charges that can be collected by source/drain diffusion nodes and alter the logic state of the node. The existence of this radiation has been known for more than 50 years, and the capacity for this radiation to create what is called *transient faults* has been studied since the 1980s. Traditionally, memories have been victims of soft errors because smaller transistors are used in this circuitry, which is the densest part of the system. Additionally, when a soft error flips a memory bit (referred to as a single event upset, or SEU), then it provokes a state change that may impact the system-level behavior. This problem is handled in memories by adopting special error-detecting codes that allow SEU detection and posterior correction, or by special hardened packages in space applications (where radiation levels can be two orders of magnitude higher [24]).

As technology scales, the total charge required to toggle a node from particle-induced charge injection is the node charge $Q = C \times V$, where both the node capacitance and voltage decrease at each technology node [55]. Recent work predicted that *soft error rates*

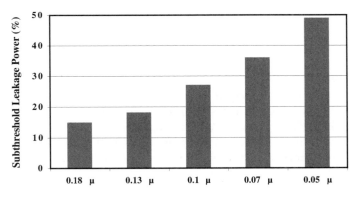

Figure 10.24. Circuit-level subthreshold leakage power contribution as percentage of total power [13].

(SER) could increase nine orders of magnitude from 1992 to 2011, and that the impact of SER on combinational logic will be comparable to that of (unprotected) memory [55]. The impact of radiation particles on combinational logic depends on the logic type (static or dynamic) and the clock speed [54]. The main reasons for this impact on combinational logic in nanometer technologies comes from the smaller critical charge required to flip a node. In addition, the smaller and faster transistors can help propagate a SE-induced glitch to a memory element and cause a permanent error. Moreover, higher operation speed of processor pipelines allows latches to cycle more frequently, with a higher probability of latching one of these SE-induced glitches.

There are no definite solutions to ensure combinational logic SE reliability, since the time or logic redundancy techniques used in memory cannot be adopted directly in random logic blocks [55]. This problem remains yet another challenge for reliability of nanometer technology ICs.

Impact of Nanometer Effects on Traditional Test Methods. The three dominant nanometer technology effects presented in the previous section pose a challenge to IC testing and verification. In the next subsection, we analyze the impact of these effects on the test techniques presented earlier in this chapter, and we present various test methodologies that have appeared in the past few years to specifically deal with these nanometer technology problems.

10.5.2 Voltage-Based Testing

Stuck-at-Fault Testing (SAF). The stuck-at-fault test is unique in that its actual measurement is reasonably unaffected by all the parametric variance of the deep-submicron technologies. Its only parametric relation to scaling is in the analog voltage setting of digital test parameters V_{OH} and V_{OL}. SAF testing is a straightforward, low-frequency, Boolean test that can be done efficiently as long as scan circuits deliver the vectors. However, SAF testing uses large amounts of tester memory in deep-submicron ICs to store the test vectors and their correct responses. The relatively slow scan delivery of test vectors in the 50–400 MHz range is a deficiency. Comparative test method studies show that SAF testing has the weakest defect detection efficiency of the major test methods (Figure 10.10) [39, 44]. In summary, we find the SAF still in common use because test patterns can be generated for scan-designed ICs. ATE has over 30 years experience in testing with the SAF. The reasons for continued dominance of SAF testing may have to do with habit, convenient test tool availability, inexperience with delay fault testing, and lack of DBT knowledge.

Delay Testing (DF). Transition delay fault testing has test limit challenges caused by parameter variations on delay, since DF testing must address problems inherent in measuring delay for a large variety of paths. Intrinsic propagation delay varies considerably for each path, and delay fault testing is not close to a capability that adjusts for a different propagation delay test limit in each path under investigation. When a single delay test limit is set, such as the operating clock period, it must then guardband against tester noise as well as die and intrinsic wafer variations. The delay fault is then limited to gross timing failure detection and not to the finer-resolution delay anomalies. To cope with these problems, several authors are investigating the adoption of statistical models for path selection, trying to maximize the probability of covering critical paths or detecting defects causing delay disturbances. Statistical methods for timing analysis have been around since the mid-1980s [62] and seem to be inevitable in the near future to predict delays in ICs affected by the statistical process variations and noise sources.

Delay fault testing should replace SAF testing, since it targets more defects, and the SAF test set is a subset of the transition delay fault test set. The availability of delay fault test generator tools is encouraging.

F_{max} Testing in Deep-Submicron ICs. F_{max} testing remains an essential IC test method now and will be in the near future. There is no replacement test that can bin the ICs for speed prior to customer delivery. Its weaknesses include no quantifying property to determine how good the F_{max} test is, and developing the large number of test vectors to reflect activity in critical paths. This is not an automated process. The F_{max} test is expensive in tester time, since the tester must do an iterative search for the minimum functional clock period.

Critical Path Timing Test. The critical paths of an IC are identified, and each can be evaluated for propagation delay. These path numbers may be on the order of 1000–2000, and their identification can be difficult. This test is quite useful, but has its own problems with parameter variation. The advanced technology ICs are reported to show die-to-die variation on the order of the critical paths due to parameter variations [45].

10.5.3 Current-Based Testing

I_{DDQ} testing has made significant progress in working around the background leakage variance and amplitude problems. At the end of the 1990s, alternative methods to single-threshold I_{DDQ} testing were investigated, and several techniques are available today that implement current-based test techniques in a production environment. We introduce these pioneer techniques and focus on those presently being used in production testing.

Single-Threshold I_{DDQ}. This test technique suffers from die parameter variation, as does delay testing, with the added disadvantage of increased intrinsic background leakage amplitude. This would not be a real problem in the absence of parameter variations, since the exact value of predicted intrinsic I_{DDQ} could be removed from the measurement. Also, the ±1% accuracy of an ATE current measurement is often not accurate enough for Ampere-level currents. It is the combination of background current amplitude increase and parameter variation that makes single-threshold I_{DDQ} unaffordable in nanometer ICs.

Current Signatures. This technique relies on a concept that does not compare measured I_{DDQ} values to a single pass/fail threshold [21, 22]. The method measures the quiescent current values for the whole test vector set (a set that contains a number of test vectors that completely exercise the circuit) and then generates a *current signature* by rank-ordering all the I_{DDQ} measurements from the smallest to the largest value. The method looks for jumps in the current signature. A smooth plot of the current value indicates a fault-free circuit, whereas significant jumps or discontinuities in the current signatures indicate faulty behavior.

This technique introduced a new concept in the adoption of quiescent current testing since a given circuit is not rejected at the first failure indication during testing, but only after all tester data are collected. The main limitation of this posttest data processing technique is related to its sensitivity to parameter variations since it requires setting a pass/fail threshold not for the absolute I_{DDQ} value, but for the magnitude of the jump at the signature. This threshold should be valid over a range of process variations.

Delta-I_{DDQ}. Thibeault proposed a similar conceptual method called delta-I_{DDQ}, in which the test observable is not the absolute I_{DDQ} value, but the differences in I_{DDQ} among successive test vectors [61]. This difference is treated probabilistically to determine if a given circuit is defective or not. Delta-I_{DDQ} was applied to commercial memory and processor circuits with a mean leakage current of 2 mA [41]. These circuits were not appropriate for single-threshold I_{DDQ} since the single-threshold I_{DDQ} 3 σ limit was 22 mA, posing a significant I_{DDQ} yield loss. The study showed that for a reduced population of samples sent to burn-in, more than 50% of the units that failed delta-I_{DDQ} but passed functional testing and single-threshold I_{DDQ} became functional failures after the experiment. Delta-I_{DDQ} has a parameter variation limitation similar to current signatures. An absolute threshold for the delta value must be set, and this can suffer from variation in the quiescent current from die to die or lot to lot.

Current Ratios. A third test technique called *current ratios* was proposed by Maxwell et. al., and uses a concept similar to current signatures with the added features of tolerating parameter variations [40]. This is done by setting a specific quiescent current limit for each die for any vector of the test set. This limit is computed individually for each part once the first minimal quiescent current vector measurement is taken. Therefore, the first I_{DDQ} value (obtained from the expected vector giving the smallest intrinsic I_{DDQ} value) establishes a range of quiescent current values that are acceptable for that part.

This technique is based on the observation that the slope of the rank-ordered quiescent current signatures for dies having significant different absolute I_{DDQ} values are similar, as shown in Figure 10.25 for two dies. The die leakages varies by about 5×, but the slopes are similar. The test limit for a given die is determined from the ratio of the maximum to min-

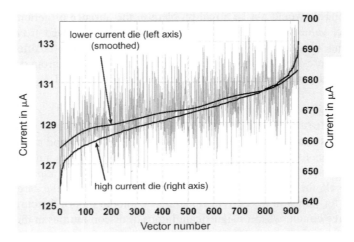

Figure 10.25. Current signatures for two dies having significantly different absolute I_{DDQ} values [40].

imum I_{DDQ} value and the slope of the rank-ordered currents. This was observed to remain reasonably constant for all the devices no matter the mean of the I_{DDQ} measurements for each die. This ratio is determined from a small population of devices having as wide a spread of current as possible, and is done through an iterative process in which the maximum current is plotted versus the minimum current for all dies. This plot is a straight line and provides a slope and an intercept value. Outliers are iteratively removed from the regression line, and the min to max ratio is computed. Once the ratio is established, the outlier regions are set to account for uncertainties in the measurement process. The minimum current is found by characterizing the response of many dies, and identifying the vector that typically gives a minimal reading for each die. The maximum current of a die under test (DUT) is computed from the measured minimum I_{DDQ} vector using the previously computed slope values.

This technique overcomes the two main limitations of other quiescent current techniques since it tolerates both high background currents and parameter variations. Another benefit of this technique is its suitability for production testing environments since the test limits are adjusted for each part once the first measurement is taken.

Nearest Neighbor Statistics. The previous techniques achieved I_{DDQ} variance reduction by considering different die patterns. Another test method reduces the I_{DDQ} variance using die nearest neighbor statistics to identify die outliers [16, 56]. This technique can be applied not only to I_{DDQ} data, but to any test data that changes continuously over the wafer or lot.

This method is based on the identification of residuals whose variance is larger than the I_{DDQ} variance. The *residual* is the difference between the actual measured I_{DDQ} value and an *estimate* of this value. For the results presented in [16] I_{DDQ} was an average of the quiescent current measured for a number of vectors on the tested die, whereas the estimate was the median of the I_{DDQ} values computed from at least eight die closest to the tested die. The median was used instead of the mean since it reduces the effect of outliers. Daasch et al. presented production data on six wafers, showing the merits of this technique, that bins the dies according to the leakage measured from neighbors [16]. Some dies that were classified as faulty from a single-threshold I_{DDQ} technique were classified as good since they belong to a high-I_{DDQ} neighborhood, whereas other dies with an I_{DDQ} value below the single-threshold limit were considered faulty as their leakage was abnormally high when compared to dies in their proximity. Nearest neighbor test assumes that defects and fabrication variation cluster on the wafer, so that dies in close proximity on the wafer will have similar average chip values of L_{eff}, V_t, I_{DDQ}, and F_{max}. The measurements of F_{max} and I_{DDQ} are referenced to the nearest eight neighboring chips to establish a level of performance consistent with nearest chips. An expression used is to distinguish "bad dies in good neighborhoods, or good dies in bad neighborhoods." More sophisticated techniques also correlate parameters in the rows and columns. The nearest neighbor test method was demonstrated using I_{DDQ}, $V_{DD_{min}}$, and F_{max}, but it has general applicability to other variables. It is a valuable tool in parametric failure detection, especially as it evaluates IC data at varying V_{DD} and temperature. Nearest neighbor testing is a production test method at LSI Logic Corp.

The method proposed in [16] was generalized in [17] in terms of neighborhood selection. The die selection used in [16] to compute the estimate was based on data from the eight neighboring dies adjacent to the die under test. This method of selecting the die by only considering position works well when the data pattern varies smoothly. In some cases, data can follow systematic stepper patterns, and in general, test data may show a combination of both smooth and step patterns. This variation, although systematic within a wafer, changes from lot to lot, making variation difficult to predict. To cope with this effect Daasch et al. presented a method to select neighborhood dies from the data itself, instead of considering physical location only [17]. This technique measured data from a 0.18 μm technology using 75 wafers and 25,000 dies. The results demonstrated significant I_{DDQ} variance reduction, and outlier detection was improved. The possibilities of applying this method to production testing were demonstrated on a single lot by considering a single I_{DDQ} measurement per die.

In summary, the nearest neighborhood, posttest statistical analysis method implemented by Daasch and Madge is a powerful demonstration of adjusting I_{DDQ} limits to the quality environment that each die sees in its immediate vicinity [17].

10.5.4 Delay Testing Using Statisitical Timing Analysis

Parameter variations and noise-induced fluctuations aggravated in nanometer technologies significantly impact the variation in the delay of signal paths from die to die and within dies. There can be some deterministic components in the parameters influencing delay, such as topological dependencies of device and interconnect processing or lithography correlation. However, their complexity in large designs, together with the impact of environmental effects (such as supply voltage fluctuations, temperature, etc.), and the existence of real random variables (random doping fluctuation) requires us to treat these variations as random. This creates the need for statistical timing models whose gate delays are modeled as random variables.

These approaches have been applied to timing analysis and delay testing to capture random effects and their sensitivity to internal signals subjected to a given type of defect or variation source. There are several approaches used to incorporate statistical methods in delay testing, Liou et al. showed how these methods are effective in selecting the most probable critical paths [34]. The problem of determining the delay pass/fail limit on the tester is still unsolved.

10.5.5 Low-V_{DD} Testing

Chapter 4 discussed the strong influence of the supply voltage on the gate delay for an inverter. The relationship applies to any CMOS gate under test and, therefore, to the overall IC speed. The need for alternative and/or complementary parametric test methods to I_{DDQ} in submicron ICs motivated the investigation of the test effectiveness in lowering the IC supply voltage to expose the presence of weak defects by correlating V_{DD} to timing [23]. The latest concept lowers the supply voltage at a given clock frequency, and measures the minimum value of the supply voltage at which the circuit will still function. This technique is referred to as V_{DD}Min [64]. The reduction of the IC supply voltage decreases the drive strength of the transistor gates, making them more sensitive to defects impacting timing, such as resistive shorts or weak opens.

The V_{DD}Min test increases the critical resistance of a bridge defect, and increases the test efficiency of voltage-based tests. Example 10.5 analyzes this observation for one circuit example. Tseng et al. showed that the V_{DD}Min value for weak chips was higher than for the fault-free ones, and that this technique might detect bridging shorts, gate oxide shorts, threshold voltage shifts, and tunneling opens [64]. An example follows that illustrates the change in R_{crit} as V_{DD} is lowered.

EXAMPLE 10.5

Calculate R_{crit} in the bridging defect example (Figure 10.26) when the circuit is tested at $V_{DD} = 1.5$ V and then at 1.0 V. $K_n = 100$ µA/V², $W/L = 2$, $V_{tn} = 0.4$ V, and $V_{TL} = 0.5\ V_{DD}$.

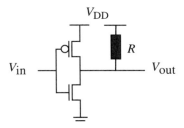

Figure 10.26.

(a) $V_{DD} = 1.5$ V and $V_{TL} = V_0 = 0.75$ V, so the *n*-channel transistor is in nonsaturation, and the *p*-channel transistor is off. Therefore,

$$I_{Dn} = (100\ \mu A)(2)[2(1.5 - 0.4)0.75 - 0.75^2] = 217.5\ \mu A$$

$$R_{crit} = \frac{(1.5 - 0.75)V}{217.5\ \mu A} = 3.45\ k\Omega$$

(b) $V_{DD} = 1.0$ V and $V_{TL} = V_0 = 0.5$ V, so the *n*-channel transistor is in nonsaturation, and the *p*-channel transistor is off. Therefore,

$$I_{Dn} = (100\ \mu A)(2)[2(1.0 - 0.4)0.5 - 0.5^2] = 70\ \mu A$$

$$R_{crit} = \frac{(1.0 - 0.5)V}{70\ \mu A} = 7.14\ k\Omega$$

The increase in critical resistance at the lower V_{DD} value approximately doubled, and that is good for voltage-based testing. However, detection of bridge defects by the voltage-based test is still not a target test. This detection relies on the chance that a sensitized path exists, and that the defect resistance lies below R_{crit}. Nevertheless, increases in test efficiencies for advanced-technology ICs are a trend reversal. ∎

The drawbacks of this technique for production testing relate to the increase in test time, and the cost to do a search test for the minimum voltage at which the die still passes the test [32]. To overcome the search-time penalty of the V_{DD}Min test, a three-step process was proposed based on V_{DD}Min to detect outliers and a posterior statistical process for outlier screening [35]. This technique uses a reduced vector set instead of the full vector set to search the V_{DD}Min value, and then applies the full set of test vectors to the die using the minimum voltage found in the previous step. Depending on the fault coverage, outliers are screened using statistical postprocessing methods. Experimental results showed that V_{DD}Min yield fallout varied from 0.2–0.8%, depending on product complexity, and approximately 20% of the outlier dies showed a MinV_{DD} positive or negative shifting. Some specific defects were found during failure analysis, including a resistive via and a tungsten stringer (bridge).

10.5.6 Multiparameter Testing

The test techniques discussed until now are based on monitoring one test observable and applying data analysis to reduce variance and set an adjustable limit, as a method to bin parts subjected to a spread of parameters. Another solution to distinguish good from bad devices in a population with large variance is to exploit the correlation of different circuit parameters owing to fundamental physical properties of devices and interconnects. These techniques started by correlating two parameters, such as speed and leakage, to establish the two-parameter techniques. This can be generalized to multiparameter techniques where more than two variables are adjusted.

F_{max} versus I_{DDQ} Two-Parameter Testing. Keshavarzi et al. showed that outliers could be pulled out of intrinsic variations by manipulating more than one variable to find a sensitive limit [29]. Single-threshold I_{DDQ} testing was based on setting a single pass/fail threshold independent of other parameters of the circuit. For a given population of circuits, high-leakage parts are intrinsically faster since their transistors have smaller V_t's and shorter L_{eff}'s. This does not necessarily correspond to a defective IC, but to a faster one, since larger I_{off} implies larger I_{Dsat} if the device is working properly. Keshavarzi et. al. observed this property and plotted the static current leakage versus F_{max} for a large number of circuits, obtaining the graph shown in Figure 10.27. Data clearly show a trend in which leaky parts run faster, whereas slower parts exhibit a smaller leakage. The technique of combining two parameters to establish pass/fail criteria is a good solution to avoid yield loss from valid parts. This test method is a low-cost alternative for saving fast, intrinsically leaky ICs while discriminating against defective ones.

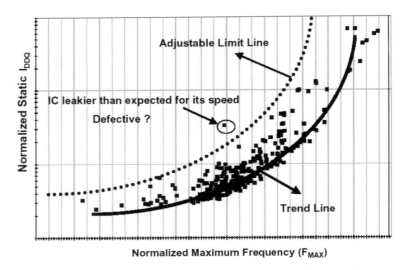

Figure 10.27. I_{DDQ} versus maximum frequency [29].

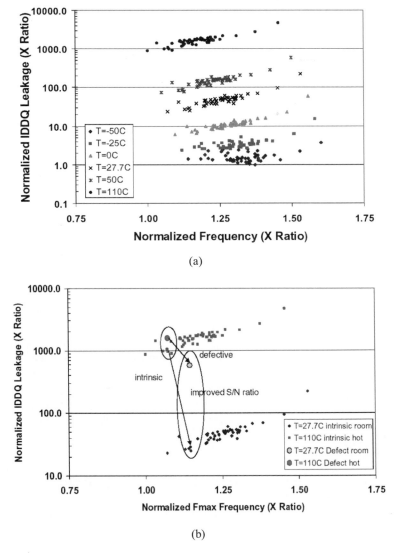

Figure 10.28. (a) Leakage current versus frequency for six different temperatures. (b) Defect detection using I_{DDQ} at different temperatures [29].

Beyond Two-Parameter Testing. The distribution of two parameters can be altered by a third or fourth parameter to extract outliers in the original distribution. Several variables can be considered for multiparameter testing, such as F_{max}, V_{DD}, V_{DD}Min, temperature, reverse body bias, I_{DDQ}, and ΔI_{DDQ}. Reverse body biasing (RBB) of the source and body terminals increases the transistor threshold voltage, decreases intrinsic I_{DDQ}, and slows the chip. Keshavarzi et al. found that temperature was more effective than RBB for distinguishing defects in a 180 nm technology [29].

Figure 10.28(a) shows normalized I_{DDQ} versus F_{max} distributions for six temperatures [29]. The normal variations at each temperature in the data do not overlap. Figure 10.28 shows the test discrimination in the presence of extrinsic leakage. A 1 MΩ resistor was inserted to emulate a low-impact bridge defect between the power rails. Without this resistance, the circled IC data point dropped 36× with temperature cooling from 110°C to 27.7°C. When the emulated defect was attached across the power lines, the drop in I_{DDQ} was only 1.6×. That value was well outside of the intrinsic distribution at room temperature, demonstrating a promising test approach to see through the terrible variance of the intrinsic deep-submicron parameters.

10.6 CONCLUSIONS

Deep-submicron and nanometer technologies challenge several areas of test and diagnosis. The complex ATE digital testers are pressed to match the timing requirements, the I/O pin counts, the high thermal environment, the power supply current demands, contain costs, and make critical measurements in the face of widening statistical variation in the normal population of ICs. Test development and actual IC test measurement costs drove acceptance of design-for-test structures for the high-performance industry. Scan circuits are mandatory to ease the pressures of the deep-submicron technologies. Defect-based testing is a deterministic technique to analyze defect electronic behavior, and then use test methods matched to those behaviors. DfT structures dictate a test pattern delivery style, and defect-based testing dictates what specific tests to use. The test strategies are compromised by the increase in transistor and part parameter variance with each technology node. These are exciting times to push the technological and human limit of computing circuits.

BIBLIOGRAPHY

M. Abramovici, M. Breuer, and A. Friedman, *Digital Systems Testing and Testable Design,* Computer IEEE Press, New York, 1993.

A. L. Crouch, *Design for Test for Digital IC's and Embedded Core Systems,* Prentice-Hall, 1999.

R. K. Gulati and C. F. Hawkins, I_{DDQ} *Testing of VLSI Circuits,* Kluwer Academic Publishers, 1993.

E. J. McCluskey, *Logic Design Principles: With Emphasis on Testable Semicustom Circuits,* Prentice-Hall, 1986.

V. P. Nelson, H. T. Nagle, B. D. Carroll, and J. D. Irwin, *Digital Logic Circuit Analysis & Design,* Prentice Hall, Inc., 1995.

F. F. Tsui, *LSI/VLSI Testability Design,* McGraw-Hill, 1987.

REFERENCES

1. 1149.1–2001 Test Access Port and Boundary Scan Architecture (IEEE), IEEE Pub., [www.computer.org/cspress/CATALOG/ST01096.htm]
2. http://grouper.ieee.org/groups/1149/4/index.html
3. http://standards.ieee.org/reading/ieee/std_public/description/testtech/1149.5–1995_desc.html

4. http://grouper.ieee.org/groups/1149/6/index.html
5. M. Abramovici, M. Breuer, and A. Friedman, *Digital Systems Testing and Testable Design,* Computer IEEE Press, New York, 1993.
6. R. Aitken, "Extending the pseudo-stuck-at fault model to provide complete I_{DDQ} coverage," in *IEEE VLSI Test Symposium,* pp.128–134, April 1999.
7. B. Alorda, M. Rosales, J. Soden, C. Hawkins, and J. Segura, "Charge based transient current testing (CBT) for submicron CMOS SRAMs," in *IEEE International Test Conference (ITC),* pp. 947–953, 2002.
8. K. Baker, G. Gronthoud, M. Lousberg, I. Schanstra, and C. Hawkins, "Defect-based delay testing of resistive vias-contacts: a critical evaluation," in *IEEE International Test Conference (ITC),* pp. 467–476, October 1999.
9. J. Beasley, H. Rammamurthy, J. Ramirez-Angulo, and M. DeYong, "IDD pulse response testing of analog and digital circuits," in *IEEE International Test Conference (ITC),* pp. 626–634, October 1993.
10. J. Beasley et al., "idd pulse response testing applied to complex CMOS ICs," in *IEEE International Test Conference (ITC),* pp. 32–39, October 1997.
11. K. Bernstein, K. Carrig, C. Durham, P. Hansen, D. Hogenmiller, E. Nowak, and N. Rohrer, *High Speed CMOS Design Styles,* Kluwer Academic Publishers, 1998.
12. H. Bleeker, P. van den Eijnden, and F. de Jong, *Boundary-Scan Test, A Practical Approach,* Kluwer Academic Press, 1993.
13. S. Borkar, "Low power design challenges for the decade," in *ASP-DAC01,* pp. 293–296, 2001.
14. A. Campbell, E. Cole, C. Henderson, and M. Taylor, "Case history: Failure analysis of a CMOS SRAM with an intermittent open circuit," in *International Symposium on Test and Failure Analysis (ISTFA),* pp. 261–269, Nov. 1991.
15. A.L. Crouch, *Design for Test for Digital IC's and Embedded Core Systems,* Prentice Hall, 1999.
16. R. Daasch, J. McNames, D. Bockelman, K. Cota, and R. Madge, "Variance reduction using wefer patterns in I_{DDQ} data," in *IEEE International Test Conference (ITC),* October 2000.
17. R. Daasch, K. Cota, J. McNames, and R. Madge, "Neighbor selection for variance reduction in I_{DDQ} and other parametric data," in *IEEE International Test Conference (ITC),* October 2001.
18. V. De and S. Borkar, "Technology and Design Challenges for Low Power and High Performance," in *International Symposium on Low Power Electron Devices,* pp. 163–168, 1999.
19. R. Degraeve, B. Kaczer, A. De Keersgieter, and G. Groeseneken, "Relation between breakdown mode and breakdown location in short channel NMOSFETs," in *IEEE International Reliability Physics Symposium (IRPS),* pp. 360–366, May 2001.
20. R. Eldred, "Test routines based on symbolic logical statements," *Association of Computing Machinery, 6,* 33–36, 1959.
21. A. Gattiker and W. Maly, "Current signatures," in *IEEE VLSI Test Symposium,* pp. 112–117, 1996.
22. A. Gattiker and W. Maly, "Current signatures: application," in *IEEE International Test Conference (ITC),* pp. 156–165, 1997.
23. H. Hao and E. McCluskey, "Very-low voltage testing for weak CMOS logic ICs," in *IEEE International Test Conference,* pp. 275–284, 1993.
24. P. Hazucha, C. Svensson, and S. Wender, "Cosmic-ray soft error rate characterization of a standard 0.6 μm CMOS process," *IEEE Journal of Solid State Circuits, 35,* 10, 1422–1429, October 2000.
25. H. Hulvershorn, "1149.5: now it's a standard, so what?," in *IEEE International Test Conference (ITC),* pp. 166–173, 1997.
26. International Technology Roadmap for Semiconductors 2001 (http://public.itrs.net/).

27. W. Jiang and B. Vinnakota, "Statistical threshold formulation for dynamic Idd test," *IEEE Transactions on Computer Aided Design of Integrated Circutis and Systems, 21,* 6, June 2002.
28. T. Karnik, B. Bloechel, K. Soumyanath, V. De, and S. Borkar, "Scaling trends of cosmic rays induced soft errors in static latches beyond 0.18 μm," in *Symposium on VLSI Circuits,* pp. 61–62, 2001.
29. A. Keshavarzi, K. Roy, M. Sachdev, C. Hawkins, K. Soumyanath, and V. De, "Multiple-parameter CMOS IC testing with increased sensitivity for I_{DDQ}," in *IEEE International Test Conference (ITC),* pp. 1051–1059, October 2000.
30. E.E. King and G.P. Nelson, "Radiation testing of an 8-bit CMOS microprocessor," *IEEE Transactions on Nuclear Science, NS-22,* 5, 2120–2123, October 1975.
31. B. Kruseman et. al., "Transient current testing of 0.25μm CMOS devices," in *IEEE International Test Conference (ITC),* pp. 47–56, October 1999.
32. B. Kruseman, S. van den Oetelaar, and J. Rius, "Comparison of I_{DDQ} testing and very-low voltage testing," in *IEEE International Test Conference (ITC),* pp. 964–973, 2002
33. M. Levi, "CMOS is MOST testable," in *IEEE International Test Conference (ITC),* pp. 217–220, October 1981.
34. J. Liou, T. Cheng, and D. Mukherjee, "Path selection for delay testing of deep sub-micron devices using statistical performance sensitivity analysis," in *IEEE VLSI Test Symposium,* pp. 97–104, 2000.
35. R. Madge, B. Goh, V. Rajagopalan, C. Macchietto, R. Daasch, C. Shuermyer, C. Taylor, and D. Turner, "Screening MinVDD outliers using feed-foward voltage testing analysis," in *IEEE International Test Conference (ITC),* pp. 673–682, 2002.
36. R. Madge, "The value of test data in manufacturing and quality," in *IEEE Defect-Based Test Workshop, Keynote Address,* Napa Valley, CA, April 2003.
37. R. Makki et al., "Transient power supply current testing of digital CMOS circuits," in *IEEE International Test Conference (ITC),* pp. 892–901, 1995.
38. Y. Malaiya and S. Su, "A new fault model and testing technique for CMOS devices," in *IEEE International Test Conference (ITC),* pp. 25–34, October 1999.
39. P. Maxwell, R. Aitken, K. Kollitz, and A. Brown, "I_{DDQ} and AC scan: The war against unmodelled defects," in *IEEE International Test Conference (ITC),* pp. 250–258, November 1982.
40. P. Maxwell, P. O'Neill, R. Aitken, R. Dudley, N. Jaarsma, M. Quach, and D. Wiseman, "Current ratios: A self-scaling technique for production I_{DDQ} testing," in *IEEE International Test Conference (ITC),* pp. 738–746, October 1999.
41. A. Miller, "I_{DDQ} testing in deep submicron integrated circuits," in *IEEE International Test Conference (ITC),* pp. 724–729, October 1999.
42. W. Needham, C. Prunty, and E. Yeoh, "High volume microprocessor test escapes: An analysis of defects that our tests are missing," in *IEEE International Test Conference (ITC),* pp. 25–34, October 1998.
43. G. F. Nelson and W. F. Boggs, "Parametric tests meet the challenge of high-density ICs," *Electronics Magazine,* 108–111, December 1975.
44. P. Nigh, D. Vallett, A. Patel, and J. Wright, "Failure analysis of I_{DDQ}-only failure from SEMATECH test methods experiment," in *IEEE International Test Conference (ITC),* pp. 43–52, October 1998.
45. M. Orshansky, L. Millor, P. Chen, K. Keutzer, and C. Hu, "Impact of spatial intrachip gate length variability on the performance of hi-speed digital circuits," *IEEE Transactions on Computer-Aided Design, 21,* 5, May 2002.
46. http://grouper.ieee.org/groups/1500/
47. I. de Paúl, M. Rosales, B. Alorda, J. Segura, C. Hawkins, and J. Soden "Defect oriented fault diagnosis for semiconductor memories using charge analysis: theory and experiments," in *IEEE VLSI Test Symposium,* April 2001.

48. G. Perry, *Fundamentals of Mixed-Signal Test,* 1999, www.soft-test.com
49. W. Riordan, R. Miller, J. Sherman, and J. Hicks, "Microprocessor reliability performance as a function of die location for a 0.25 μm, five layer metal CMOS logic process," in *International Reliability Physics Symposium (IRPS),* pp. 1–9, April 1999.
50. R. Rodríguez and J. Figueras, "I_{DDQ} detectability of bridges in CMOS sequential circuits," *IEEE Electronics Letters, 30,* 1, 30–31, 1994.
51. M. Sachdev, P. Janssen, and V. Zieren, "Defect detection with transient current testing and its potential for deep sub-micron CMOS ICs," in *IEEE International Test Conference (ITC),* pp. 204–213, October 1998.
52. J. Saxena, K. Butler, and L. Whetsel, "An analysis of power reduction techniques in scan testing," in *IEEE International Test Conference (ITC),* pp. 670–677, October 2001.
53. J. Segura, I. De Paúl, M. Roca, E. Isern, and C. Hawkins, "Experimental analysis of transient current testing based on charge observation," *Electronics Letters, 35,* 6, March 1999.
54. N. Seifert, X. Zhu, and L. Massengill, "Impact of scaling on soft-error rates in commercial microprocessors," *IEEE Transactions on Nuclear Science, 49,* 6, Dec. 2002.
55. P. Shivakumar, M. Kisler, S. Keckler, D. Burger, and L. Alvisi, , "Modeling the effect of technology trends on the soft error rate of combinational logic," in *IEEE International Conference on Dependable Systems and Networks (DSN'02),* 2002.
56. A. Singh, "A comprehensive wafer oriented test evaluation (WOTE) scheme for the IDDQ testing of deep submicron technologies," in *IEEE International Workshop on IDDQ Testing,* pp. 40–43, November 1997.
57. A. Singh, J. Plusquellic, and A. Gattiker, "Power supply transient signal analysis under real process and test hardware models" in *IEEE VLSI Test Symposium,* pp. 367–362, 2002.
58. J. Soden, R. Treece, M. Taylor, and C. Hawkins, "CMOS stuck-open electrical effects and design considerations," in *IEEE International Test Conference (ITC),* pp. 423–430, October 1989.
59. J. Soden and C. Hawkins, "I_{DDQ} Testing: A Review," *Journal of Electronic Testing: Theory and Applications (JETTA), 3,* 4, December 1992.
60. J. Suehle, "Ultrathin gate oxide reliability: Physical models, statistics, and characterization," *IEEE Transactions on Electron Devices, 49,* 6, 958–971, June 2002.
61. C. Thibeault, "An histogram based procedure for current testing of active defects," in *IEEE International Test Conference (ITC),* pp. 714–723, October 1999.
62. D. Tryon, F. Armstrong, and R. Reiter, "Statistical failure analysis of system timing," *IBM Journal of Reserach and Development, 28*(4), 340–355, July 1984.
63. C. Tseng, E. McCluskey, X. Shao, J. Wu, and D. Wu, "Cold delay defect screening," in *IEEE VLSI Test Symposium,* April 2000.
64. C. Tseng, R. Chen, P. Nigh, and E. McCluskey, "MINVDD testing for weak CMOS ICs," in *IEEE VLSI Test Symposium,* pp. 339–344, 2001.
65. http://tab.computer.org/tttc/Standards/index.html
66. A.J. van de Goor, *Testing Semiconductor Memories,* Wiley, 1991.
67. K. Wallquist, "Achieving I_{DDQ}/I_{SSQ} production testing with QuiC-Mon," *IEEE Design & Test of Computers,* 62–69, Fall Issue, 1995.
68. R. Wadsack, "Fault modeling and logic simulation of CMOS and MOS integrated circuits," *Bell Systems Technical Journal,* 1449–1488, May-June 1978.
69. J. Walls, "Stress-voiding in tungsten-plug interconnect systems induced by high-temperature processing," *IEEE Electron Devices Letters, 16,* 10, October 1995.
70. B. Weir et al., "Ultra-thin dielectrics: they break down, but do they fail?," in *IEEE International Electron Device Physics Symposium (IEDM),* pp. 41–44, December 1997.
71. M.J.Y. Williams and J.B. Angel, "Enhancing the testability of large-scale integrated circuits via test points and additional logic," *IEEE Transactions on Computers, C-22,* 46–60, January 1973.

72. J. Altet, A. Rubio, J. Rossello, and J. Segura, Structural RFIC device testing through built-in thermal monitoring," *IEEE Communications Magazine, 41,* 9, 98–104, September 2004.

EXERCISES

10.1. Example 10.1 uses a 1K memory to illustrate the futility of functional testing. Therefore, should we eliminate functional testing from our IC test methods?

10.2. Can a functional test set replace a stuck-at-fault test set?

10.3. What does it mean to sensitize a logic signal path?

10.4. An exclusive OR gate (XOR) is shown in Figure 10.29. The output is a logic-1 only when a single logic-1 exists in AB. XOR = logic-1 if AB = 10 or 01. Derive the stuck-at-fault vectors to detect nodes x-SA0 and y-SA1.

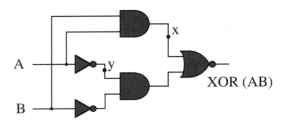

Figure 10.29.

10.5. The circuit in Figure 10.30 is a 2-bit input to 4-bit decoder. The four logic states of A_0A_1 can activate the output of the four output lines with a logic-0. Derive the test vectors that detect x-SA1 and y-SA0.

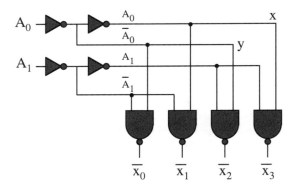

Figure 10.30.

10.6. List three advantages and three disadvantages of the stuck-at-fault test method.

10.7. Derive the delay fault vectors that detects nodes x slow to rise and y slow to fall (Figure 10.31).

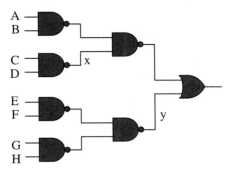

Figure 10.31.

10.8. Derive the delay fault vectors that detect nodes x slow to rise and y slow to fall (Figure 10.32).

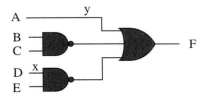

Figure 10.32.

10.9. List three advantages and three disadvantages of the delay fault (DF) test method.

10.10. List four advantages and four disadvantages of the I_{DDQ} test method.

10.11. Derive an I_{DDQ} test vector to detect the transistor bridge defects in Gate–3 of Figure 10.33.

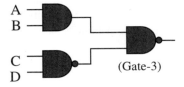

Figure 10.33.

10.12. Derive an I_{DDQ} test vector to detect a bridge defect across nodes x and y in Figure 10.34.

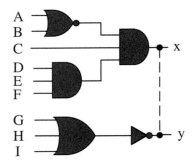

Figure 10.34.

10.13. Scan DfT makes sequential circuit (flip-flop) testing simple; you simply send logic ones and zeros through the chain, and measure the result. Describe the sequential test problem if scan DfT is not used.

10.14. A bridge defect may cause a DC leakage path from V_{DD} to GND. The I_{DDQ} going through the bridge causes a current elevation. For this circuit, you measured an I_{DDQ} strongly in excess of the current in the bridge defect. Explain.

10.15. What is the thermal problem associated with testing ICs using scan circuits?

10.16. LBIST testing requires that all flip-flops in LFSR initially be set to logic zero. Why?

10.17. If an LBIST LFSR generated all possible logic combinations driving a subcircuit, describe the defects that can quantifiably be detected. Assume that the all-zeros state of the LFSR is generated by a separate test structure, and that the circuit is clocked at the operational speed of the IC.

10.18. Assume that an LFSR drives all inputs of the circuit in Figure 10.35.
(a) Is the probability high or low that a single random vector will detect node-A as a stuck-at one?
(b) Is the probability high or low that a single random vector will detect node-A as a stuck-at zero?

10.19. Table 10.4 lists an electromigration reliability open circuit as rarely causing stuck-at-fault behavior. Is this conclusion a contradiction since we studied that hard open circuit defects can cause stuck-at-fault behavior in two of the five combinational logic defect classes?

10.20. The DC transfer curve and the timing results in Figure 10.18 should match node voltages when the quiescent logic state is reached. Do they? Why?

10.21. Why does Figure 10.18(c) show a peak for a defect resistance between 1 kΩ and 2 kΩ?

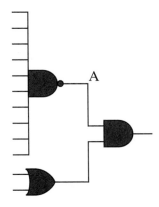

Figure 10.35.

10.22. Is there a single test method that can detect all open circuit defect behavior classes? Explain.

10.23. The Venn diagram in Figure 10.10(a) obtained by Maxwell et al. [39] was obtained with a stuck-at fault coverage of 99%. The functional test set was fault-simulated and had a 83.7% stuck-at-fault coverage. Why did the functional test set in Figure 10.10(a) detect more defects?

10.24. Resistive vias grew to be significant for IC test problems in the 1990s. Why is the present test solution to resistive vias a near-perfect example of defect-based testing?

10.25. The development of deep-submicron ICs introduced serious noise into the test process that was insignificant in older technologies. Describe the noise that affects test methods in deep-submicron technologies.

APPENDIX A

SOLUTIONS TO SELF-EXERCISES

A.1 CHAPTER 1. ELECTRICAL CIRCUIT ANALYSIS

Self-Exercise 1.1
$V_0 = 1.65$ V
$V_P = 52.33$ V

Self-Exercise 1.2
$R_{eq} = 1200 \parallel 950 \parallel R3$
$250 = [1/1200 + 1/950 + 1/R3]^{-1}$
$1/R_3 = 1/250 - 1/1200 - 1/950 \Rightarrow R_3 = 473.0 \ \Omega$

Self-Exercise 1.3
$R_{in} = 8$ kΩ
$I_{BB} = 1.25$ mA
$V_0 = 7.5$ V

Self-Exercise 1.4
(a) $R_{in} = 667 \ \Omega$
(b) $R_{in} = 2$ kΩ
(c) $R_{in} = 990.1 \ \Omega$
(d) $R_{in} = 999 \ \Omega$

CMOS Electronics: How It Works, How It Fails. By Jaume Segura and Charles F. Hawkins
ISBN 0-471-47669-2 © 2004 Institute of Electrical and Electronics Engineers

Self-Exercise 1.5
$R_{in} = 11.5 \text{ k}\Omega$
$I_{BB} = 86.96 \text{ µA}$
$V_0 = 652.2 \text{ mV}$

Self-Exercise 1.6
(a) $I_3 = 100 \text{ µA} - 50 \text{ µA} - 10 \text{ µA} = 40 \text{ µA}$
(b) $V_{R3} = 40 \text{ µA} \times 50 \text{ k}\Omega = 2.0 \text{ V}$, so that
$R_1 = 2.0 \text{ V}/50 \text{ µA} = 40 \text{ k}\Omega$
$R_2 = 2.0 \text{ V}/10 \text{ µA} = 200 \text{ k}\Omega$

Self-Exercise 1.7
$V_{R3} = V_0 = 200 \text{ µA} \times 8 \text{ k}\Omega = 1.6 \text{ V}$
$R_1 = [3.3 - 1.6]/650 \text{ µA} = 2.615 \text{ k}\Omega$
$I_2 = 650 \text{ µA} - 200 \text{ µA} = 450 \text{ µA}$
$R_2 = 1.6/450 \text{ µA} = 3.556 \text{ k}\Omega$

Self-Exercise 1.8
$R_1 = 7.125 \text{ k}\Omega$

Self-Exercise 1.9
(a) $R_{eq} = 1 \text{ M}\Omega \parallel 2.3 \text{ M}\Omega$; that is, $R_{eq} = 697.0 \text{ k}\Omega$
(b) $R_{eq} = 75 \text{ k}\Omega \parallel [150 \text{ k}\Omega + 35 \text{ k}\Omega]$; that is, $R_{eq} = 53.37 \text{ k}\Omega$

Self-Exercise 1.10
$R_{eq} = R_1 + R_3 \parallel (R_2 + R_4)$
$R_{eq} = R_1 + R_3 \parallel (R_2 + R_4) + R_5$
$R_{eq} = R_1 \parallel [R_2 + R_4 \parallel (R_3 + R_5)]$

Self-Exercise 1.11
(a) $R_{eq} = 31.98 \text{ k}\Omega$
(b) $R_{eq} = 36.98 \text{ k}\Omega$
(c) $R_{eq} = 10.32 \text{ k}\Omega$

Self-Exercise 1.12
$V_0 = (12k \parallel 20k)/[4k + (12k \parallel 20k)] \times 1 \text{ V} = 0.652 \text{ V}$
$V_{4k} = (4k)/[4k + (12k \parallel 20k)] \times 1 \text{ V} = 0.348 \text{ V}$
$V_{BB} = 0.652 + 0.348 = 1.00 \text{ V}$
$R_{in} = 4k + (12k \parallel 20k) = 11.5 \text{ k}\Omega$
$I_{BB} = V_{BB}/R_{in} = 87.0 \text{ µA}$

Self-Exercise 1.13
$R_{in} = 45k \parallel [8k + (18k \parallel 30k)] = 13.48 \text{ k}\Omega$
$V_0 = (18k \parallel 30k)/[8k + (18k \parallel 30k)] \times 2.5 \text{ V} = 1.461 \text{ V}$
$I_{BB} = 185.4 \text{ µA}$

Self-Exercise 1.14
$I_{12k} = [20/(20 + 12)] \times 87.0 \text{ µA} = 54.38 \text{ µA}$
$I_{20k} = [12/(20 + 12)] \times 87.0 \text{ µA} = 32.63 \text{ µA}$
(Notice that the KCL is satisfied: 87.0 µA = 54.38 µA + 32.63 µA.)

Self-Exercise 1.15
$I_{45k} = [8k + (18k \| 30k)]/[45k + 8k + (18k \| 30k)] \times 185.4 \, \mu A = 55.55 \, \mu A$
$I_{8k} = (45k)/[45k + 8k + (18k \| 30k)] \times 185.4 \, \mu A = 129.9 \, \mu A$
(Notice that I_{8k} is more easily obtained from KCL, where $I_{8k} = 185.4 \, \mu A - 55.55 \, \mu A = 129.9 \, \mu A$.)
$I_{18k} = (30k)/(30k + 18k) \times 129.9 \, \mu A = 81.16 \, \mu A$
$I_{30k} = (18k)/(30k + 18k) \times 129.9 \, \mu A = 48.69 \, \mu A$
$V_{BB} = 2.5 \, V$

Self-Exercise 1.16
(a) $I_{10k} = 46.15 \, \mu A$
  $I_{15k} = 30.77 \, \mu A$
  $I_{20k} = 23.08 \, \mu A$
(b) $R_{20k} \Rightarrow 114 \, k\Omega$

Self-Exercise 1.17
(a) $V_0 = 3.60 \, mV$
(b) $I_2 = 714.3 \, \mu A$, $I_9 = 400 \, \mu A$

Self-Exercise 1.18
$R_{in} = 250 + [4k + (3k \| 1k)] \| [2k \| (750 + 1.5k)] + 250 = 1.366 \, k\Omega$
$I_{1.5k} = [5 \, V/R_{in}] \times [2k \| (4k + (3k \, 1k))]/[2k \| (4k + (3k \| 1k) + (750 + 1.5k)]$
$= 1.409 \, mA$

Self-Exercise 1.19
(a) $C_{eq} = 38.1 \, nF$
(b) $C_{eq} = 25.9 \, fF$

Self-Exercise 1.20
$$V_1 = \frac{C_2}{C_1 + C_2} \times 100 \, mV$$

$$= \frac{75}{45 + 75} \times 100 \, mV = 62.5 \, mV$$

$$V_2 = \frac{45}{45 + 75} \times 100 \, mV = 37.5 \, mV$$

Self-Exercise 1.21
$$V_2 = \frac{C_1}{C_1 + C_2} \times V_D$$

$$= \frac{30}{30 + 25} \times V_D = 0.7 \, V$$

$$V_D = \frac{55}{30} \times 0.7 \, V = 1.28 \, V$$

Self-Exercise 1.22

(a) $V_D = \left(\dfrac{86.17\ \mu eV/K}{1\ eV}\right)(298\ K) \ln\left(\dfrac{200\ nA}{1\ nA} + 1\right) = 136.2\ mV$

(b) $I_D = 1\ nA(e^{400/26}) = 4.80\ mA$

Self-Exercise 1.23
$I_D = 185\ \mu A$
$V_D = 315.3\ mV$

Self-Exercise 1.24
(a) 5.179 V, –0.179 V
(b) 5 V

Self-Exercise 1.25
$V_{D1} = 18.02\ mV$
$V_0 = 4.982\ V$

Self-Exercise 1.26
$I_D = 26\ mV/1\ \mu A \Rightarrow 26\ k\Omega$
$I_D = 26\ mV/100\ \mu A \Rightarrow 260\ \Omega$
$I_D = 26\ mV/1\ mA \Rightarrow 26\ \Omega$
$I_D = 26\ mV/10\ mA \Rightarrow 2.6\ \Omega$
Dynamic diode resistance can vary over a wide range, depending upon the bias current.

A.2 CHAPTER 3. MOSFET TRANSISTORS

Self-Exercise 3.1
(a) Ohmic
(b) Ohmic
(c) Saturated
(Hint: watch your terminals.) The source terminal always has a lower or equal voltage than the drain in a conducting nMOS transistor.

Self-Exercise 3.2
$I_D = 12.5\ \mu A$ using saturated state model.
$V_D = 9.375\ V$

Self-Exercise 3.3
$I_D = 184.8\ \mu A$
$V_D = 0.761\ V$

Self-Exercise 3.4
$V_{GS} = 3.33\ V$
M1 in saturation

Self-Exercise 3.5
$R_1 = 180\ k\Omega$

Self-Exercise 3.6
$R_0 = 537.6\ \Omega$

Self-Exercise 3.7
(a) Saturated state
(b) Ohmic state
(c) Boundary point of both saturated and ohmic state

Self-Exercise 3.8
$I_D = 29.4$ μA, $V_0 = 2.94$ V

Self-Exercise 3.9
$I_D = 48.46$ μA, $V_0 = 4.846$ V

Self-Exercise 3.10
$I_D = 153.6$ μA, $V_0 = -1.848$ V

Self-Exercise 3.11
R = 239.0 kΩ

Self-Exercise 3.12
$W/L > 6024$

Self-Exercise 3.13
$I_{SAT} = 6$ mA

Self-Exercise 3.14
$I_{SAT} = 0.66$ mA

A.3 CHAPTER 4. CMOS BASIC GATES

Self-Exercise 4.1
$W_p/W_n = 3.36$

Self-Exercise 4.2
(a) 61.4%
(b) 71.7%; you must derive the equation for this from the V_{in} equation in Example 4.2.
(c) 10.4%

Self-Exercise 4.3
$\Delta V_{out}/\Delta V_{in} \approx -17$

Self-Exercise 4.4
$\tau_D = 103.1$ ps at $V_{DD} = 2.5$ V
$\tau_D = 186.0$ ps at $V_{DD} = 1.8$ V
The extra delay is 83.0 ps, which represents an increase of 80.4%.

Self-Exercise 4.5
(a) $I_1I_2I_3I_4 = 11X0$
(b) $I_1I_2I_3I_4 = 1X00$

Self-Exercise 4.6
(a) $I_1I_2I_3I_4 = 1100$
(b) $I_1I_2I_4I_5 = 1101$
(c) $I_2I_3I_4I_5 = 1001$

A.4 CHAPTER 5. CMOS BASIC CIRCUITS

Self-Exercise 5.1

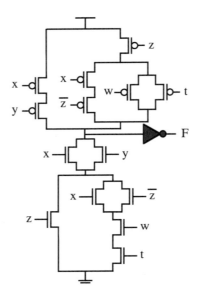

Self-Exercise 5.2
The gate level design of a D-latch shown in Figure 5.16(a) has four transistors in each NOR gate (thus, a total of 16 devices), plus two transistors for the inverter, resulting in 18 transistors. A transistor-level design of the circuit in Figure 5.16(b) is

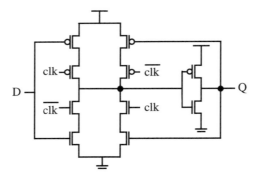

This design uses 10 transistors shown plus two more to invert the clock for a total of 12 transistors. The circuit loads data when clk is high, and holds that data to output Q when the clk goes negative.

A.5 CHAPTER 6. FAILURE MECHANISMS IN CMOS ICs

Self-Exercise 6.1
(a) 29.7% Reduction in current density
(b) 50.7% Reduction in lifetime at the higher temperature

Self-Exercise 6.2
(a) $P = 1.2$ W/cm^2
(b) $P = 12$ kW/cm^2. Since the passivation material is highly insulating, you can get a hot region. 10 mA is typical for electromigration test structure experiments and 100 μA is typical for IC operation.
(c) The 100 μA current drops 300 μV and the 10 mA current drops 30 mV. These drops would typically not alter logic function.

Self-Exercise 6.3
250 μm

Self-Exercise 6.4
(a) 8.658×10^{-6} lb, 3.936×10^{-6} kg
(b) 0.1329 μm = 132.9 nm

A.6 CHAPTER 7. BRIDGING DEFECTS

Self-Exercise 7.1
(a) $R_{crit} = 3.246$ kΩ
(b) $K_p/K_n = 1$ and both transistor drains are at the logic threshold voltage.

Self-Exercise 7.2
$R_{crit} = 1.433$ kΩ – 1.722 kΩ

Self-Exercise 7.3
$V_G = 2.30$ V, so gate passes function (barely).

A.7 CHAPTER 8. OPEN DEFECTS

Self-Exercise 8.1
(a) $V_2 = 1$ V
(b) $V_{DD} = 2.513$ V

A.8 CHAPTER 10. DEFECT BASED TESTING

Self-Exercise 10.1
Node-c SA1 has three possible test vectors:
01 11 11
01 11 10
01 11 01
Node-m SA0 has nine possible test vectors including:
11 00 10
11 00 01
11 00 11

INDEX

Acceptor atom, 44, 48
AHI (*see* Anode hole injection model)
AHR (*see* Anode hydrogen release model)
Anode hole injection model (AHI), 182
Anode hydrogen release model (AHR), 182
Antennas, 191
Aspect ratio
   via, 177, 253–254, 306
   impact on gate threshold voltage, 102
   impact on parameter variations, 247
   symmetric transfer characteristic, 103
ATE (*see* Automatic test equipment)
ATPG (*see* Automatic test pattern generation)
Automatic test equipment (ATE), 282, 296, 299, 310, 314–315, 322
Automatic test pattern generation (ATPG) (*see also* Test), 207, 214, 281, 284, 292, 296
Avalanche
   weak, 93

Band bending, 45
Band gap, 38
BCC (*see* Metal Body-centered cubic)
Bi-directional (bipolar) current, 165–166
Bi-directional pin, 146, 147
Bipolar current (*see* Bi-directional current)
Bipolar transistor, 53, 207
   parasitic structure from gate-oxide short, 211–212, 214
   role in latchup, 145–146
BIST (*see* Built-in self test)
Bit line, 141, 142
Black's law, 161, 162
Blech Effect, 165–168, 176, 196
Body effect (*see also* Transistor threshold voltage)
   constant, 78, 93, 94, 97, 145
Bohr, Neils, 37
Boltzmann constant, 24, 41, 158, 160, 162, 186
Bounce (*see* Noise)
Bridging defect
   control loops, 217–218
   floating state, 217–218
   forced state, 217
   voltage based testing of, 302
   feedback, 214–215
   inverted, 216
   nonfeedback, 214
   noninverted, 216
Built-in electric field (*see* Diode and Transistor)
Built-in self test (BIST), 293, 296
   combinational logic test (LBIST), 209, 296, 298, 328

Built-in self test (BIST) *(continued)*
　embedded core memory test (MBIST), 296
　linear feedback shift register (LFSR), 296–298, 328
　multiple input shift register (MISR), 296–298

Capacitor voltage divider. 20–22, 224–226, 233
Capacitor
　parallel connection, 18, 19
　series connection, 19, 21, 81, 225
Carrier transport, 44–46
Catastrophic defect (*see* defect)
CD (*see* Critical dimension)
Channel length (*see* Transistor Channel length)
Channel length modulation, 83
Channel width (*see* Transistor Channel width)
Charge based testing, 323
Charge sharing, 134–35, 142
Chemical mechanical planarization (CMP), 177, 244, 259, 275
Chemical vapor deposition (CVD), 177
CMP (*see* Chemical mechanical planarization)
Column decoder, 140, 296
Comparative test methods, 292–293, 314
Complementary static gates, 127–128, 133
Compound domino logic, 135, -137, 148
Conduction band, 38–43, 44, 45
Conduction channel (*see* Transistor)
Conduction,
　diffusion, 45–47, 49
　drift, 40, 45–46, 52
　in diodes (*see also* Diode), 48–50
　in metal voids, 165
　in solids (*see also* Semiconductor), 37–41
　in transistors (*see also* Transistor), 56, 60
　trap-assisted (*see* Trap), 181–182
　tunneling, 229–230, 236–237
Conductivity, 37, 156, 254
Coulomb, 4, 18, 40
Critical dimension (CD), 244, 310
Critical path, 265
　global interconnect, 259
　impact of noise on, 269–270
　timing test, 315, 318
　V_t adjustment of, 104
Critical resistance, 199–207, 284, 287, 301–302
　impact of technology scaling on, 218
　variation with V_{DD}, 318–319
Cross-talk (*see also* Noise), 17, 243, 253, 255, 308–309
　in Cu-interconnect, 177
　parametric failures, 243, 245–246
　impact on delay, 259, 266–268
　increase with interconnect scaling, 255
Current based testing, 283, 289–293
　of bridge defect, 301–302
　current ratios test, 316
　current signatures, 315–316
　current-based test, 218, 283, 289–290, 293, 301–302, 304–305, 311, 315
　delta-IDDQ, 316
　I_{DDQ} test (see I_{DDQ} test)
　of hard opens, 304–305
　of nanometer ICs, 311, 315–318
　of semiconductor memories, 218
　transient, 292
　transient current test, 283, 292
Current divider, 14–15, 17
Current drive strength (*see also* Drive strength), 109, 142, 201, 249–250
　impact on bridge critical resistance, 205, 214
　impact on inverter characteristics, 103
　of memory transistors, 304
　of NAND gate, 120
Current leakage (*see* Transistor subthreshold leakage and Diode)
Current source, 6–7, 12, 60, 110
Current
　Conduction current, 18
　Displacement current, 18
CVD (*see* Chemical vapor deposition)

Damage
　thermal current, 182, 270
　trap (*see also* Trap), 93, 129, 180, 182, 185–187, 191
Dangling bonds, 178–179
DBT (*see* Defect-based testing)
DCVS (*see* Differential cascode voltage switch)
Defect activation, 201
Defect clustering, 317
Defect location, 107
Defect
　bridge (*see* Bridging defect),
　catastrophic, 244
　delay (*see* Parametric defects)
　environmentally sensitive, 243
　gate oxide short (*see* Gate oxide short)
　in crystals, 156
　open (*see* Open defect)
　parametric (*see* Parametric defects)
　random, 272, 282
　short (*see* Bridging defect)

INDEX   **341**

Defect-based testing (DBT), 281
Delay
   impact of threshold voltage, 110, 112
   signal slow-down, 265–266, 268
   signal speed-up, 255, 260, 268
Delay fault (DF), 305–306, 309, 312, 314–315, 327
Delay fault test, 266, 283, 288
   comparative analysis, 292–293
   in nanometer ICs, 314
   pattern generation, 289, 296
   using Statistical timing analysis, 318
Depletion region,
   diode, 47–50
   in channel length modulation, 83
   light emission from, 67
   transistor channel, 57
   transistor drain (*see* Drain depletion region)
Depletion, transistor operation region (*see* Weak inversion)
Design for test (DfT)
   scan design, 293
Design verification, 281, 284
DF (*see* Delay fault)
DfT (*see* Design for test)
DIBL (*see* Drain induced barrier lowering)
Die
   identification code, 307
   per wafer, 81
Differential cascode voltage switch (DCVS), 136
Diffusion current, 46, 47, 49
Diffusivity, 158, 160, 175, 177
Diode resistor circuits, 25–28
Diode
   built-in electric field, 48–49, 51, 57
   depletion region (*see* Depletion region)
   dynamic resistance, 28–29
   I-V characteristics, 24
   leakage current (*see* Diode reverse bias saturated current)
   parasitics, 50–51
   reverse bias saturated current (Is), 24
   under equilibrium, 46–48
   under forward bias, 49
   under reverse bias, 49
Displacement current, 18
Domino logic, 135–136
Domino logic, Self Resetting (SRCMOS), 135
Donor atom, 43, 47–48
Doping, 42–44, 51
   in *pn* junction, 46–48,
   role in parameter variation, 247–248
   role in substrate noise, 264
   role in temperature variation, 255
Drain bias mobility reduction, 84
Drain depletion region, 52, 59, 67, 79–81, 83, 107
   in gate oxide short, 210–211
   reduction in short-channel transistors, 248
   role in hot carrier injection, 185–189
Drain induced barrier lowering (DIBL), 82, 91, 93, 98, 248
DRAM (*see* Dynamic random access memory)
Drift current, 45
Drive strength (*see also* Current drive strength), 103, 120, 214, 221, 318
   impact on bridge critical resistance, 199, 202, 204–205
   impact on IC performance, 252
   of transistor (K), 61
   variation with V_{DD} lowering, 318
Dual damascene process, 177–178, 196
Duality property, 41, 229
Duty cycle, 169, 190
Dynamic logic, 127, 135, 147, 262
Dynamic power, 113, 131, 312
Dynamic random access memory (DRAM), 142, 150, 218, 310

E^{-1} model, 185
E-model, 185
Electrical overstress (EOS), 143, 150
Electromigration (EM), 159–161
   failure process, 162–165
   failure time, 161–162
   flux, 177
   weak spot, 178
Electron, 37–38
Electron tunneling, 183, 229–230, 236
Electron-hole pair creation, 40–41, 43–44, 52, 93, 186
Electron-hole pair recombination, 40–41
Electrostatic discharge (ESD), 143–144, 150
EM (*see* Electromigration)
Energy band model, 37, 44
Energy gap, 39, 41
EOS (*see* Electrical overstress)
ESD (*see* Electrostatic discharge)
ESD protection circuit, 143–144, 150
Evaluation phase, 134
Evaluation transistor, 134
Extrinsic failures, 223, 243–244, 271–272, 275, 322
Extrinsic material, 154, 243
Extrinsic semiconductor, 42

Farads (F), 18
Fault (*see* Fault model)
Fault equivalence, 286
Fault model
   biased voting, 207–208
   delay (*see* Delay fault),
   logic wired AND/OR, 207–208
   pseudo stuck-at, 207, 291
   stuck-at (*see* Stuck-at)
   voting, 207–208
FCC (*see* Metal Face-centered cubic)
Feedback transistor, 135
Fermi–Dirac, 41
Fick's law, 46, 158
Field emission, 185
Flip-flop, 137–139, 265, 302, 304
   bridges in, 217–218,
   in scan design, 294–295
   opens in, 239,
F_{MAX} (*see* Maximum operating frequency)
Forward bias, 25, 28, 49–52, 211
Functional failure, 202
Functional test, 283

Gate bias mobility reduction, 85
Gate induced leakage current (GIDL), 186
Gate overdrive, 103, 115, 256
Gate oxide short (GOS), 181, 183, 191, 199,
      208–214, 271, 301–302, 318
   in semiconductor memories, 218
   soft, 181, 213
Gate threshold voltage (V_{thr} or V_{TL}), 101–103
GIDL (*see* Gate induced leakage current)
GOS (*see* Gate oxide short)
Ground, 4
Ground bounce (*see* Noise)
Guardband, 188, 314

Hard opens, 271–272, 275, 304, 306, 328
HBD (*see* Oxide Hard breakdown)
HCI (*see* Hot carrier injection)
High-Z, 131–132, 134, 224
Hole, 39–49, 52, 57–58, 60, 70–71, 146,
      179–180, 186, 189, 191
Horizontal mobility, 84, 93
Hot carrier effects, 82, 91, 93, 116
Hot carrier injection (HCI), 124, 178, 185,
      188–189, 192, 196, 232, 262
   lifetime, 190
Hot electrons, 146, 186

I_{DDQ} test, 283, 288, 290–292, 302, 304–305,
      309, 311, 315, 320, 327

IDL (*see* Interlayer dielectrics)
IEEE Test standard 1149.1, 299–300
IEEE Test standard 1149.4, 299–300
IEEE Test standard 1149.5, 300
IEEE Test standard 1149.6, 300
IEEE Test standard 1450, 300
IEEE Test standard 1532, 300
IEEE Test standard P1500, 300
IEEE Test standard P1581, 300
IEEE Test standards, 300
Inductor, 3, 53, 262
Input high voltage (V_{IH}), 102, 123
Input low voltage (V_{IL}), 102, 123
Insulator, 18, 37–39, 52, 54, 56–57, 79, 248,
      261
Interconnect line
   capacitance model, 257
   RC model, 258, 279
   RLC model, 258, 279
Interconnect wire
   capacitive coupling, 261
   Copper reliability, 176–179
   crystal structure, 154–158
   defects (*see* Defects)
   delay variation, 266
   electrical models, 257–259
   electromigration (*see* Electromigration)
   inductive coupling, 261
   material in ICs, 154–155
   parameter variation (*see* Parameter variation)
   resistance dependence with temperature,
      256
   via-structure, 163–164
Interlayer dielectrics (IDL), 244
Intrinsic failures, 243–246, 256, 270, 275, 311,
      314–316, 320, 322
Intrinsic material, 154, 249–253
Intrinsic semiconductor, 42
Inverter logic threshold (*see* Gate threshold
   voltage)

KCL (*see* Kirchhoff Current law)
Kirchhoff Current law (KCL), 4, 7, 29
Kirchhoff Voltage Law (KVL), 4, 5, 29
KVL (*see* Kirchhoff Voltage Law)

Latch, 135, 137–139, 147, 150, 238–239, 265,
   314, 336
Latchup, 145–146, 150, 211–212,
LBIST (*see* Built-in self test)
Leakage power, 312–313
LFSR (*see* Built-in Self test, Linear feedback
   shift register)

Logic threshold voltage (*see* Gate threshold voltage)
Low V_{DD} testing (*see* Voltage based testing)

Master-slave flip flop (MSFF), 138–139
Maximum operating frequency (FMAX), 186, 188, 245, 250, 264–266, 268–269, 280, 304–306, 312, 315, 317, 321–322
   test, 288
   test in nanometer ICs, 315
MBIST (*see* Built-in Self test, Embedded core memory test)
MCM (*see* Multichip modules)
Metal
   bamboo grains, 157, 161, 176–177
   body-centered cubic (BCC), 155
   Bravais lattices, 154
   close-packed plane, 155
   edge dislocation, 156–157
   face-centered cubic (FCC), 155
   flux divergence, 162–164
   grain boundaries, 154, 156–159, 161–162, 175, 195
   grains, 154, 161, 177
   interstitial, 156
   mousebites, 271–275, 282
   slip plane, 156–157
   sliver defect, 199–200, 221
   substitutionals, 156, 195
   triple point, 156–157
   vacancy, 156, 159
Min V_{DD} (*see* Voltage based testing), 317–319, 321
MISR (*see* Built-in self test, Multiple input shift register)
Mobility, 45–46, 83
   of electron, 61, 103
   reduction, 84, 86
   reduction, horizontal, 84, 93
   reduction, vertical, 85, 93
   of hole, 103
MODL (*see* Multiple output domino logic)
Moore's law, 309
MSFF (*see* Master-slave flip flop)
Multichip modules (MCM), 299
Multiparameter testing, 320–321
Multiple output domino logic (MODL), 135

n-type semiconductor, 42–43, 46
NAND logic gate, 99, 118–120, 129, 227–228, 237
   path sensitization, 285
   with input-output short, 302

NBTI (*see* Negative bias temperature instability)
Nearest neighbor test, 317–318
Negative bias temperature instability (NBTI), 178, 191
NM (*see* Noise margin)
NMH (*see* Noise margin high)
NML (*see* Noise margin low)
Noise margin (NM), 100, 102, 127, 184, 232, 247, 257, 259, 261, 264, 275,
Noise margin high (NMH), 102, 123
Noise margin low (NML), 102, 123
Noise
   coupling noise, 259–260
   crosstalk noise (*see also* Cross-talk), 260–261, 279
   Delta-I noise, 259, 262, 264
   ground bounce, 253, 263, 275, 279, 308
   IR drop noise, 259, 261–262
   substrate noise, 64
   supply bounce, 253, 263, 308
   switching noise, 243, 245, 253, 264
Noncontrolling logic state, 118–121, 137, 237, 285, 287
NOR logic gate, 99, 120–121, 286–287
   path sensitization, 285

Off-state, 54–55
   current (*see* Transistor subthreshold leakage)
Ohm's Law, 4–8, 28–29
Ohm, 4
On-state, 54–55, 60, 262
OPC (*see* Optical proximity correction)
Open defect class,
   delay, 232, 236–237
   sequential, 232, 238–239
   transistor pair-on, 232, 234–235
   transistor pair-on/off, 232, 234–236
   transistor-off (memory), 232, 237–238
   transistor-on, 232–234
Open defect
   impact of surrounding lines, 226
   large opens, 224–226
   tunneling opens, 229
Optical proximity correction (OPC), 244
Outer energy band, 40
Output high voltage (V_{OH}), 102, 123, 314
Output low voltage (V_{OL}), 102, 123, 315
Overshoot, 107–110
Oxide hard breakdown (HBD), 181, 183–184, 213–214
Oxide soft breakdown (SBD), 181, 183–184, 213

Oxide ultrathin, 181–185, 213–214, 221, 248, 271
Oxide wearout, 178–183, 185, 191–192, 213, 244, 301, 308, 312

p-type semiconductor, 44, 47
Parallel plate capacitor, 18, 50, 52, 79, 80, 258, 275
Parallel resistance, 8–11
Parameter variations
    channel length variation, 246–247, 250, 266
    channel width variation, 246–247
    die to die, 244
    extrinsic, 243–244, 271–275
    impact on delay, 265
    in interconnect, 244, 252–255
    intrinsic, 243–253, 257, 261, 264–266, 268, 270, 275, 279–280, 288, 306, 308–309, 315
    lot to lot, 244
    nMOS to pMOS length ratio, 244, 246–247
    random, 245, 317–318
    random doping fluctuation, 248318
    systematic, 245, 317–318
    wafer to wafer, 244
    within die, 244
Parametric defects, (see also Parameter variations)
    delay, 264–270
    extrinsic metal mousebites, 272
    extrinsic metal slivers, 275
    extrinsic resistive vias/contacts, 271
    extrinsic weak opens, 271
Parasitics (see Transistor Parasitic and Diode Parasitic)
Pascal (Pa), 171
Pass gate, 127, 132–133, 217
Pass transistor logic, 127, 132
Pass transistor, 122, 127, 132–133, 141–142, 144, 147, 256
Pauli exclusion principle, 41
PCB (see Printed circuit board)
Percolation model, 182–183, 196, 214
Permittivity, 18, 79–80
Phase shift masks (PSM), 244
PICA (see Timing path analyzer)
Pinchoff (see Transistor pinchoff), 59, 61, 83, 93
Pitch, 244, 253, 310
Plasma vapor deposition (PVD), 177
pn junction, (see also Diode) 46, 50, 52, 56, 113, 177, 209, 212, 214, 248

POH (see Power-on hours)
Positive current convention, 4
Power consumption, 113
    leakage component, 113
    shortcircuit component, 113
    transient component, 113
Power grid, 262–264
Power-on hours (POH), 190–191
Precharge, 134
Precharge transistor, 131–136, 141–142
Printed circuit board (PCB), 142, 145
Printed wire boards (PWB), 299
Process corners, 253, 269
PSM (see Phase shift masks)
Pull-down (see Transistor Pull-down)
Pull-down network, 128–131
Pull-up (see Transistor Pull-up)
Pull-up network, 128–131, 148
PVD (see Plasma vapor deposition)
PWB (see Printed wire boards)

Quic-Mon I_{DDQ} circuit, 291

RAM (see Random access memory)
Random access memory (RAM), 139
RBB (see Reverse body biasing)
Redundant faults (nodes), 216, 287, 291
Register, 137, 139
    linear feedback shift register (LFSR), 296–297
    multiple input shift register (MISR), 298
    scan, 294–296
Resistive contacts, 306
Resistive via, F_{max} testing of, 288
Resistive vias, 237, 271–272, 275, 288, 305–307, 309, 320, 329
Reverse bias saturation current, 25, 27, 49–50, 52, 103
Reverse bias, 49–50, 52, 56–57, 79, 83, 191, 233
Reverse body biasing (RBB), 321
Ring oscillator, 187, 213
Row decoder, 238

Saddle point, 159
SAF (see Stuck-at)
Sakurai model, 88
Saturation current (see Transistor saturation and Diode saturation)
SBD (see Oxide Soft breakdown)
Scan design for test, 293–296, 298–299, 304–305, 309–310, 314, 322, 328
Scan flip-flop, 295

INDEX **345**

Schmitt trigger, 144, 150
SE (*see* Soft errors)
Self-resetting domino logic (SRCMOS), 135
Semiconductor
    doping, 42–44
    extrinsic, 42
    intrinsic, 42
    pure, 39
SER (*see* Soft error rates)
Series elements, 5
Series resistance, 6, 14
Shallow trench isolation (STI), 244, 247, 250
Sheet resistivity, 273
Silicon (*see also* Semiconductor)
SIV (*see* Stress induced Voiding)
Smiconductor, conduction in, 44–46
Soft breakdown (*see* Gate oxide short)
Soft error rates (SER), 313–314
Soft errors (SE), 312
Soft gate oxide rupture (*see* Gate oxide short)
Speed binning, 288
SRAM (*see* Static random access memory)
SRCMOS (*see* Self-Reseting CMOS)
Static random access memory (SRAM), 140
Statistical timing analysis, 315, 318
STI (*see* Shallow trench isolation)
Stress gradient, 157, 160
Stress induced Voiding (SIV), 169
Stress voiding (SV), 157, 169–178
Stuck-at
   comparative analysis, 291–293,
   fault (SAF), 207, 283, 284, 308
   test, 284
   test for nanometer ICs, 314
   test Pattern Generation, 284–288
   weak, 234–235
Stuck-open, fault, 237
Subthreshold current, 92
Subthreshold slope, 92
    degradation, 186
Supply bounce (*see* Noise)
Supply
    current, 6
    voltage, 5
SV (*see* Stress voiding)
Switching noise (*see* Noise)

Tapered buffers, 117
TCE (*see* Thermal coefficient of expansion)
Test pattern, 282
Test vector, 282
Test
   current based (*see* Current based testing)

delay fault (*see* Delay fault test)
functional (*see* Functional test)
I_{DDQ} (*see* I_{DDQ} test)
stuck-at fault (*see* Stuck-at)
voltage-based (*see* Voltage based testing)
weak defect (*see* Weak defects)
Thermal coefficient of expansion (TCE), 169
Thermal voltage, 24
Threshold voltage (*see* Transistor threshold
    voltage, and/or Gate threshold voltage)
Timing path analyzer (PICA), 107
Transient faults, 313
Transistor lifetime (t), 190
Transistor pull-down (*see also* Evaluate
    transistor), 103
Transistor pull-up (*see also* Precharge
    transistor), 103, 141
Transistor threshold voltage (V_t), 54, 58
   body effect, 77
   graphical determination, 105
   impact of technology scaling on, 93
   impact on delay, 110, 112
   $n$mos (V_{tn}), 54, 58
   pMOS (V_{tp}), 54, 58
   scaling, 92
   variation with V_{sb}, 77
   zero bias (V_{t0}), 78
Transistor
   built in diodes, 56–57
   bulk, 54–56
   channel length, 56
   channel width, 56
   drain region, 54, 56
   drive strength (*see* Drive strength)
   family of curves, 60, 70
   gate oxide, 55
   internal diodes, 56–57
   inversion channel, 58
   junction leakage current, 113
   linear operation region (*see* Transistor ohmic
     operation region)
   nonsaturated operation region (*see* Transistor
     ohmic operation region)
   ohmic operation region, 59, 61–63, 70, 86,
     89
   parasitic capacitors, 79–81
   pinchoff, 59, 83
   pinchoff voltage, 61
   saturation operation region, 59, 70, 86, 88
   source region, 54, 56
   subthreshold leakage scaling with technology,
     312
   subthreshold leakage (I_{off}), 92, 252

Transistor *(continued)*
  subthreshold leakage stress induced degradation, 188
Transmission gate, 122
  weak voltage in, 132
Transmission line effects, 259, 261
Trap, 93, 179
  assisted conduction, 181
  charge, 191, 232
  damage, 182
  distribution, 183
  interface, 180, 186
  plasma-induced, 232
  state, 179
Tri-state gate, 131
Tri-state level, 122

Undershoot, 107–108
Unipolar current, 165–166

Vacancy
  defect, 156
  of electron, 39–40, 44
Valence band, 38–40
Velocity saturation, 45–46, 83–85, 93
  with voltage scaling, 115–116
Vertical mobility reduction *(see* Mobility)
V_{IH} *(see* Input high voltage)
V_{IL} *(see* Input low voltage)
V_{OH} *(see* Output high voltage)
V_{OL} *(see* Output low voltage)
Voltage based testing (VBT), 282–283
  critical path timing test *(see* Critical path)
  delay fault test *(see* Delay fault)
  FMAX test *(see* Maximum Operating Frequency)
  for nanometerIC, 314
  low V_{DD} test (Min V_{DD}), 318
  of bridge defects *(see* Bridging defect)
  speed testing, 288
  stuck-at fault test *(see* Stuck-at)

Voltage divider, 14-16, 20–22
  due to open defect, 224–226, 233
Voltage source, 5–6
Voltage, 3–5
  weak *(see* Weak voltage)
V_t *(see* Transistor threshold voltage)
V_{thr} *(see* Gate threshold voltage)
V_{TL} *(see* Gate threshold voltage)
V_{tn} *(see* Transistor threshold votlage)
V_{t0} *(see* Transistor threshold voltage)
V_{tp} *(see* Transistor threshold voltage)

Wafer hot chuck, 307
Wafer-level reliability (WLR), 189
Wafer probing, 307, 310
Watt (W), 5
Weak bond
  in Cu, 177
  in oxide, 179, 182
Weak defect
  bridge,
  open, 223, 271–272, 304–306
  test for, 318
Weak inversion, 79–80
Weak logic gate, 264
Weak logic value *(see* Weak voltage)
Weak oxide, 181
  defect due to, 210
Weak Stuck-at *(see* Stuck-at)
Weak transistor, 205
Weak voltage, 102, 107, 132, 144, 184, 201, 202, 205, 211
  defect-induced, 213, 219, 239, 304
Well, 77
  single technology, 145–146, 211
Wired logic, 208

XOR gate, 128, 132, 133, 297, 298, 326
  weak voltage in, 122

Young's modulus, 171, 176

ABOUT THE AUTHORS

Jaume Segura is an associate professor of Electronic Technology in the Electronic Technology Group at the Balearic Islands University, Spain. He teaches graduate courses in microelectronic analog and digital design, microprocessor-based design, test engineering, and reliability. He also teaches short courses on these topics in Europe, the United States, and South America. Professor Segura has done research in these topics for more than 10 years, establishing research collaborations with several universities and companies. He has been a visiting researcher at Philips Semiconductors in New Mexico and Intel Corporation in Oregon.

Dr. Segura has been a guest editor for the *IEEE Design & Test of Computers* and was the general chair of the IEEE International On-Line Testing Workshop in 2000. Dr. Segura is a member of the executive committee of the Design Automation and Test in Europe (DATE) conference and a member of the organizing committee of the IEEE VLSI Test Symposium (VTS). He was director of the IEEE-TTTC Summer Courses on Design and Test Topics from 1999–2002. He also served as program committee member for several international conferences including the IEEE International Test Conference (ITC), IEEE VLSI Test symposium (VTS), the IEEE International On-Line Testing Symposium (IOLTS), the IEEE European Test Symposium (ETS), and the IEEE Workshop on Defect Based Testing (DBT). He is the chairman of the IEEE-CAS Spanish Chapter. He received his Ph.D. in Electronic Technology from the Polytechnic University of Catalonia in 1992, and a Physics degree from the Balearic Islands University in 1989. In 1994, he was a visiting researcher at the University of New Mexico.

Charles F. Hawkins is a professor in the Electrical and Computer Engineering Department at the University of New Mexico, Albuquerque. He teaches graduate courses in Microelectronics Design, Reliability, Test Engineering, and Failure Analysis. For more than

20 years, Professor Hawkins has also taught industry short courses on these topics in the United States, Canada, Europe, and Australia. His research in these topics includes a 20-year research collaboration with the Microelectronics Group at Sandia National Laboratories in Albuquerque, New Mexico; a four-year collaboration with the AMD Corporation and Sandia Labs in failure analysis of timing paths in high-speed microprocessors; and a four-month sabbatical with Intel Corporation in Rio Rancho, New Mexico.

He was the editor of the *ASM Electron Device Failure Analysis Magazine (EDFA)* from 1999–2003 and the general and program chair of the International Test Conference (ITC) in 1996 and 1994, respectively. He co-shared eight Best or Outstanding Paper conference awards with colleagues at Sandia Labs, Intel, and AMD Corp at ITC and at the International Symposium on Test & Failure Analysis (ISTFA). In addition to this book, he has co-authored two books on electronics. He received his Ph.D. in Bioengineering from the University of Michigan, a Master degree in Electrical Engineering from Northeastern University, and a BEE from the University of Florida. He was the associate dean of the School of Engineering at the University of New Mexico from 1980–1982. He held summer faculty appointments at the University of the Balearic Islands, Spain from 1998–2002.